Climate Change Management and Social Innovations for Sustainable Global Organization

Sumita Dave
Amity University, Raipur, India

Monica Sainy
Amity University, Raipur, India

Disha Sharma
Amity University, Raipur, India

Anupam Singh
University of Stirling, UK

IGI Global
PUBLISHER of TIMELY KNOWLEDGE

A volume in the Advances in Logistics, Operations, and Management Science (ALOMS) Book Series

Published in the United States of America by
IGI Global
Engineering Science Reference (an imprint of IGI Global)
701 E. Chocolate Avenue
Hershey PA, USA 17033
Tel: 717-533-8845
Fax: 717-533-8661
E-mail: cust@igi-global.com
Web site: http://www.igi-global.com

Library of Congress Cataloging-in-Publication Data

Names: Dave, Sumita, 1972- editor. | Sainy, Monica, 1975- editor. | Sharma, Disha, 1991- editor.
Title: Climate change management and social innovations for sustainable global organization / edited by Sumita Dave, Monica Sainy, Disha Sharma, Anupam Singh.
Description: Hershey, PA : Engineering Science Reference, [2023] | Includes bibliographical references and index. | Summary: "This book is envisioned as a manuscript that presents chapters, and case studies exploring the nature and impact of climate change on business"-- Provided by publisher.
Identifiers: LCCN 2023041811 (print) | LCCN 2023041812 (ebook) | ISBN 9781668495032 (hardcover) | ISBN 9781668495049 (paperback) | ISBN 9781668495056 (ebook)
Subjects: LCSH: Climatic changes--Economic aspects. | Business enterprises--Environmental aspects. | Sustainable development--Economic aspects.
Classification: LCC QC903 .C5526 2023 (print) | LCC QC903 (ebook) | DDC 363.7--dc23/eng/20230912
LC record available at https://lccn.loc.gov/2023041811
LC ebook record available at https://lccn.loc.gov/2023041812

This book is published in the IGI Global book series Advances in Logistics, Operations, and Management Science (ALOMS) (ISSN: 2327-350X; eISSN: 2327-3518)

British Cataloguing in Publication Data
A Cataloguing in Publication record for this book is available from the British Library.

For electronic access to this publication, please contact: eresources@igi-global.com.

Advances in Logistics, Operations, and Management Science (ALOMS) Book Series

John Wang
Montclair State University, USA

ISSN:2327-350X
EISSN:2327-3518

Mission

Operations research and management science continue to influence business processes, administration, and management information systems, particularly in covering the application methods for decision-making processes. New case studies and applications on management science, operations management, social sciences, and other behavioral sciences have been incorporated into business and organizations real-world objectives.

The **Advances in Logistics, Operations, and Management Science** (ALOMS) Book Series provides a collection of reference publications on the current trends, applications, theories, and practices in the management science field. Providing relevant and current research, this series and its individual publications would be useful for academics, researchers, scholars, and practitioners interested in improving decision making models and business functions.

Coverage

- Marketing engineering
- Operations Management
- Production Management
- Information Management
- Computing and information technologies
- Decision analysis and decision support
- Networks
- Organizational Behavior
- Risk management
- Services management

IGI Global is currently accepting manuscripts for publication within this series. To submit a proposal for a volume in this series, please contact our Acquisition Editors at Acquisitions@igi-global.com or visit: http://www.igi-global.com/publish/.

The Advances in Logistics, Operations, and Management Science (ALOMS) Book Series (ISSN 2327-350X) is published by IGI Global, 701 E. Chocolate Avenue, Hershey, PA 17033-1240, USA, www.igi-global.com. This series is composed of titles available for purchase individually; each title is edited to be contextually exclusive from any other title within the series. For pricing and ordering information please visit http://www.igi-global.com/book-series/advances-logistics-operations-management-science/37170. Postmaster: Send all address changes to above address.

Titles in this Series

For a list of additional titles in this series, please visit: www.igi-global.com/book-series

Black Women's Formal and Informal Ways of Leadership Actualizing the Vision of a More Equitable Workplace
Audrey D. Bowser (Arkansas State University, USA) Kimberley M. Davis (University of Arkansas at Pine Bluff, USA) and Natalie A. Johnson-Leslie (Arkansas State University, USA)
Information Science Reference • copyright 2023 • 340pp • H/C (ISBN: 9781668438275) • US $215.00 (our price)

Multidisciplinary Approaches to Organizational Governance During Health Crises
Carla Sofia Vicente Negrão (CeBER, University of Coimbra, Portugal) Isabel Guerreiro Pimentel Maia (Centre for Business and Economics Research, University of Coimbra, Portugal) and João António Furtado Brito (University of Cape Verde, Cape Verde)
Business Science Reference • copyright 2023 • 300pp • H/C (ISBN: 9781799892137) • US $250.00 (our price)

New Perspectives and Possibilities in Strategic Management in the 21st Century Between Tradition and Modernity
Javier Martínez-Falcó (University of Alicante, Spain & Stellenbosch University, South Africa) Bartolomé Marco-Lajara (University of Alicante, Spain) Eduardo Sánchez-García (University of Alicante, Spain) and Luis A. Millan-Tudela (University of Alicante, Spain)
Business Science Reference • copyright 2023 • 514pp • H/C (ISBN: 9781668492611) • US $315.00 (our price)

Government Impact on Sustainable and Responsible Supply Chain Management
Atour Taghipour (Normandy University, France)
Information Science Reference • copyright 2023 • 354pp • H/C (ISBN: 9781668490624) • US $240.00 (our price)

Transport and Logistics Planning and Optimization
Jaouad Boukachour (University of Le Havre, France) and Abdelhamid Benaini (University of Le Havre, France)
Business Science Reference • copyright 2023 • 317pp • H/C (ISBN: 9781668484746) • US $250.00 (our price)

Considerations and Characteristics of the Millennial Leader in Today's Global Society
Tymon M. Graham (TMGraham Consulting, LLC, USA)
Business Science Reference • copyright 2023 • 300pp • H/C (ISBN: 9781668447116) • US $240.00 (our price)

The Economics of Talent Management and Human Capital
Sharif Uddin Ahmed Rana (Paris Graduate School and World Talent Economy Forum, Malaysia) and Adrian David Cheok (Imagineering Institute, Malaysia)
Business Science Reference • copyright 2023 • 300pp • H/C (ISBN: 9781668466414) • US $235.00 (our price)

Entrepreneurship Ecosystems and Their Opportunities and Challenges

IGI Global
PUBLISHER of TIMELY KNOWLEDGE

701 East Chocolate Avenue, Hershey, PA 17033, USA
Tel: 717-533-8845 x100 • Fax: 717-533-8661
E-Mail: cust@igi-global.com • www.igi-global.com

Editorial Advisory Board

Table of Contents

Foreword xvi

Preface xviii

Chapter 1
Understanding the Role of Green HRM as a Contributor Towards Sustainable Global Organization ... 1
Anushka Lydia Issac, Westford University College, UAE

Chapter 2
How Do Equity Market Performance and ICT Influence Clean Energy Consumption in High-Income Countries? 15
Pınar Avci, Tekirdağ Namık Kemal University, Turkey

Chapter 3
Does Economic Growth Increase CO2 Emissions in BRICS Countries? Question on Sustainability . 33
Esha Sunil Bharadwaj, Jain University (Deemed), India
Pratik Mukherjee, Jain University (Deemed), India
Suvam Sivam, Jain University (Deemed), India
Ashish Kumar, Jain University (Deemed), India
Susobhan Maiti, Jain University (Deemed), India

Chapter 4
Empirical Evidence on the Relation Between Progress on Sustainable Development Goals and Adaptation of Islamic Finance 47
Dogus Emin, Social Sciences University of Ankara, Turkey
Aysegul Aytac Emin, Social Sciences University of Ankara, Turkey

Chapter 5
The Effects of Climate Change on Human Health and the Healthcare Sector 64
Ranjit Barua, Omdayal Group of Institutions, India
Sumit Bhowmik, Omdayal Group of Institutions, India
Debasish Banerjee, Omdayal Group of Institution, India
Sudipto Datta, Indian Institute of Science, Bangalore, India

Chapter 6
Green Marketing: Sustainable Strategies for Success 76
Shriya Tripathi, Ganpat University, India
Disha Sharma, Amity University, Raipur, India

Chapter 7
Development of Quaternary Concrete 93
Sagar Dhengare, Yeshwantrao Chavan College of Engineering, Nagpur, India
R. L. Sharma, Lovely Professional University, India

Chapter 8
The Anthropology and Anthropocene Urban Socio-Ecology Resilience Approach to Resilience Planning and Green Innovations 105
José G. Vargas-Hernandez, Tecnológico Superior oe Jalisco Mario Molina, Zapopan, Mexico
Carlos A. Rodriguez-Maillard, Universidad Cristòbal Colòn, Mexico
Omar C. Vargas-González, Tecnológico Nacional de México, Ciudad Guzmán, Mexico

Chapter 9
Green Technology: Exploring Its Impact on Indian Society 127
Ankit Mishra, Amity University, Raipur, India

Chapter 10
Green Knowledge in Green Roofs and Organizational Green Innovation 134
José G. Vargas-Hernandez, Tecnológico Superior de Jalisco Mario Molina Pasquel y Henríquez, Unidad Académica Zapopan, Mexico
Elsa Patricia Orozco Quijano, Laurentian University, Canada
Carlos A. Rodriguez-Maillard, Universidad Cristobal Colón, Mexico

Chapter 11
Green Marketing Among Z-Generation (Z-Gen.) 157
Abhinav Agrawal, Agrasen Mahavidyalaya, India
Minal Shah, Bhilai Institute of Technology, India
Dharmendra Kumar Gangeshwer, Bhilai Institute of Technology, India

Chapter 12
Gross National Happiness: Bhutan's Model for Sustainable Development – Its Relevancy in Current Times 175
Bhuwan Shrivastava, Royal Thimphu College, Bhutan & Amity University, Raipur, India
Sumita Dave, Amity Business School, Amity University, India

Chapter 13
Green Cement: A Step Towards Sustainable Development 195
Sushma Singh, Bhilai Institute of Technology, Durg, India
Urvashi Shrivastava, Bhilai Institute of Technology, Durg, India

Chapter 14
Characterising a Green Supply Chain Using Multi-Dimensional Construct: Evidence From an Emerging Economy 212
Chetan V. Hiremath, Kirloskar Institute of Management, India
Chandan Chavadi, Presidency Business School, India
Sudhansu Sekhar Nanda, Kirloskar Institute of Management, India
Shankargouda C. Patil, Rani Channamma University, India

Chapter 15
Impact of Digital Culture Integration on Performance of Infosys Ltd. A Pre- and Post-Merger Analysis 232
Abhay Kant, Agra Public Group of Institutions, India
Renu Bharti, Dayalbagh Educational Instiution, India
Shobhit Sagar, Mewar Institute of Management, Ghaziabad, India

Chapter 16
The Mediating Role of Job Satisfaction Between Work-Life Balance and Employee Retention 241
Soni Rathi, Chandigarh University, India
Praveen Kumar, Chandigarh University, India

Compilation of References 263

About the Contributors 306

Index 311

Detailed Table of Contents

Foreword xvi

Preface xviii

Chapter 1
Understanding the Role of Green HRM as a Contributor Towards Sustainable Global Organization... 1
Anushka Lydia Issac, Westford University College, UAE

This chapter examines the role of green human resource management (HRM) in promoting sustainable global organisations. The chapter adopts a case study methodology to explore how organizations like McDonald's, Interface, Unilever, Patagonia, Toyota, and the Co-Operative Group have successfully integrated green HRM practices into their operations. The chapter also identifies critical issues surrounding green HRM, such as the lack of standardisation, the need for employee buy-in, and the potential for green-washing. To address these issues, the chapter proposes solutions such as developing industry-wide standards, creating a culture of environmental responsibility, and adopting transparent reporting practices. It is hoped that this chapter will contribute to a deeper understanding of the role of green HRM in promoting sustainable practices in organisations. Overall, this chapter aims to contribute to the ongoing discourse on sustainability and the role of HRM in promoting sustainable practices in organisations.

Chapter 2
How Do Equity Market Performance and ICT Influence Clean Energy Consumption in High-Income Countries? 15
Pınar Avci, Tekirdağ Namık Kemal University, Turkey

The aim of this study is to examine whether equity market development and ICT affect clean energy consumption in 18 high-income countries during the period 1990-2015. In the analysis, firstly, the stationarity properties of the variables are tested with the CADF and CIPS tests. In the second stage, cointegration between the variables in the model is investigated by Westerlund, Kao, and Pedroni methods. In the next step, the long-run elasticities of the variables are estimated using the PCSE and FGLS methods. The causality between the variables is also determined by applying the Dumitrescu-Hurlin panel causality method. According to findings, cointegration between clean energy consumption and equity market development, ICT, economic growth, and CO2 emissions variables, which integrated at I(1) level, is proven. The long-term findings showed that the equity market, ICT, and economic growth encourage clean energy consumption of 18 high-income countries, and CO2 emissions also prevent them. In addition, there is bidirectional causality between clean energy consumption and other variables.

Chapter 3
Does Economic Growth Increase CO2 Emissions in BRICS Countries? Question on Sustainability . 33
Esha Sunil Bharadwaj, Jain University (Deemed), India
Pratik Mukherjee, Jain University (Deemed), India
Suvam Sivam, Jain University (Deemed), India
Ashish Kumar, Jain University (Deemed), India
Susobhan Maiti, Jain University (Deemed), India

Sustainability is an increasingly important issue in the modern world and an essential concept that is increasingly gaining importance in today's world. It refers to the responsible use of natural resources while ensuring that future generations can meet their needs. The impact of human activities on the environment is becoming more apparent, with climate change and biodiversity loss being some of the major challenges that we face today. Considering the foregoing discussion, the present study comprises two main objectives: firstly, to find out the relationship between economic growth, which is measured by the growth of GDP per capita and CO2 emissions per capita in the BRICS nations, and to find out unidirectional or bidirectional causality between carbon emissions and proxies of economic growth based on the data collected from the world development indicator for the period 1990 to 2022. The results suggest that a bi-directionally causality exists between carbon emissions per capita and economic growth per capita in the BRICS countries.

Chapter 4
Empirical Evidence on the Relation Between Progress on Sustainable Development Goals and Adaptation of Islamic Finance 47
Dogus Emin, Social Sciences University of Ankara, Turkey
Aysegul Aytac Emin, Social Sciences University of Ankara, Turkey

In this study, the authors aim to investigate the dynamic correlation and lead/lag relation between the progress of the countries on the Sustainable Development Goals (SDGs) and adaptation of Islamic finance. As the proxy of progress level on SDGs, SDG Index of the Bertelsmann Stiftung and the Sustainable Development Solutions Network (SDSN) are used. To be able to determine the adaptation level of Islamic finance, Islamic Finance Adaptation (IFA) Index is introduced. The data set is formed with top eight countries according to Islamic finance assets and top eight countries according to the SDG Index for the period of 2012 and 2021. To assess the relation between progress level on Sustainable Development Goals (SDG Index) and Islamic finance adaptation level (IFA Index), the Dynamic Conditional Correlation Model (DCC Model) of Engle and Sheppard and Geweke Feedback Measure of Geweke are used.

Chapter 5
The Effects of Climate Change on Human Health and the Healthcare Sector 64
Ranjit Barua, Omdayal Group of Institutions, India
Sumit Bhowmik, Omdayal Group of Institutions, India
Debasish Banerjee, Omdayal Group of Institution, India
Sudipto Datta, Indian Institute of Science, Bangalore, India

Just a few of the ways that climate change has influenced human lives and health over the past several decades include changes in extreme weather and climatic events, changes in the labor force, changes in food security, and changes in the prevalence and geographic distribution of infectious diseases. Indicators for climate change and health were developed as useful tools to track the complex network of interdependent relationships that connect climate change and human health. While the weather has an impact on the timing and severity of outbreaks, climate primarily influences the range of infectious diseases. This chapter aims to offer an outline of the impacts of climate change on human health in analysis, research potential, and applications in the healthcare sector.

Chapter 6
Green Marketing: Sustainable Strategies for Success 76
Shriya Tripathi, Ganpat University, India
Disha Sharma, Amity University, Raipur, India

Consumers in the 21st century are more informed and discerning than ever about the products they buy. This is positively obvious with regards to assessing the ecological effect and manageable acts of the brands they support. That is why many of the world's most significant and effective organizations are seeking green marketing drives. Green marketing has become an essential business strategy as consumers' awareness of environmental issues has grown. Purchasers are progressively worried about the natural effect of their utilization decisions, and they are bound to purchase items or administrations that line up with their ecological qualities. Businesses can draw in these environmentally conscious customers, set themselves apart from their rivals, and enhance their brand image by employing green marketing strategies.

Chapter 7
Development of Quaternary Concrete 93
Sagar Dhengare, Yeshwantrao Chavan College of Engineering, Nagpur, India
R. L. Sharma, Lovely Professional University, India

The amount of unregulated trash produced by industry, particularly in developing countries, has resulted in an increase in environmental challenges. The cultivation of this type of quaternary concrete material is both a sustainable and cost-effective solution to pollution. From the standpoint of producing quaternary cement concrete products from waste materials, the chapter provides a quick overview of which materials are suitable for use in construction as cement replacements. A number of writers and academics conduct a variety of physic-mechanical tests. Various approaches are employed in various studies. As a result, resource-efficient cements with appropriate qualities can be produced to strike the necessary balance between the industry's objective for high-performance concrete and increasingly stringent environmental laws. On the other side, the increase in construction activity in the country caused a scarcity of most supplies used in making concrete, and consequently, cement costs will rise as a result.

Chapter 8
The Anthropology and Anthropocene Urban Socio-Ecology Resilience Approach to Resilience Planning and Green Innovations 105
José G. Vargas-Hernandez, Tecnológico Superior oe Jalisco Mario Molina, Zapopan, Mexico
Carlos A. Rodriguez-Maillard, Universidad Cristòbal Colòn, Mexico
Omar C. Vargas-González, Tecnológico Nacional de México, Ciudad Guzmán, Mexico

This study aims to analyze some of the existing implications between urban anthropology and Anthropocene urban socio-ecology planning resilience. Beginning from the assumption that urban anthropology gives support to create and develop any urban planning based on the Anthropocene urban socio-ecology resilience, the methods employed are the analytical-descriptive based on an ethnographic interpretation and reflection of the theoretical and empirical literature review. The analysis concludes that urban anthropology fundamentals give support to strengthen the Anthropocene orientation of the urban socioecological planning resilience.

Chapter 9
Green Technology: Exploring Its Impact on Indian Society 127
Ankit Mishra, Amity University, Raipur, India

In the domain of language, the term "green tech" encompasses a wide range, referring to innovative creations meticulously designed with environmental considerations woven into every stage, from inception to utilization. The swift depletion of Earth's natural resources, surpassing their regeneration, has spurred an unrelenting quest to refine such technology, making it more harmonious with our delicate ecology. The urgency to lessen reliance on finite resources, notably fossil fuels, aligns with the need to heal ecological wounds. These imperatives define green technology's core mission. Sectors like waste management, energy, and transportation ardently embrace this mission. Despite its benefits, green technology faces substantial hurdles before becoming mainstream. Over time, significant employment growth is observed in green tech, underscoring its growing relevance in our contemporary era.

Chapter 10
Green Knowledge in Green Roofs and Organizational Green Innovation 134
José G. Vargas-Hernandez, Tecnológico Superior de Jalisco Mario Molina Pasquel y Henríquez, Unidad Académica Zapopan, Mexico
Elsa Patricia Orozco Quijano, Laurentian University, Canada
Carlos A. Rodriguez-Maillard, Universidad Cristobal Colón, Mexico

This study aims to analyze the implications of green knowledge and technology in organizational green innovation, urban green innovation, and green roofs. Green roofs can be an effective tool for cities to improve the thermal environment, save energy, and combat climate change, and are an appropriate method of saving energy. The analysis is supported by the assumption that green technology is basic to organizational green innovation and urban green innovation areas practices, operations, and activities. Making the balance between urban developments and environmental issues, in consideration to sustainable development principle and its innovative green solutions, the methods employed are based on the analytical-reflective and descriptive supported with the review of theoretical and empirical literature. The analysis concludes that green knowledge sharing is relevant to create and develop the green technology with positive implications for organizational green innovation, urban green innovation areas, and green roofs.

Chapter 11
Green Marketing Among Z-Generation (Z-Gen.)... 157
Abhinav Agrawal, Agrasen Mahavidyalaya, India
Minal Shah, Bhilai Institute of Technology, India
Dharmendra Kumar Gangeshwer, Bhilai Institute of Technology, India

Environmental problems have become prime global challenges, and in developing countries like India, there is a need for high attention to curb these problems. These problems bring unsustainable economic and environmental development, and global sustainable growth has been a major concern of G-20 countries. For this purpose, in every economy, the business organizations, producers, and consumers can play their roles to protect the environment through their business and buying activities. Manufactures and businesses can play vital roles through production of green product and green marketing activities, and buyers can play their role through demanding such product as these products are proven to be good for the environment as well as for the health for consumers and also attract maximum profit for the business organizations. Marketers can target young consumers aged between 11-26 called "Z-Generation" consumers because they are kind of consumers who are more aware of the environmental issues and express their will to protect mother nature.

Chapter 12
Gross National Happiness: Bhutan's Model for Sustainable Development – Its Relevancy in Current Times 175
Bhuwan Shrivastava, Royal Thimphu College, Bhutan & Amity University, Raipur, India
Sumita Dave, Amity Business School, Amity University, India

Bhutan, known to the outside world as the land of happiness, has attracted attention due to its unique way of transforming the lives of its citizen through its approach towards Gross National Happiness and Sustainable Development Goals. This chapter explores Bhutan's way to treat the concept of Gross National Happiness. This chapter provides an overview of GNH philosophy by examining its different pillars. This chapter also takes into consideration the criticism and shortcoming which GNH has over the most common indicator of growth, the GDP. Furthermore, the chapter also explores the development of concept of GNH and its presence and relevancy in current times. It tries to understand the attainment of Sustainable Development Goals via GNH thereby understanding further Bhutan's model of SDG attainment and how it acts as a model for other nations aspiring for Sustainable Development Goals. Overall, this chapter is an attempt to understand, explore, and present Bhutan's way to reach the target of SDG.

Chapter 13
Green Cement: A Step Towards Sustainable Development.. 195
Sushma Singh, Bhilai Institute of Technology, Durg, India
Urvashi Shrivastava, Bhilai Institute of Technology, Durg, India

With the focus on sustainable development of the global arena in terms of construction and infrastructure, it is necessary to evaluate the impact of one of the ingredient "cements" on the environment. Therefore, this chapter investigates the literature available in the manufacturing process of cements, alternatives to it, and in particular, the emergence of "green cement," which assumes to be environmentally friendly. The action taken by the cement companies is clean and green, with profitability and sustainability as the prime focus by lesser CO2 emission. By innovating new technologies and adopting modernizations, the companies are focusing on developing such concrete, which is an environmentally friendly and economically sustainable. The future of green cement holds the promise of reduced carbon, enhanced durability, and improved performance. As environmental regulations become more stringent, the demand for green cement is expected to rise significantly. This presents a significant opportunity for the construction industry to embrace greener practices and contribute to a more sustainable environment.

Chapter 14
Characterising a Green Supply Chain Using Multi-Dimensional Construct: Evidence From an Emerging Economy..212
Chetan V. Hiremath, Kirloskar Institute of Management, India
Chandan Chavadi, Presidency Business School, India
Sudhansu Sekhar Nanda, Kirloskar Institute of Management, India
Shankargouda C. Patil, Rani Channamma University, India

The study proposed a construct for a green supply chain, which aims to analyze the different elements that contribute to a network being considered "green." The study found that there are many types of research done on green products, but very few on green corporations and supply chains. Also, there is a lack of consensus in definitions and previous studies have not examined the inter-relations among the variables that contribute to a supply chain being considered green. The review of the literature reveals the range of factors, from the procurement of raw materials to end customer perceptions which make the supply chain green. In a simple random sampling, 125 responses were collected from Karnataka to explore the interrelations between the factors. The analysis evaluates whether the underlying structure of a green supply chain is uni-dimensional or multi-dimensional. Whether the factors have a hierarchical relationship and are factors independent or correlated, the analysis found that there are four correlated factors without any higher order.

Chapter 15
Impact of Digital Culture Integration on Performance of Infosys Ltd. A Pre- and Post-Merger Analysis...232
Abhay Kant, Agra Public Group of Institutions, India
Renu Bharti, Dayalbagh Educational Instiution, India
Shobhit Sagar, Mewar Institute of Management, Ghaziabad, India

The customer is an important resource for achieving the digitalisation capabilities. Organisations mainly emphasise the customer taste and preference because the customer is the king of the market as well as a help in assessing the internal as well as external process that can simplify the various points that are useful to make a balance between customers and life cycle of products and services. The main aim of the chapter is to know the digital culture transformation and its impact on financial performance of Infosys. The present study utilizes secondary data sources. The secondary data sources primarily consist of published studies in various international and national journals, magazines, and conference proceedings. The statistical tools such as the ratio analysis and sample test were used to test impact of digital culture transformation. The study found that electronic payment growth (EPG) did not significantly impact four important financial metrics, namely net interest margin (NIM), current ratio, advance growth, and capital adequacy ratio (CAR).

Chapter 16
The Mediating Role of Job Satisfaction Between Work-Life Balance and Employee Retention 241
Soni Rathi, Chandigarh University, India
Praveen Kumar, Chandigarh University, India

Work-life balance is a well-researched area, and it does not only escalate the effectiveness and retention of employees, but also reforms satisfaction in both the professional as well as personal lives. The current study highlights the impact of WLB on employee retention and JS and the influence of job satisfaction on employee retention. It further explores the sub-factors of work-life balance and job satisfaction of distinct sectors. The chapter deals with surveying the past literature reviews on work-life balance, job satisfaction, and employee retention and discovering the association among them in various sectors of industries in India and other nations. It is found via this study that work-life balance has an influence on the employee retention and it also influences job satisfaction of employees. Further, there is a relationship also depicted between job satisfaction and retention of employees at the workplace. The limitations and future directions are also discussed.

Compilation of References 263

About the Contributors 306

Index 311

Foreword

There is a pressing need for the creation and application of original and creative strategies because of a global society dealing with the more complicated difficulties brought on by climate change. The inadequacy of traditional approaches and tactics is becoming clearer as the effects of human activity on the environment grow more pronounced. To create successful global organizations, it is essential to incorporate fresh perspectives and innovative methodologies. This important book, *Climate Change Management and Social Innovations for Sustainable Global Organization*, is something I'm proud to present on behalf of the Amity Universe.

This book takes readers on a transformative journey that aims to close the gap between business leadership and climate responsibility. The investigation digs into the areas of technological advancement and corporate practices, ultimately inspiring substantial reforms that may help to create a more promising and sustainable future. The complexity of climate change necessitates the adoption of group initiatives involving not only governmental bodies but also international businesses and NGOs. It is crucial to recognize that to successfully navigate the upcoming difficulties, a fundamental change in our cognitive framework and operational procedures is required. Social innovations are essential for bringing about systemic change and fostering productive teamwork.

In the current work, the authors' perspectives on managing climate change are covered throughout the course of 18 chapters. They participate in a lively study of the potential for social innovations to serve as powerful tools for advancing sustainability through a thorough investigation. The writers have offered insightful and useful suggestions and tactics for transforming multinational corporations into role models of good stewardship.

The fundamental elements of climate change are thoroughly examined in this literary work, along with its effects on the environment, the economy, and society. However, this project's scope goes beyond merely increasing awareness. Instead, it provides readers with innovative and useful tactics that encourage flexibility, reduce carbon emissions, and start positive changes both locally and globally. The subsequent chapters explain how organizations of any size or industry can adopt sustainability as a fundamental principle and incorporate it into their essential structure. The writers deftly negotiate the complex terrain of sustainability issues while providing readers with a thorough manual for building a wealthy and inclusive future. There is an increasing need for leaders who are prepared to adopt progressive ideologies and implement revolutionary solutions as the visible effects of climate change worsen.

The search for sustainability is a group effort. The active involvement and steadfast dedication of all parties concerned, from the highest levels of decision-making to the grassroots level of the community, are essential to the successful completion of this project. It is sincerely encouraged to actively join in the group effort to generate positive development as one begins the journey through the contents of this literary work. Utilizing social innovations gives us the chance to jointly tap into their potential to further the growth of sustainable international organizations and leave a legacy that future generations will value highly. Let's work together to create a global society that thrives in the face of climate change, one that appreciates and protects our Earth and its many inhabitants, and to fulfil the demands of the current situation. 'Climate Change Management and Social Innovations for Sustainable Global Organizations' is a promising initiative in our shared search for a better future, and I'm happy to promote it.

Piyush Kant Pandey
Amity University, Raipur, India

Preface

Welcome to *Climate Change Management and Social Innovations for Sustainable Global Organization*. In the face of an ever-evolving world, where the impacts of climate change are becoming increasingly evident, the need for a comprehensive and actionable guide has never been more urgent. As editors, we are delighted to present this edited reference book, which delves into the critical intersection of climate change, sustainable development, and organizational practices.

In the following pages, we embark on a journey to dissect the multifaceted challenges that climate change poses to organizations on a global scale. The undeniable reality of our current production and consumption patterns has ushered in an era of unprecedented strain on our environment, affecting the very foundations upon which businesses operate. From the availability and quality of essential resources such as energy, water, and skilled labor, to the resilience of sectors spanning agriculture, production, and services, the impact of climate change reverberates across the entire organizational landscape.

In the midst of these challenges, organizations are striving for growth and prosperity, but often at the expense of sustainability. This book is dedicated to exploring innovative pathways that lead to a harmonious coexistence of growth and ecological equilibrium. We firmly believe that social and technological innovations hold the key to steering organizations toward a Net Zero economy, a pressing need for our times.

The compilation of chapters and case studies presented within this manuscript embodies our commitment to shedding light on the intricate relationship between climate change and business. Our esteemed contributors have meticulously formulated strategies, technologies, and initiatives that aim not only to mitigate the impacts of climate change but also to catalyze positive transformations within organizations.

This book addresses an array of dimensions within the realm of climate change management, ranging from the examination of sustainable practices in agriculture and circular economies, to the exploration of green technologies, innovative human resource management, and the integration of environmental sustainability into marketing and finance. Additionally, legal, social, and cultural perspectives provide a comprehensive understanding of climate change's intricate web.

We have curated this book with a diverse readership in mind. Whether you are an academician, a researcher, a policy maker, a corporate leader, an environmentalist, or a student, you will find valuable insights and actionable knowledge within these pages. For educators, this book can serve as a valuable companion in fostering discussions around the critical juncture of sustainable development and climate change.

As the urgency of addressing climate change looms ever larger, we must remember that collective action is paramount. We express our heartfelt gratitude to all the contributors who have shared their expertise, experiences, and ideas within this volume, enriching the discourse surrounding climate change management and social innovations. May this book inspire and guide individuals and organizations alike toward a future that is both prosperous and ecologically sound.

CHAPTER OVERVIEW

Welcome to the diverse world of *Climate Change Management and Social Innovations for Sustainable Global Organization*. In this collection of chapters, we delve into the intricate tapestry of sustainable practices, innovative strategies, and their implications for organizations in the face of climate change challenges. Let's embark on a journey through the insights offered by our esteemed contributors.

Chapter 1: Understanding the Role of Green HRM as a Contributor Towards Sustainable Global Organization

Anushka Issac explores the pivotal role of Green Human Resource Management (HRM) in fostering sustainable practices within organizations. Through insightful case studies, the chapter showcases successful integration of Green HRM by renowned organizations. Critical issues and potential solutions are examined, highlighting the importance of industry-wide standards, environmental responsibility, and transparent reporting practices.

Chapter 2: How Do Equity Market Performance and ICT Influence Clean Energy Consumption in High-Income Countries?

Pinar Avci investigates the intricate relationship between equity market development, information, and communication technology (ICT), and clean energy consumption in high-income countries. Employing a robust methodology, the chapter reveals the interplay between these variables, shedding light on the long-term impact on clean energy consumption and emissions.

Chapter 3: Does Economic Growth Increase CO2 Emissions in BRICS Countries? Question on Sustainability

Esha Bharadwaj, Pratik Mukherjee, Suvam Sivam, Ashish Kumar, and Susobhan Maiti delve into the complex relationship between economic growth and CO2 emissions in BRICS nations. Through meticulous analysis, the chapter uncovers unidirectional and bidirectional causality, offering insights into the dynamics between economic growth and environmental impact.

Chapter 4: An Empirical Evidence on the Relation Between Progress on Sustainable Development Goals and Adaptation of Islamic Finance

Dogus Emin and Aysegul Aytac Emin navigate the realm of Sustainable Development Goals (SDGs) and Islamic finance adaptation. Through innovative indices and methods, the chapter establishes a dynamic correlation between SDG progress and Islamic finance adaptation, contributing to a nuanced understanding of their interrelation.

Chapter 5: The Effects of Climate Change on Human Health and Healthcare Sector

Ranjit Barua, Sumit Bhowmik, Debasish Banerjee, and Sudipto Datta dissect the multifaceted impacts of climate change on human health and healthcare. Exploring factors ranging from extreme weather to infectious diseases, the chapter emphasizes the intricate web of climate-related health challenges and potential solutions.

Chapter 6: Green Marketing – Sustainable Strategies for Success

Shriya Tripathi and Disha Sharma delve into the realm of green marketing, unveiling its pivotal role in engaging environmentally conscious consumers. The chapter elucidates how businesses can align with consumer values through sustainable strategies, fostering a harmonious relationship between commerce and ecological well-being.

Chapter 7: Development of Quaternary Concrete

Sagar Dhengare and R. Sharma present an innovative approach to sustainable construction through the development of quaternary concrete. By repurposing waste materials, this chapter proposes a resource-efficient solution to the environmental challenges posed by traditional construction methods.

Chapter 8: The Anthropology and Anthropocene Urban Socio-Ecology Resilience Approach to Resilience Planning and Green Innovations

José Vargas-Hernandez, Carlos Rodriguez-Maillard, and Omar Vargas-González explore the fascinating intersection of urban anthropology, Anthropocene urban socio-ecology, and resilience planning. The chapter unveils the crucial role of anthropology in shaping sustainable urban development and green innovations.

Chapter 9: Green Technology: Exploring Its Impact on Indian Society

José Vargas-Hernandez, Elsa Orozco Quijano, and Carlos Rodriguez-Maillard delve into the synergy between green knowledge, green technology, and organizational innovation. Focusing on green roofs, this chapter uncovers how knowledge sharing contributes to the development of sustainable practices in organizations and urban environments.

Chapter 10: Green Knowledge in Green Roofs and Organizational Green Innovation

Abhinav Agrawal, Minal Shah, and Dharmendra Gangeshwer navigate the world of green marketing targeted at the environmentally-conscious Z-Generation. The chapter explores how this tech-savvy cohort's values and preferences align with green products, shaping new paradigms in marketing strategies.

Chapter 11: Green Marketing Among Z-Generation (Z-Gen.)

Diwakar Kadam offers a comprehensive exploration of the Green Economy concept and its significance for sustainable development. The chapter delves into strategies and case studies that showcase how a Green Economy can harmonize economic growth, social equity, and environmental well-being.

Chapter 12: Gross National Happiness: Bhutan's Model for Sustainable Development – Its Relevancy in Current Times

Sushma Singh and Urvashi Shrivastava unravel the journey towards greener construction practices through the concept of green cement. The chapter examines manufacturing processes, alternative materials, and the emergence of eco-friendly cement, ushering in a more sustainable era in construction.

Chapter 13: Green Cement: A Step Towards Sustainable Development

Chetan Hiremath, Chandan Chavadi, Sudhansu Nanda, and Shankargouda Patil probe the complexities of green supply chains, unraveling the interconnected elements that contribute to a network's eco-friendliness. The chapter explores multi-dimensional constructs, hierarchies, and correlations, shedding light on achieving sustainability in supply chain operations.

Chapter 14: Characterizing a Green Supply Chain Using Multi-Dimensional Construct: Evidence From an Emerging Economy

Abhay Kant, Renu Bharti, and Shobhit Sagar delve into the digital culture's impact on organizational performance through the lens of Infosys Ltd. The chapter examines the merger's influence on digitalization capabilities and its subsequent effect on financial performance, unveiling insights into the changing landscape of business operations.

Chapter 15: Impact of Digital Culture Integration on Performance of Infosys Ltd. A Pre- and Post-Merger Analysis

Soni Rathi and Praveen Kumar delve into the intricate relationship between work-life balance, job satisfaction, and employee retention. Through a comprehensive exploration of past literature and sector-specific sub-factors, the chapter uncovers the nuanced interplay between these essential dimensions of organizational dynamics.

With this broad spectrum of chapters, *Climate Change Management and Social Innovations for Sustainable Global Organization* offers a wealth of perspectives, strategies, and insights to empower organizations, researchers, policymakers, and individuals in their pursuit of sustainability in a rapidly changing world.

Chapter 16: The Mediating Role of Job Satisfaction Between Work-Life Balance and Employee Retention

In today's fast-paced world, achieving a balance between work and personal life has become a critical concern. This chapter delves into the extensive research conducted on work-life balance, revealing its multifaceted significance beyond mere temporal adjustments. The authors assert that work-life balance not only enhances employee effectiveness and retention but also reshapes overall satisfaction, both within the professional realm and in personal life.

The primary focus of this chapter revolves around the intricate relationship between work-life balance and employee retention, with a specific emphasis on the mediating role played by job satisfaction. Through empirical analysis and review of literature, the authors unveil how work-life balance significantly influences job satisfaction, further impacting the propensity of employees to remain within an organization.

IN SUMMARY

As we bring *Climate Change Management and Social Innovations for Sustainable Global Organization* to a close, we find ourselves at the intersection of knowledge, innovation, and action. Throughout this reference book, our contributors have meticulously explored the myriad challenges and opportunities that climate change presents to organizations worldwide. From the intricate dynamics of Green HRM to the transformative power of sustainable technologies, these chapters have offered a panoramic view of the evolving landscape of sustainability.

The journey we embarked upon through these pages has allowed us to traverse diverse terrains, from the boardrooms of corporations to the fertile fields of agricultural practices. We've delved into the mediating role of job satisfaction, the profound impacts of green marketing, and the complex relationship between economic growth and CO2 emissions. Each chapter has provided a unique lens through which to examine the intricate web of interactions that underpin the global efforts towards sustainable development.

In the ever-changing tapestry of our world, the relevance of climate change management and social innovations cannot be understated. Our contributors have skillfully elucidated the strategies, challenges, and potentials that organizations encounter as they navigate this complex terrain. From harnessing green technologies to embracing green supply chains, the chapters have illuminated the diverse paths toward a more sustainable future.

Furthermore, the comprehensive exploration of climate change's impact on various sectors and aspects of society reflects our commitment to addressing this pressing challenge from a holistic perspective. The fusion of academia, research, and practical insights woven throughout these pages showcases the collective effort required to drive meaningful change.

As editors, we are deeply grateful to our contributors for their dedication and expertise in shaping this compilation. We are equally appreciative of our readers, whose engagement with this work demonstrates a shared commitment to fostering sustainable practices and embracing innovation. The transformative journey towards a sustainable global organization demands the collective dedication of scholars, researchers, policymakers, and practitioners alike. With this book, we hope to spark conversations, inspire solutions, and propel the mission of sustainability forward.

In the face of climate change, every step towards a sustainable future matters. As we reflect on the myriad perspectives presented in this reference book, let us remember that the choices we make today will resonate through generations to come. The journey continues, and as we step into this shared endeavor, may we walk hand in hand towards a world of resilience, innovation, and lasting change.

Thank you for joining us on this enlightening voyage.

Sumita Dave
Amity University, Raipur, India

Monica Sainy
Amity University, Raipur, India

Disha Sharma
Amity University, Raipur, India

Anupam Singh
University of Stirling, UK

Chapter 1
Understanding the Role of Green HRM as a Contributor Towards Sustainable Global Organization

Anushka Lydia Issac
https://orcid.org/0000-0002-1493-1047
Westford University College, UAE

ABSTRACT

This chapter examines the role of green human resource management (HRM) in promoting sustainable global organisations. The chapter adopts a case study methodology to explore how organizations like McDonald's, Interface, Unilever, Patagonia, Toyota, and the Co-Operative Group have successfully integrated green HRM practices into their operations. The chapter also identifies critical issues surrounding green HRM, such as the lack of standardisation, the need for employee buy-in, and the potential for green-washing. To address these issues, the chapter proposes solutions such as developing industry-wide standards, creating a culture of environmental responsibility, and adopting transparent reporting practices. It is hoped that this chapter will contribute to a deeper understanding of the role of green HRM in promoting sustainable practices in organisations. Overall, this chapter aims to contribute to the ongoing discourse on sustainability and the role of HRM in promoting sustainable practices in organisations.

1. INTRODUCTION

In recent years, there has been growing recognition of the need for organizations to address the challenges of climate change and adopt sustainable practices. As a result, the concept of Green Human Resource Management (HRM) has gained importance as a way to promote sustainability within organizations. Green HRM involves integrating environmental sustainability into HRM practices, such as recruitment, training, performance management, and employee engagement. Green HRM is important and relevant

DOI: 10.4018/978-1-6684-9503-2.ch001

in today's time due to the urgent need for organizations to address climate change and adopt sustainable practices.

Climate change and its impact on business have emerged as one of the most pressing concerns of our times. The adverse effects of climate change have far-reaching implications for the survival of businesses and the global economy. As organizations grapple with the challenge of mitigating the impact of climate change, Green HRM has emerged as a critical tool in promoting sustainability and positive impact. This has led to growing concerns among stakeholders, including employees, customers, investors, and regulators, about the environmental impact of organizations.

Green HRM can help organizations address these concerns by promoting sustainability in all aspects of their operations. By incorporating environmental sustainability into HRM practices, organizations can reduce their carbon footprint, conserve resources, and promote social responsibility. Green HRM practices such as employee training, performance management, and sustainability teams can help organizations achieve their sustainability goals while also enhancing their reputation as socially responsible entities. In today's competitive business environment, organizations that adopt Green HRM practices are more likely to attract and retain employees, customers, and investors who are committed to sustainability. Green HRM can also help organizations reduce costs, increase efficiency, and promote innovation by encouraging employees to identify and implement sustainability initiatives.

In this chapter, we will explore the concept of Green HRM and its importance and relevance in promoting sustainability within organizations. We will draw on academic journals, scholarly work, and corporate reports to provide insights into the role of Green HRM in addressing the challenges of climate change and promoting social innovations for sustainable global organizations. Moreover, this chapter will explore the challenges and opportunities that organizations face in implementing Green HRM practices. While there is growing awareness of the importance of Green HRM, many organizations struggle to integrate environmental sustainability into their HRM practices. This may be due to a lack of understanding, resources, or organizational support. This book chapter aims to explore the role of Green HRM in promoting sustainable global organizations. By examining the theoretical foundations of Green HRM and its practical applications, this chapter seeks to provide insights into the potential of Green HRM to drive positive change. Through an analysis of case studies, this chapter also highlights successful examples of Green HRM in action and the impact it has had on promoting sustainability in organizations. The chapter concludes by emphasizing the importance of Green HRM as a critical component of organizational strategy in the context of climate change and the need to develop innovative and effective practices to achieve a net-zero future.

In recent years, sustainability has emerged as a critical concern for organizations worldwide. With the growing awareness of the environmental and social impact of business operations, firms are increasingly recognizing the need to adopt sustainable practices that promote environmental protection, social responsibility, and economic development. In this context, Human Resource Management (HRM) has emerged as a key area of interest for scholars and practitioners seeking to understand how organizations can achieve sustainable development.

Green HRM, also known as sustainable HRM, is a new and rapidly evolving concept that focuses on integrating environmental sustainability into HRM practices. Green HRM involves the development and implementation of policies, practices, and strategies that promote environmentally sustainable practices, reduce carbon emissions, and enhance social responsibility. The concept is gaining importance in the context of global organizations as a means of achieving long-term sustainability goals while also contributing to business success.

The purpose of this chapter is to explore the role of Green HRM as a contributor towards sustainable global organizations. The chapter will begin by providing an overview of the concept of Green HRM, its importance, and its relevance in the context of global organizations. It will then examine the key dimensions of Green HRM, including recruitment, training and development, performance management, and employee engagement. The chapter will also discuss the challenges and opportunities associated with the adoption of Green HRM practices, as well as the implications for organizational performance and sustainability. Overall, this chapter seeks to contribute to the growing body of literature on Green HRM and its role in promoting sustainability in global organizations. By examining the key dimensions and challenges of Green HRM, the chapter will provide insights into how organizations can effectively integrate environmental sustainability into their HRM practices and contribute towards sustainable development.

2. UNDERSTANDING GREEN HUMAN RESOURCE MANAGEMENT

2.1 Definition of Green HRM and Its Key Dimensions

Green Human Resource Management (HRM) is a concept that emerged in the late 1990s, as organizations began to recognize the importance of sustainability and environmental protection in their operations. Green HRM refers to the integration of environmental sustainability into HRM practices, with the aim of promoting sustainability and reducing the carbon footprint of organizations. In this critical report, Green HRM will be defined, its key dimensions examined, and insights provided into its importance and implications for organizational performance and sustainability. Green HRM is a relatively new and rapidly evolving concept, and there is no universally accepted definition of the term. However, scholars have provided various definitions of Green HRM. According to (Renwick et al., 2013), Green HRM involves the integration of environmental concerns into HRM practices, such as recruitment, selection, training and development, performance management, and compensation. Similarly, (Kramar & Syed, 2012) define Green HRM as "a set of policies, practices, and systems that are designed to facilitate the development of an environmentally sustainable workforce and organizational culture."

2.2 Key Dimensions of Green HRM

Green HRM involves several key dimensions, including recruitment, training and development, performance management, and employee engagement. In terms of recruitment, organizations can adopt Green HRM practices by incorporating environmental sustainability into their job descriptions and selection criteria (Renwick et al., 2013). For example, companies can require job applicants to have knowledge of sustainability issues or experience in environmental management. In terms of training and development, organizations can promote environmental awareness and sustainability through training programs and workshops for employees (Kramar & Syed, 2012). Such training programs can help employees understand the importance of sustainability and how they can contribute towards environmental protection through their work. Performance management is another key dimension of Green HRM. Organizations can incorporate environmental sustainability into their performance evaluation criteria and reward employees who demonstrate a commitment to sustainability (Renwick et al., 2013). This can help promote a culture of sustainability within the organization and encourage employees to adopt sustainable practices in their work. Employee engagement is also an essential dimension of Green HRM. Organizations can involve employees in sustainability initiatives, such as recycling programs or carbon reduction initiatives, and provide opportunities for employees to participate in decision-making processes related to environmental sustainability (Kramar & Syed, 2012). This can help create a sense of ownership and commitment among employees towards sustainability.

2.3 An Overview of Green HRM and Its History in Brief

Green Human Resource Management (Green HRM) is a relatively new concept that has emerged as a response to the increasing environmental challenges and the need for sustainability in organizational practices. Green HRM involves integrating environmental concerns and sustainability practices into the traditional HRM practices such as recruitment, selection, training, performance management, and compensation. This approach is intended to align the HRM function with the organization's sustainability goals and promote sustainable business practices. The origins of Green HRM can be traced back to the 1970s and 1980s when the environmental movement gained momentum, and companies began to adopt environmentally friendly practices. However, the integration of environmental concerns into HRM practices did not gain significant attention until the late 1990s and early 2000s, when sustainability emerged as a key organizational issue. The concept of Green HRM has been extensively discussed in academic literature, with numerous studies focusing on the integration of environmental sustainability into HRM practices. Scholars have identified a range of factors that have led to the development of Green HRM, including increasing awareness of environmental issues, changes in societal values, and the emergence of environmental regulations. Despite the growing attention and interest in Green HRM, it remains a relatively new and underdeveloped field. While there have been some successful examples of Green HRM practices in organizations, there is still much to be done to fully integrate environmental concerns into HRM practices. Additionally, the effectiveness of Green HRM practices in promoting sustainability and organizational performance is still under debate. Green HRM has emerged as a promising approach to address environmental challenges and promote sustainability in organizations. However, it requires significant attention, resources, and commitment from organizations to fully integrate environmental concerns into HRM practices and to reap the benefits of a sustainable future.

2.4 Green HRM in the Contemporary Era

Green HRM, also known as environmental or sustainable HRM, has become an increasingly important aspect of human resource management in the contemporary era. The focus on sustainability and environmental responsibility has brought attention to the need for organizations to adopt environmentally friendly practices, and Green HRM plays a crucial role in achieving this goal. Green HRM practices encompass a wide range of activities, including recruitment and selection, training and development, compensation and rewards, and employee engagement. Organizations that adopt Green HRM practices aim to reduce their carbon footprint, promote eco-friendliness, and support sustainable development. The contemporary era has seen a significant shift in the global business environment, with sustainability and environmental responsibility becoming more prominent concerns. Many organizations have realized that their operations have a significant impact on the environment, and that they have a responsibility to mitigate their negative effects. Green HRM has emerged as an effective way for organizations to manage their environmental impact while also promoting sustainable development. The adoption of Green HRM practices has numerous benefits for organizations. It can improve their reputation and brand image, which can attract customers who prioritize environmental responsibility. It can also improve employee engagement and motivation, as employees are more likely to be committed to an organization that shares their values. Additionally, Green HRM practices can lead to cost savings through more efficient resource usage and reduction in waste. However, the adoption of Green HRM practices also presents challenges for organizations. It requires a significant shift in organizational culture, as well as investments in new technologies and training programs. Additionally, organizations must ensure that their Green HRM practices are consistent with their overall business strategy, as well as local laws and regulations. Green HRM has become an important aspect of human resource management in the contemporary era. Organizations that adopt Green HRM practices can reap numerous benefits, including improved reputation, employee engagement, and cost savings. However, the adoption of Green HRM practices also presents challenges, and organizations must carefully consider their implications before implementing them.

2.5 Importance and Implications of Green HRM

Green HRM is becoming increasingly important in the context of global organizations, as companies are recognizing the need to adopt sustainable practices and reduce their carbon footprint. According to a report by the (United Nations Global Impact, 2013), "the integration of environmental, social, and governance (ESG) issues into core business practices is essential to long-term value creation." Green HRM can help organizations achieve this goal by promoting sustainability and reducing their environmental impact. Moreover, Green HRM can have several implications for organizational performance and sustainability. Research has shown that the adoption of Green HRM practices can improve employee motivation, job satisfaction, and organizational commitment (Renwick et al., 2013). This can lead to increased productivity and improved organizational performance. Additionally, Green HRM practices can help reduce the environmental impact of organizations, promote social responsibility, and enhance the organization's reputation as a socially responsible entity (Kramar & Syed, 2012).

However, the adoption of Green HRM practices also poses several challenges for organizations. For example, the implementation of Green HRM practices may require significant investments in training, development, and infrastructure, which can impact the financial performance of the organization. Moreover, Green HRM practices may also face resistance from employees who may perceive them as an additional burden or disruption to their daily work routine. The success of Green HRM practices, therefore, requires effective communication, leadership, and change management skills to ensure buy-in from all stakeholders.

While Green HRM presents significant opportunities for organizations to promote sustainability and positive impact, it also poses several challenges that require careful consideration and planning. The adoption of Green HRM practices requires a holistic approach that integrates social, environmental, and economic objectives and considers the unique context and culture of the organization.

2.6 Critiquing Green HRM: Challenges Involved

Green HRM, as a relatively new field, faces several challenges that hinder its effectiveness in promoting sustainable practices within organizations. These challenges can be categorized into three main areas: institutional, managerial, and employee-related challenges. The first challenge relates to institutional barriers such as a lack of government policies or regulations that promote green practices. In many countries, there are no laws that require organizations to implement environmental policies or reduce their carbon footprint. This lack of regulatory support can make it difficult for organizations to prioritize green initiatives, including those related to HRM. The second challenge is managerial in nature, where managers and HR professionals may not have the necessary knowledge, skills, or resources to effectively implement green practices. This may lead to a lack of commitment or investment in green HRM initiatives, resulting in limited progress towards sustainability goals. Finally, employee-related challenges can occur when employees do not fully embrace green practices or lack awareness and understanding of their importance. Resistance to change, lack of training or awareness programs, and misaligned incentives can all hinder the successful implementation of green HRM initiatives. In addition to these challenges, there is also a potential for green HRM to become superficial or tokenistic, with organizations implementing symbolic gestures without addressing underlying environmental issues. This can lead to a lack of credibility and trust among stakeholders, ultimately hindering progress towards sustainability. To address these challenges, organizations must prioritize the integration of green HRM into their overall business strategy, providing sufficient resources and support for HR professionals to implement and maintain green initiatives. Organizations should also foster a culture of sustainability, encouraging employee engagement and participation in green practices. This can be achieved through training, education, and communication initiatives that promote environmental awareness and understanding. While green HRM presents a promising approach to promoting sustainability within organizations, there are several challenges that need to be addressed to maximize its potential. It requires a holistic approach that involves institutional support, managerial commitment, and employee engagement to overcome these challenges and drive real change towards a sustainable future.

3. RESEARCH DESIGN

3.1 Current Gap in Literature

Despite the increasing attention given to Green HRM, there is still a significant gap in the literature related to its practical implementation and effectiveness in achieving sustainable outcomes. While there are many theoretical models and conceptual frameworks available, there is a lack of empirical research on the actual implementation of Green HRM practices and their impact on environmental sustainability and organizational performance. Additionally, there is a lack of standardization and measurement tools, which makes it difficult to compare and evaluate the effectiveness of different Green HRM practices. Another significant gap in the literature is the limited attention given to the role of employees in driving and implementing Green HRM practices. While there is some research on the importance of employee engagement and participation in environmental initiatives, there is a need for more empirical research on the mechanisms that drive employee participation and how it impacts organizational outcomes. Furthermore, there is a lack of research on the challenges faced by organizations in implementing Green HRM practices, particularly in developing countries. The literature tends to focus on developed countries, where Green HRM practices are more prevalent and easier to implement due to the availability of resources and infrastructure. However, there is a need to understand the challenges faced by organizations in developing countries, where resources and infrastructure may be limited, and the cultural context may be different. Overall, there is a need for more empirical research on the practical implementation and effectiveness of Green HRM practices, the role of employees in driving sustainability outcomes, and the challenges faced by organizations in implementing these practices, particularly in developing countries. Addressing these gaps can provide valuable insights for organizations and policymakers looking to promote sustainable practices and improve environmental outcomes.

3.2 Methodology

Case study research method is used in this chapter to explore the role of Green HRM as a contributor towards sustainable global organization. Case study research is commonly used in social sciences to investigate complex phenomena, such as organizational behavior, with a holistic approach (Yin, 2018). It allows the researcher to analyze real-life situations and practices in their natural settings (Stake, 2006). The use of case study method enables us to study Green HRM practices and their contribution towards sustainability in different organizations. In this chapter, six organizations – McDonald's, Interface, Unilever, Patagonia, Toyota, and the Co-operative Group – are selected as case studies. These organizations were chosen based on their notable efforts towards implementing Green HRM practices and achieving sustainability. The benefits of using the case study method in this chapter are many. Firstly, it provides a comprehensive and in-depth understanding of the relationship between Green HRM practices and sustainability. Secondly, it enables us to examine the practical application of Green HRM practices in real-world situations. Thirdly, it helps to identify the challenges and barriers that organizations face when implementing Green HRM practices. Fourthly, it allows for the comparison of Green HRM practices across different organizations and industries. The key impact that this topic could bring about is a better understanding of the role of Green HRM in promoting sustainability in organizations. This can help organizations to implement effective Green HRM practices that contribute towards their sustainability goals. Additionally, this chapter can serve as a guide for organizations that are seeking to integrate Green HRM practices into their operations. To conduct the case study research, data will be collected through various sources, such as published reports, company websites, academic articles, and interviews with key personnel. The collected data will be analyzed using a qualitative approach to identify the common themes and patterns in Green HRM practices and their contribution towards sustainability. The limitations and challenges of the case study method will also be acknowledged and addressed.

4. GREEN HRM AND SUSTAINABILITY: A GLOBAL PERSPECTIVE

4.1 The Role of Green HRM as a Contributing Factor Towards a Sustainable Global Organization

Green HRM plays a crucial role in promoting sustainability and contributing to a sustainable global organization. By integrating sustainable practices into human resource management, organizations can reduce their carbon footprint, minimize waste, and promote sustainable behaviors among employees. For instance, Green HRM practices such as eco-friendly recruitment, training, and development can help organizations attract and retain employees who share the organization's sustainability values and ensure that sustainability is integrated into the organization's culture.

Furthermore, Green HRM can also contribute to the development of sustainable leaders who can effectively drive change and inspire others to adopt sustainable practices. Sustainable leaders are individuals who possess a deep understanding of the social, economic, and environmental issues facing the organization and can lead by example to promote sustainability. In addition, Green HRM can also help organizations develop a sustainable supply chain by partnering with suppliers who share the same sustainability values and promote sustainable practices in their operations. By doing so, organizations can reduce their environmental impact, improve their reputation, and contribute to positive social and environmental outcomes. However, the successful implementation of Green HRM practices requires a strategic and holistic approach that integrates social, environmental, and economic objectives. The adoption of Green HRM practices may require significant investments in training, development, and infrastructure, which can impact the financial performance of the organization. Moreover, effective communication, leadership, and change management skills are critical to ensure buy-in from all stakeholders. Green HRM plays a crucial role in promoting sustainability and contributing to a sustainable global organization. However, the adoption of Green HRM practices requires careful planning and a strategic approach that considers the unique context and culture of the organization. By adopting Green HRM practices, organizations can contribute to a sustainable future while also achieving their business objectives.

4.2 The Impact of Green HRM to Promote Positive Impact in Global Organizations

Green HRM has the potential to promote positive impact in global organizations by integrating sustainable practices into human resource management processes and policies. Through the implementation of Green HRM practices, organizations can reduce their environmental impact, promote social responsibility, and improve their overall sustainability performance. This can lead to a range of benefits, such as increased employee engagement and productivity, reduced costs, improved reputation with stakeholders, and compliance with regulatory requirements. Several studies have shown the positive impact of Green HRM practices on organizational performance. For instance, a study by (Thite & Russell, 2010) found that organizations that implement Green HRM practices have higher levels of environmental performance and financial performance compared to those that do not. Additionally, a study by (Renwick et al., 2013) found that Green HRM practices can lead to reduced energy consumption, waste generation, and greenhouse gas emissions. Furthermore, Green HRM can also promote positive impact in global organizations by fostering a culture of sustainability and social responsibility. By integrating sustainability into human resource management processes and policies, organizations can communicate their commitment to sustainability to employees, stakeholders, and the wider community. This can lead to increased trust, loyalty, and support from stakeholders, as well as enhanced reputation and competitive advantage in the market. However, while Green HRM has the potential to promote positive impact in global organizations, there are challenges to its implementation. One challenge is the lack of awareness and understanding of the importance of sustainability and Green HRM practices among employees and managers. Another challenge is the need for a shift in organizational culture and values towards sustainability and social responsibility, which requires a commitment from top management and a willingness to invest in the necessary resources, such as training and technology, to support Green HRM practices. Thus, the impact of Green HRM to promote positive impact in global organizations is significant. By integrating sustainable practices into human resource management processes and policies, organizations can reduce their environmental impact, promote social responsibility, and improve their overall sustainability performance. However, to realize the full potential of Green HRM, organizations must overcome the challenges of implementation and foster a culture of sustainability and social responsibility.

4.3 Roles of Green HRM in Promoting Sustainability

Green HRM can play a vital role in promoting sustainability by incorporating environmental sustainability into recruitment practices. According to (Kramar & Syed, 2012), companies can require job applicants to have knowledge of sustainability issues or experience in environmental management. This can help create a workforce that is committed to sustainability and capable of contributing towards environmental protection through their work. Green HRM can also promote sustainability through training and development programs for employees. These programs can help employees understand the importance of sustainability and how they can contribute towards environmental protection through their work (Kramar & Syed, 2012). According to a report by (Deloitte, 2017), companies that invest in employee training and development in sustainability are more likely to achieve their sustainability goals. Green HRM can promote sustainability by incorporating environmental sustainability into performance evaluation criteria and rewarding employees who demonstrate a commitment to sustainability (Renwick et al., 2013). This can help create a culture of sustainability within the organization and encourage employees to adopt sustainable practices in their work. Green HRM can promote sustainability by involving employees in sustainability initiatives and providing opportunities for employees to participate in decision-making processes related to environmental sustainability ((Kramar & Syed, 2012). This can help create a sense of ownership and commitment among employees towards sustainability.

4.4 Implications of Green HRM for Organizational Performance and Sustainability

Green HRM has significant implications for organizational performance and sustainability. Through implementing green HRM practices, organizations can reduce their environmental footprint, enhance their reputation as socially responsible entities, and attract and retain environmentally conscious employees. Studies have shown that green HRM practices can lead to increased employee satisfaction and motivation, which in turn can lead to improved performance and productivity. For example, a study conducted by Saeed and Azmi (2017) found that green HRM practices positively influenced employee job satisfaction and organizational commitment, which in turn resulted in improved organizational performance. Moreover, green HRM practices can contribute to the achievement of sustainability goals. By incorporating sustainability into their HR practices, organizations can promote a culture of environmental responsibility and encourage sustainable behaviors among employees. This can lead to reductions in resource consumption, waste generation, and greenhouse gas emissions, contributing to a more sustainable future. However, implementing green HRM practices may also pose challenges for organizations. For example, it may require additional resources and investment in employee training and development. It may also require changes in organizational culture and employee attitudes towards environmental sustainability. Research has shown that the adoption of Green HRM practices can improve employee motivation, job satisfaction, and organizational commitment (Renwick et al., 2013). This can lead to increased productivity and improved organizational performance. Green HRM practices can help reduce the environmental impact of organizations by promoting sustainable practices and reducing the carbon footprint of organizations (Kramar & Syed, 2012). Green HRM practices can help promote social responsibility by creating a culture of sustainability within the organization and encouraging employees to adopt sustainable practices in their work (Deloitte, 2017). Green HRM practices can enhance the organization's reputation as a socially responsible entity by demonstrating the organization's commitment towards environmental sustainability and social responsibility (Kramar and Syed, 2012). While the adoption of green HRM practices may present some challenges, the benefits for organizational performance and sustainability make it a worthwhile endeavor. Organizations that prioritize green HRM practices can reap the benefits of improved performance, employee satisfaction, and a more sustainable future.

5. CASE STUDIES

5.1 Critical Case Studies Evidencing Green HRM as a Contributor to Sustainability in Organizations

Green Human Resource Management (HRM) is increasingly being recognized as a key factor in promoting sustainability within organizations. The concept of Green HRM involves integrating environmental sustainability into HRM practices, such as recruitment, training, performance management, and employee engagement. In this report, critical and successful case studies of Green HRM as a contributor to sustainability in organizations shall be analyzed.

5.1.1 Critical Case Study: McDonald's

One critical case study of Green HRM in promoting sustainability is McDonald's. In 2011, McDonald's launched its "Best of Green" program, which aimed to reduce the company's carbon footprint and promote sustainability in all aspects of its operations (McDonald, 2012). The program included several Green HRM practices, such as employee engagement, training, and performance management. As part of the program, McDonald's implemented a training program for all employees, which included modules on environmental sustainability and waste reduction. The company also developed a Green Team, consisting of employees who were trained to identify and implement sustainability initiatives within the organization. Additionally, McDonald's incorporated environmental sustainability into its performance management system, with employees being evaluated based on their contribution towards environmental sustainability. The "Best of Green" program was successful in reducing McDonald's carbon footprint by 7% and achieving a 30% reduction in water usage (McDonald, 2012). The program also led to increased employee engagement and commitment towards sustainability, with the Green Team implementing several sustainability initiatives across the organization.

5.1.2 Successful Case Study: Interface

One successful case study of Green HRM in promoting sustainability is Interface, a global flooring company. Interface has been recognized for its innovative approach to sustainability, with its Mission Zero program aiming to achieve zero negative impact on the environment by 2020 (Interface, 2019). The program includes several Green HRM practices, such as employee engagement, training, and performance management. Interface implemented a training program for all employees, which included modules on environmental sustainability and the company's Mission Zero program. The company also developed a Green Team, consisting of employees who were trained to identify and implement sustainability initiatives within the organization. Additionally, Interface incorporated environmental sustainability into its performance management system, with employees being evaluated based on their contribution towards environmental sustainability. The Mission Zero program was successful in reducing Interface's carbon footprint by 96% and achieving a 91% reduction in waste sent to landfill (Interface, 2019). The program also led to increased employee engagement and commitment towards sustainability, with the Green Team implementing several sustainability initiatives across the organization.

5.1.3 The Case of Unilever

Unilever is a multinational consumer goods company that has been implementing various Green HRM practices to contribute to sustainability in their organization. The global consumer goods' company has implemented various green HRM practices to promote sustainability. The company has set ambitious targets to reduce its environmental impact, and its HR policies have played a critical role in achieving these targets. One of the most significant initiatives they have taken is the Unilever Sustainable Living Plan, which includes a series of ambitious targets aimed at reducing the company's environmental footprint while increasing social impact. The plan includes goals such as reducing greenhouse gas emissions, improving water management, and enhancing the livelihoods of millions of people in their supply chain. To achieve these goals, Unilever has implemented various Green HRM practices, including eco-friendly workplace policies, employee engagement programs, and sustainability training. For instance, Unilever's eco-friendly workplace policies include initiatives such as reducing paper usage, promoting energy-efficient practices, and encouraging sustainable transportation options. These policies have not only reduced the company's environmental impact but have also contributed to a more engaged and motivated workforce. Additionally, Unilever has implemented various employee engagement programs to encourage employees to participate in sustainability initiatives actively. For example, they launched a 'Sustainable Living Ambassador' program, where employees volunteer to promote sustainable living in their communities. This program has not only contributed to Unilever's sustainability goals but also helped to create a sense of purpose and pride among employees.

Unilever also provides sustainability training to its employees to enhance their knowledge and skills on environmental and social issues. The company has developed a comprehensive sustainability training program called the 'Sustainable Living Plan Academy.' This program provides employees with the knowledge and skills needed to integrate sustainability into their work practices and make a positive impact on the environment and society. Unilever's Green HRM practices have contributed significantly to the company's sustainability efforts. By implementing eco-friendly workplace policies, employee engagement programs, and sustainability training, Unilever has not only reduced its environmental impact but also contributed to a more engaged and motivated workforce. The company's commitment to sustainability has not only benefited the environment and society but has also helped to improve their brand reputation and financial performance.

5.1.4 The Case of Patagonia

Patagonia is a well-known outdoor apparel company that has been recognized for its sustainability efforts, including its use of Green HRM practices. One of the ways Patagonia has demonstrated its commitment to sustainability is through its "Footprint Chronicles," which is a comprehensive online platform that provides transparency into the company's supply chain and environmental impact. This platform includes information about the materials used in their products, the factories where the products are produced, and the transportation methods used to distribute the products. In terms of Green HRM practices, Patagonia has implemented several initiatives to reduce its environmental impact, such as offering employees incentives for using public transportation, reducing travel-related emissions by using video conferencing, and implementing a waste reduction program. Additionally, the company offers an on-site organic cafe for employees and a wellness program that encourages exercise and healthy living. Patagonia's sustainability efforts are not only beneficial for the environment but also for its business performance. The company's CEO, Rose Marcario, has stated that "sustainability is not only the right thing to do, but also good for business." Patagonia's sustainable practices have helped the company attract and retain customers who value sustainability and have also helped to reduce costs by decreasing waste and improving efficiency. Overall, Patagonia's use of Green HRM practices has been instrumental in contributing to the company's sustainability and its success as a business. By implementing initiatives to reduce its environmental impact and offering incentives for employees to do the same, Patagonia has demonstrated that it is possible for companies to be profitable while also being environmentally responsible.

5.1.5 The Case of Toyota

Toyota, the multinational automobile manufacturer, has been recognized for its commitment to sustainability and green HRM practices. The company has implemented several initiatives to reduce its carbon footprint and improve environmental performance. One of the most notable initiatives is the Toyota Environmental Challenge 2050, which aims to reduce carbon emissions and achieve zero carbon dioxide emissions from Toyota vehicles and operations by 2050. Green HRM practices are also evident in Toyota's employee engagement and development programs. The company has implemented training programs for employees to raise awareness about environmental issues and encourage the adoption of sustainable practices. Toyota also encourages its employees to develop innovative solutions that reduce the environmental impact of its products and operations. This approach has resulted in the creation of new hybrid and electric vehicles, such as the Toyota Prius, which has become a symbol of sustainable transportation. Moreover, Toyota has also implemented sustainable supply chain practices, such as reducing packaging waste, promoting sustainable forestry practices, and encouraging suppliers to adopt sustainable practices. This has not only helped Toyota reduce its environmental impact but has also created a ripple effect throughout its supply chain, encouraging other companies to adopt sustainable practices. Overall, Toyota's green HRM practices have contributed significantly to the company's sustainability efforts, making it a role model for other organizations in the automobile industry and beyond.

5.1.6 The Co-Operative Group

The Co-operative Group is a UK-based consumer co-operative, which is owned and controlled by its members. The organization operates in various sectors, including food retail, funeral care, legal services, insurance, and energy. The Co-operative Group has been recognized for its commitment to sustainability, and its Green HRM practices play a vital role in this regard. One of the Co-operative Group's flagship initiatives is its Ethical Plan, which sets out a comprehensive sustainability strategy. The plan includes a range of ambitious targets, such as sourcing 100% of the palm oil used in its products sustainably by 2015 and achieving carbon neutrality by 2017. The Co-operative Group's Green HRM practices are critical in achieving these targets. For example, the organization has implemented a Green Rewards scheme, which incentivizes employees to adopt more sustainable behaviors. Through the scheme, employees can earn rewards for actions such as using public transport, cycling to work, and reducing energy consumption at home. The Co-operative Group also recognizes the importance of sustainability in its supply chain and has implemented various initiatives to promote sustainability among its suppliers. For instance, the organization's Ethical Trading Initiative (ETI) Base Code sets out minimum labor standards that suppliers must meet, and the Co-operative Group works closely with suppliers to help them improve their sustainability practices. The organization also encourages its suppliers to adopt more sustainable practices, such as reducing their carbon footprint and promoting sustainable agriculture. The Co-operative Group's Green HRM practices are essential in achieving the organization's sustainability goals. The Green Rewards scheme incentivizes employees to adopt more sustainable behaviors, and the organization works closely with its suppliers to promote sustainability throughout its supply chain.

6. CONCLUSION

In conclusion, this chapter has highlighted the important role that Green Human Resource Management (HRM) plays in contributing towards sustainable global organizations. The chapter has explored the concept of Green HRM and its implications for organizational sustainability, employee engagement, and social responsibility. Through a review of academic literature and case studies of successful organizations, it is evident that Green HRM practices can significantly reduce the environmental impact of organizations while promoting employee engagement and commitment towards sustainability. Green HRM practices such as employee training, performance management, and sustainability teams can help organizations achieve their sustainability goals while also enhancing their reputation as socially responsible entities. As organizations continue to face increasing pressure to operate sustainably and responsibly, the role of Green HRM in promoting sustainability will become increasingly critical. It is essential for organizations to adopt Green HRM practices and incorporate environmental sustainability into their HRM practices to ensure that they are contributing towards sustainable global organizations. Overall, this chapter highlights the need for further research and action in the area of Green HRM to support the transition towards a more sustainable future.

REFERENCES

Deloitte. (2017). *Rewriting the rules for the digital age 2017 Deloitte Global Human Capital Trends*. www.deloitte.com

Interface. (2019). *2019 Annual Report*. Author.

Kramar, R., & Syed, J. (2012). *Human Resource Management in a Global Context: A Critical Approach Making critical sense of human resource management in a globalized world*. https://www.researchgate.net/publication/295867671

McDonald. (2012). *2012 Annual Report*. Author.

Renwick, D. W. S., Redman, T., & Maguire, S. (2013). Green Human Resource Management: A Review and Research Agenda. *International Journal of Management Reviews*, *15*(1), 1–14. doi:10.1111/j.1468-2370.2011.00328.x

Thite, M., & Russell, B. (2010). Work organization, human resource practices and employee retention in Indian call centers. *Asia Pacific Journal of Human Resources, 48*(3), 356–374. doi:10.1177/1038411110381623

United Nations Global Impact. (2013). *Global Corporate Sustainability Report 2013 2*. www.unglobalcompact.org

Chapter 2
How Do Equity Market Performance and ICT Influence Clean Energy Consumption in High-Income Countries?

Pınar Avci
https://orcid.org/0000-0001-9480-8016
Tekirdağ Namık Kemal University, Turkey

ABSTRACT

The aim of this study is to examine whether equity market development and ICT affect clean energy consumption in 18 high-income countries during the period 1990-2015. In the analysis, firstly, the stationarity properties of the variables are tested with the CADF and CIPS tests. In the second stage, cointegration between the variables in the model is investigated by Westerlund, Kao, and Pedroni methods. In the next step, the long-run elasticities of the variables are estimated using the PCSE and FGLS methods. The causality between the variables is also determined by applying the Dumitrescu-Hurlin panel causality method. According to findings, cointegration between clean energy consumption and equity market development, ICT, economic growth, and CO2 emissions variables, which integrated at I(1) level, is proven. The long-term findings showed that the equity market, ICT, and economic growth encourage clean energy consumption of 18 high-income countries, and CO2 emissions also prevent them. In addition, there is bidirectional causality between clean energy consumption and other variables.

INTRODUCTION

Scientists agree that carbon dioxide emissions causing global warming are brought about by the use of fossil fuels such as coal, oil and gas as energy sources. Reducing the use of fossil fuel energy is necessary to mitigate the effects of global warming and climate change, and this is one of the most urgent demands of the international community to protect the global environment (Chang et al., 2020). Therefore, in recent years, many researchers and policy makers have focused on the use of renewable energy and its

DOI: 10.4018/978-1-6684-9503-2.ch002

subunits of clean/green energy rather than traditional energy sources (such as fossil resources) in the production process. Clean energy has many advantages over fossil fuels, meeting the increasing energy demand and reducing carbon dioxide emissions. International organizations are pushing countries by focusing on the use of clean energy at the Kyoto Protocol Summit in 1997. Developed and developing countries also promote clean energy production and use (Paramati et al., 2016). In this context, it has been determined that clean energy sources are needed in the energy system in order to reduce the carbon dioxide emission level in the atmosphere from past to present (Alam et al., 2021).

Equity market development plays an important role in the entire energy system in general, especially in the use of clean energy (Alam et al., 2021). In addition, equity market developments enable to create a healthy business environment for both companies and investors. This is because the equity market provides additional funds for listed companies, which offers alternative investment opportunities for both national and international investors. Equity market developments lead to increase energy demand by increasing economic activities in countries (Paramati et al., 2016). According to some researchers in the literature, equity market development has been found to reduce carbon dioxide emissions (Paramati et al., 2017b; Habiba et al., 2021; Younis et al., 2021) while according to some, equity market development has increased clean energy consumption (Paramati et al., 2016; Alam et al., 2021). At the same time, the clean/renewable energy consumption in 28 EU countries is positively affected by the banking sector and bond market development for the period of 1990–2015 but it is not affected by the stock market. In addition, the developments in the capital market have an impact on clean energy consumption, some studies (Talan et al., 2022; Yu et al., 2023) investigate whether information, communication and technology (ICT) have an impact on clean energy consumption. ICT is seen as a fundamental tool that contributes to easing human life, helping economic development and improving the energy efficiency of any country equally (Bano et al., 2022).

In this context, the rapid developments in energy efficiency and clean energy infrastructures of high-income countries (Taşbaşı, 2022) and the development of ICT and equity market for clean energy consumption cause high-income countries to be accepted as a case in this study. Since the 1990s, large ICT investments of high-income countries have become important (Öztürk, 2002). Likewise, stock markets in such countries are experiencing a rapid development in terms of volume and efficiency (Alam et al., 2021). Thus, ICT and equity market development improve renewable energy generation and increase its consumption (Yu et al., 2023; Qiu et al., 2023). It is, therefore, of vital importance to examine the relationship between ICT and stock market development and clean energy consumption in the case of high-income countries.

Therefore, in line with the above evaluations, the lack of the studies based on the relationship between equity market development, ICT and clean energy consumption is a motivated tool of this study. The aim of this study is to inspect the impact of equity market development and ICT on clean energy consumption of 18 high-income countries in the period 1990-2015 by integrating the economic growth and carbon dioxide (CO2) emissions. In order to attain this purpose, the studies conducted by Paramati et al. (2016) for developing countries, Alam et al. (2021) for OECD countries, Tiwari et al. (2022) for Asian countries, Talan et al. (2023) for G7 countries and Yu et al. (2023) for 35 countries are examined and considering the advantages of their studies, the empirical model of this study is obtained. In the study, firstly, after testing the presence of cross-sectional dependence and slope homogeneity, the stationarity of the variables is investigated with the second generation unit root tests, CADF and CIPS unit root tests. In the second stage, cointegration test is performed by Westerlund, Kao and Pedroni methods. In the next step, the long-run coefficients of the variables are estimated using second generation panel data techniques such as Panel Corrected Standard Errors (PCSE) and Feasible Generalized Least Squares (FGLS) estimation methods. In the last stage, Dumitrescu-Hurlin panel causality method is applied to determine causality among the variables in the model.

The findings obtained from the study analyses provide important contributions to the literature. First, the clean energy consumption of 18 high-income countries appears to be promoted by equity market development, ICT and economic growth, and carbon dioxide emissions reduce clean energy consumption. Finally, the long-term coefficients of the variables are determined by PCSE and FGLS methods, which are the second generation panel data techniques.

The rest of the study is designed as follows. In the second part, literature review is presented. In the third part, the study model and data are explained. In the fourth part, the study method is mentioned. In the fifth part, empirical findings of applied second generation tests are evaluated and discussed with previous study findings in literature. In the last part, the results of the study are explained.

LITERATURE REVIEW

In general, although it has been obtained that the effects of equity market development, financial development, ICT, economic growth and CO2 emissions on energy consumption have been determined in the literature (Sadorsky, 2010;2011; Zhang et al., 2011; Saud et al., 2019; Chang et al., 2020; Khan et al., 2022), the effects of these and other factors on renewable energy and its sub-sectors clean-green energy consumption have been examined by very few studies (Paramati et al., 2017a; Shahbaz et al., 2018; Razmi et al., 2020; Liu et al., 2022). For example, Sadorsky (2010;2011) obtains that stock market affects the energy consumption for both emerging economies and Central and Eastern European frontier economies and Zhang et al. (2011) conclude that the stock market scale causes the energy consumption while the energy consumption causes the stock market efficiency in China. At the same time, Paramati et al. (2016) determine that the equity market developments play an important role for clean/renewable energy investments and projects and they provide the funds and the liquidity to obtain more returns from these projects and investments. Al-Mulali et al. (2015) reveal the causality between financial development and renewable energy in 23 European countries over the 1990-2013 period while Kim and Park (2016), who examine the relationship between the financial market development and renewable energy, find that the financial markets of 30 countries promote renewable energy over the period of 2000-2013.

On the other hand, Burakov and Freidin (2017) conclude that there is not a significant causality between the financial development and renewable/clean energy in Russian for the period of 1990-2014 while Hassine and Harrathi (2017) obtain that renewable energy consumption causes financial development Gulf Cooperation Council (GCC) over the period 1980-2012.

Paramati et al. (2017a) investigate whether both the domestic stock market and foreign capital flows affect clean energy use among EU, G20 and OECD countries in the 1993-2012 period. Empirical findings show that both stock market development and foreign capital flows increase the use of clean energy. Also, Cai et al. (2018) view the relationship between clean energy consumption, economic growth and CO2 emissions of G-7 countries, and the findings show unidirectional causality from clean energy consumption to CO2 emissions for Germany and bidirectional causality between clean energy consumption and CO2 emissions for the USA.

Examining the impact of foreign capital, economic growth, and financial development on clean energy consumption and environmental degradation, Shahbaz et al. (2018) find out that economic growth and foreign capital increase clean energy consumption in N-11 countries during the 1992-2016 period while CO2 emissions decrease it. In addition, it is determined that while economic growth and financial development reduce clean energy consumption in BRICS countries, CO2 emissions prevent it. Yilanci et al. (2019) reveal that trade openness in Russia, China and South Africa negatively affects clean energy consumption by applying the Fourier Toda-Yamamoto and Asymmetric Fourier Toda Yamamoto causality tests in the 1985-2017 period, and foreign direct investments in Russia positively affect clean energy use.

Ji and Zhang (2019) aim to explore how financial development contributes to renewable/clean energy development in China. The researchers detect that stock market development, credit market growth and growth of international investments as financial development indicators contributed 42.42% to renewable/clean energy development in the 1992-2013 period. Razmi et al. (2020) seek the linkage among Iran's renewable/clean energy consumption, capital market development and economic growth in the period 1990-2014 utilizing the ARDL model. The findings specify that capital market development and economic growth can encourage renewable/clean energy in Iran. Similarly, economic growth promotes clean energy consumption for 27 cities in the central region of the Yangtze River Delta in China (Liu et al., 2020) and in Pakistan. Sohail et al. (2022) and Sbia et al. (2014) detect that between the economic growth and clean energy of the UAE sample over the period 1975Q1-2011Q4 are Granger causes of each other. On the other hand, Anton and Nucu (2020) probe the effect of the financial sector on renewable/clean energy consumption of 28 countries between 1990 and 2015 and find that financial development, banking sector development and stock market development encourage renewable/clean energy consumption.

However, Raza et al. (2020) reveal a negative link between renewable/clean energy consumption and financial development, stock market development and banking sector development using annual panel data of the top 15 countries where renewable energy was consumed in the 1997-2017 period. At the same time, Tiwari et al. (2022) obtain that there is an insignificant relationship between equity market development and renewable/clean energy consumption among 16 Asian countries over the period 1990-2019.

Liu et al. (2022) seek the effect of education and training expenditures, population, GDP per capita and research and development on clean energy consumption and CO2 emissions in BRICS countries during the period of 1991-2019. Findings attain that the economic growth, which is GDP per capita of BRICS countries, increases the population and R&D clean energy consumption. Yu et al. (2023) also examine the relationship between ICT financial development and renewable/clean energy consumption applying panel data from 35 countries for the period 1996-2018. The findings demonstrate that both ICT and financial development boost renewable/clean energy consumption, and economic development, trade openness and urbanization are also positively affected.

As a result, when the studies in the literature are examined, it is determined that there are many factors (such as economic growth, foreign direct investments, equity market development, banking sector development, financial sector development, ICT, R&D, education expenditures, CO2 emissions) affecting renewable/clean energy consumption. These evaluations are motivating in determining the purpose of this study.

MODEL AND DATA

The aim of the study is to investigate the impact of equity market development and ICT on clean energy consumption by including the explanatory variables of economic growth and CO2 emissions. The series of 18 high-income countries over the period 1990-2015 are analysed. In this context, we consider the studies by Paramati et al. (2016) for developing countries, Alam et al. (2021) for OECD countries, Tiwari et al. (2022) for Asian countries, Talan et al. (2023) for G7 countries and Yu et al. (2023) for 35 countries and the following empirical regression model is developed. Variables are added to the model by transforming them into logarithmic form (Bano et al., 2022).

$$\ln ANE_{it} = \gamma_0 + \gamma_1 \ln SMP_{it} + \gamma_2 \ln ICT_{it} + \gamma_3 \ln CO2_{it} + \gamma_4 \ln GDP_{it} + \varepsilon_{it} \quad (1)$$

$\gamma 0$ is the constant coefficient of model; $\gamma 1, \gamma 2, \gamma 3$ and $\gamma 4$ are the exponents of equity market performance, ICT, CO2 emissions and economic growth that are independent variables; ε point outs the error term. t is time and i shows country. ANE is used to show clean-energy consumption measured by the alternative and nuclear energy (% of total energy use) (Sohail et al., 2022; Liu et al., 2022). SMP is equity market performance that is indicated by market capitalization of listed domestic companies (% of GDP) (Carp, 2012; Qamruzzaman & Wei, 2018). ICT is information and communication technology that is measured with fixed telephone subscriptions (per 100 people) (Pradhan et al., 2017; Raza et al., 2023). CO2 emissions are per capita CO2 emissions (Saud et al., 2019) and GDP presents economic growth detected by per capita GDP (Paramati et al. 2018). The study covers from 1990 to 2015.

The alternative and nuclear energy (% of total energy use), market capitalization of listed domestic companies (% of GDP), fixed telephone subscriptions (per 100 people), per capita GDP and per capita CO2 emissions data are collected from the World Development Indicators (WDI) database published by the World Bank. The explanations, measurements, sources and expected signs of the variables in the model are summarized in Table 1.

Table 1. Variable descriptions

Variables	Symbol	Measure	Source	Expected Sign
Clean-Energy consumption	ANE	Alternative and nuclear energy (% of total energy use)	WDI	
Equity market performance	SMP	Market capitalization of listed domestic companies (% of GDP)	WDI	(+) Paramati et al. (2017a); Alam et al. (2021)
Information and communication technology	ICT	Fixed telephone subscriptions (per 100 people)	WDI	(+) Talan et al. (2023)
CO2 emissions	CO2	CO2 emissions (metric tons per capita)	WDI	(-) Shahbaz et al. (2018)
Economic growth	GDP	GDP per capita (current US$)	WDI	(+) Jiang and Chang (2022)

Table 2 reports the summary statistics of the variables selected for high-income countries for the period 1990-2015. As can be seen, the average alternative and nuclear energy consumption of 18 high-income countries is 2,143. The minimum value of alternative and nuclear energy use is -2.950, the maximum value is 3.903 and the standard deviation is 1.425. This indicates a high divergence among 18 high-income countries. The ratio of the average market capitalization of listed domestic companies (% of GDP), average fixed phone subscriptions (per 100 people), average CO2 emissions per capita and average per capita GDP are 26,715, 3,840, 2,230 and 10,312, respectively. For high-income countries, the minimum and maximum values of the market capitalization of listed domestic companies (% of GDP) with 1.697 standard deviations are 21.678 and 30.901.

Similarly, in logarithmic form, fixed telephone subscriptions (per 100 people) have a standard deviation of 0.316, with the lowest value of 1.873 and the highest value of 4.317. Moreover, the minimum and maximum values of CO2 emissions per capita with 0.465 standard deviations are 0.784 and 3.415 while the lowest and highest values of GDP per capita with 0.621 standard deviations are 7,821 and 11,725. On the other hand, Table 3 summarizes the correlation matrix of the variables in the study. According to the results, there is a positive correlation between the alternative and nuclear energy use of 18 high-income countries over the 1990-2015 period and the market capitalization of listed domestic companies (% of GDP), fixed telephone memberships (per 100 people), and CO2 emissions per capita while there is a negative correlation between the alternative and nuclear energy use and GDP per capital.

Table 2. Summary statistics

	lnANE	lnSMP	lnICT	lnCO2	lnGDP
Mean	2.143	26.715	3.840	2.230	10.312
Median	2.576	26.801	3.886	2.207	10.359
Std. dev.	1.425	1.697	0.316	0.465	0.621
Min.	-2.950	21.678	1.873	0.784	7.821
Max.	3.903	30.901	4.317	3.415	11.725
Skewness	-1.484	0.013	-2.498	-0.057	-0.838
Kurtosis	4.944	2.532	12.106	3.330	4.869
Obs.	468	468	468	468	468

Table 3. Correlation matrix

Variables	lnANE	lnSMP	lnICT	lnCO2	lnGDP
lnANE	1.000				
lnSMP	0.164	1.000			
lnICT	0.174	0.304	1.000		
lnCO2	-0.324	0.145	0.032	1.000	
lnGDP	0.026	0.283	0.525	-0.138	1.000

METHODOLOGY

In the study, the estimation process based on second-generation econometric approaches consists of five stages since 18 high-income countries are interrelated. In this context, the methodology of the study includes the latest econometric methods applied in chronological order: (1) Breusch-Pagan LM, Pesaran CD, Pesaran scaled LM and bias-corrected scaled LM tests to confirm cross-section dependence, and homogeneity test to test slope homogeneity of panel data are used; (2) CIPS and CADF panel unit root tests probe the stationarity properties of the variables; (3) Westerlund (2005, 2007), Kao and Pedroni panel cointegration tests are used to examine the cointegration between the variables selected in the model; (4) PCSE and FGLS estimators are applied in order to long-term estimates of variables; and (5) Dumitrescu-Hurlin panel causality test is utilized to test the causal effect between the selected variables.

In the first stage, before testing the stationarity of the variables, by applying Breusch-Pagan LM, Pesaran CSD, Pesaran scaled LM and bias-corrected scaled LM tests suggested by Breusch and Pagan (1980), Pesaran (2004), Pesaran, Ullah and Yamagata (2008) and Baltagi, Feng and Kao (2012) cross-section dependencies (CSD) are examined. The null hypothesis of the CSD test is that there is cross-section independence between the sample countries, and the alternative hypothesis is that there is cross-sectional dependence between the sample countries. In addition, slope-homogeneity test proposed by Pesaran and Yamagato (2008) is applied to test the slope homogeneity of the panel data, and the null hypothesis states that the slope of the panel data sets is homogeneous. This test will not be appropriate when regression errors are heteroscedastic and serially related. Therefore, the varying variance and autocorrelation consistent (HAC) version of the Blomquist and Westerlund (2013) Δ tests are applied.

In the second stage, since the presence of CSD in panel data is proven by CSD test, second generation panel unit root tests are applied rather than first generation in this study, and the results of second generation panel unit root tests are more consistent and reliable (Saud et al., 2019). Cross-sectional augmented Dickey-Fuller (CADF) and cross-sectional augmented Im-Pesaran-Shin (CIPS) tests proposed by Pesaran (2007) are used to detect the integration of variables in the model.

Following the second generation unit root tests, in the third stage, cointegration analysis is focused. In the study, in order to estimate the statistically significant coefficient of the variables, the variables should be stationary or cointegrated at the I(1) level (Saud et al., 2020). Therefore, Kao (1999), Pedroni (2004) and Westerlund (2005;2007) cointegration tests examine whether the selected variables are cointegrated. These tests analyse the null hypothesis that there is no cointegration in the model, and the rejection of the null hypothesis points out the existence of a long-term relationship between the variables in the model (Khan et al., 2021).

After determining that the variables selected for the model in the study have a balanced relationship, the long-term coefficients of the variables are estimated in the fourth step. Panel Corrected Standard Errors (PCSE) proposed by Beck and Katz (1995) and Feasible Generalized Least Squares (FGLS) estimators developed by Parks (1967) are used to measure the long-term elasticity of variables. The PCSE method is less sensitive to outlier estimates and provides an independent estimate of serial correlation, is used when working with time series and cross-section data, and produces accurate standard error estimates (Gozgor et al., 2020).

Forecasters such as PCSE and FGLS provide information about the long-term elasticity of the variables, but do not provide information about the causal relationship between the variables (Çeştepe et al., 2022). There are some suitable Granger causality approaches in the applied economic literature and some previous studies (Rahman et al., 2019; Ahmet et al., 2021) are based on this study. Therefore, in the last step of the methodology, causality is investigated among the selected variables using the Dumitrescu–Hurlin panel Granger causality test proposed by Dumitrescu and Hurlin (2012). This test is the latest and advanced version of Granger's (1969) causality test for panel data and includes Wbar statistics (representing the mean of Wald statistics) and Zbar statistics (calculated from the mean and variance of Wald statistics). At the same time, the Dumitrescu–Hurlin panel Granger causality test does not impose any restrictions on T and N and it is also consistent with the characteristics of our dataset (Rahman et al., 2019; Ahmet et al., 2021).

FINDINGS AND DISCUSSION

The econometric analysis for panel data begins with the examination of cross-sectional dependence, and the findings from the CSD test are summarized in Table 4. Findings reject the null hypothesis that there is no cross-sectional dependence between countries and point the cross-sectional dependence at the 1% significance level for all variables. This implies that the shock in one of the highest-income countries has some spillover effect that has been observed in the entire sample.

The findings of the slope homogeneity test proposed by Blomquist and Westerlund (2013) are presented in Table 5 and according to the findings, the null hypothesis of slope homogeneity cannot be rejected. This means that the slope coefficient of the model is homogeneous and the slope does not diversify among the highest income countries.

Since the selected highest-income countries are cross-section dependent, the stationarity properties of the variables are tested with the second generation unit root tests, CADF, CIPS, and the findings are reported in Table 5. According to two findings, the null hypothesis stating that the variables do not

Table 4. CSD tests

Variables	Breush-Pagan LM Statistic	*p*-Value	Pesaran Scaled LM STATISTIC	*p*-Value	Bias-Corrected Scaled LM	*p*-Value	Pesaran CD	*p*-Value
lnANE	778.371***	0.000	34.721***	0.000	34.361***	0.000	0.270	0.786
lnSMP	2533.877***	0.000	135.076***	0.000	134.716***	0.000	49.129***	0.000
lnICT	1389.420***	0.000	69.652***	0.000	69.292***	0.000	25.390***	0.000
lnCO2	1243.063***	0.000	61.285***	0.000	60.925***	0.000	12.698***	0.000
lnGDP	3332.563***	0.000	180.734***	0.000	180.374***	0.000	57.311***	0.000

Note: *** denotes significance at %1 level.

Table 5. Slope homogeneity test results

Test Statistics	t-Statistics	*p*-Value
$\tilde{\Delta}$	0.909	0.363
$\tilde{\Delta}_{adjusted}$	1.086	0.277

Table 6. Unit root tests

	CADF Test Statistic				CIP S Test Statistic			
	Level		First Difference		Level		First Difference	
Variables	Constant	Constant & Trend	Constant	Constant & Trend	Constant	Constant & Trend	Constant	Constant & Trend
lnANE	-1.197	-2.047	-3.314***	-3.383***	2.417	1.200	-6.842***	-4.896***
lnSMP	-1.959	-1.983	-3.422***	-3.441***	-0.913	1.491	-7.312***	-5.160***
lnICT	-1.126	-1.641	-2.436***	-2.757**	2.729	3.053	-2.999***	-2.040**
lnCO2	-1.730	-2.463	-3.526***	-2.821**	0.989	-0.696	-3.955***	-2.332**
lnGDP	-1.875	-2.316	-2.751***	-2.767**	-0.546	-0.028	-4.377***	-2.087**

Note: ***, ** and * denote significance at %1, %5 and %10 level, respectively.

contain a unit root cannot be rejected at the level. However, once the first differences of the variables are taken, the null hypothesis of all variables is rejected at the first difference level, which indicates that all variables are integrated at the I(1) level. In other words, the stationarity properties of clean energy consumption, equity market development, ICT, economic growth and CO2 emissions variables at the I(1) level are proven.

After the stationarity properties of the variables are determined, the cointegration between the selected variables in the model is examined and the findings of the Westerlund (2005; 2007), Kao and Pedroni cointegration tests are summarized in Table 7. When the findings obtained from the Westerlund (2007) test are evaluated, the null hypothesis that there is no cointegration between the variables in the model is rejected by the Gt and Ga statistics and findings indicate that there is a long-term relationship between the variables. At the same time, the reliability of the findings from the Westerlund (2007) test is re-examined using Westerlund (2005), Kao and Pedroni approaches, and the findings in Table 7 prove the long-term equilibrium relationship among clean energy consumption and equity market development, ICT, economic growth and CO2 emissions. The cointegration relationship between these determined variables also provides an opportunity to define the long-term effect of each variable on the clean energy consumption of high-income countries.

After confirming the cointegration relationship, the long-term effects of independent variables on clean energy consumption of high-income countries are estimated by PCSE and FGLS methods. The findings of the two estimators are presented in Table 8 and these findings are similar with each other. In this context, when the findings of the FGLS estimator are examined, the coefficient of equity market development is negative at the level of 1% significance. In a study sample of 18 high-income countries, if equity market development is increased by 1%, clean energy consumption will increase by 0.110%. This expresses us that the development of the equity market encourages clean energy consumption. In other words, equity market development plays an important role in creating a healthy environment for both companies and investors (Paramati et al, 2016). This finding is in line with the findings of Alam et al. (2021), who detect that stock market development of 30 OECD countries increases clean energy consumption and environmental protection in the 1996-2013 period. At the same time, it is also supported by previous studies in the literature such as Paramati et al. (2016) and Razmi et al. (2020).

Table 7. Cointegration tests results

Westerlund (2007) Cointegration			
Statistics	Value	Z-value	*p*-value
Gt	-3.316**	-3.870	0.024
Ga	-5.807*	3.959	0.064
Pt	-4.657	4.115	0.858
Pa	-2.249	3.949	0.930
Pedroni cointegration		test statistic	*p*-value
Modified Phillips-Perron t		2.036**	0.020
Phillips-Perron t		-3.695***	0.000
Augmented Dickey-Fuller t		-0.673	0.250
Kao cointegration			
Modified Dickey-Fuller t		1.292*	0.098
Dickey-Fuller t		2.047**	0.020
Augmented Dickey-Fuller t		-0.243	0.403
Unadjusted modified Dickey-Fuller t		0.794	0.213
Unadjusted Dickey-Fuller t		1.564*	0.058
Westerlund (2005)			
Variance Ratio		-1.621*	0.052

Note: ***, ** and * denote significance at %1, %5 and %10 level, respectively.

In addition, it is stated that ICT is positively associated with clean energy consumption at the 1% significance level. A 1% increase in ICT increases the clean energy consumption of high-income countries by 1.935%. Therefore, the increase in ICT infrastructure brings a big increase in clean energy consumption. Murshed (2020) states that ICT applications are important in smartening and greening traditional electricity branches where renewable energy sources will be used to generate small-scale electricity outputs. Therefore, the renewable energy production and consumption of sample countries are promoted by ICT investments (Lee et al., 2023). This finding of the study is similar to the finding of Talan et al. (2023), which revealed that the G-7 countries played a vital role in renewable energy consumption in the 1990-2018 period. These findings are consistent with Yu et al. (2023). However, Bano et al. (2022) conclude that ICT prevented renewable/clean energy consumption in the BRICS countries in the 2000-2017 period.

The FGLS findings in Table 8 indicate that the coefficient of CO2 emissions is negative at the 1% significance level. A 1% increase in CO2 emissions reduces clean energy consumption by 1.771%. This means that the increase in CO2 emissions of 18 high-income countries hinders clean energy consumption. This finding of the study reveals that there is a negative relationship between CO2 emissions and clean energy consumption. This finding of the study is supported by the findings of Shahbaz et al. (2018), which revealed a negative relationship between CO2 emissions and clean energy consumption.

Table 8. PCSE and FGLS estimates

Estimators	PCSE		FGLS	
Dependent Variable: lnANE	**Coefficient**	**Standard Error**	**Coefficient**	**Standard Error**
lnSMP	0.111***	0.011	0.110***	0.001
lnICT	1.938***	0.142	1.935***	0.009
lnCO2	-1.778***	0.071	-1.771***	0.007
lnGDP	0.037	0.044	0.038***	0.005
Constant	-4.700***	0.504	-4.684***	0.042
Wald chi2(5)		719.24		104876.30
p-value		0.000***		0.000***
Observation	468			
Number of countries	18			

Note: *** denote significance at %1 level.

Finally, the long-term forecaster findings show that the economic growth elasticity is positive at the 1% significance level. It is obtained that 1% increase in economic growth encourages clean energy consumption by 0.037%, in other words, there is a positive correlation between economic growth and clean energy consumption. This means that economic growth is important in providing renewable energy resources, which are necessary for sustainable development. (Pao & Fu, 2013). This finding of the study is confirmed by Sohail et al. (2022), who conclude that Pakistan's economic growth in the 1970-2019 period had an increasing effect on clean energy use.

After long-term estimations are made, the causality relationship between the series chosen in the model is examined with the Dumitrescu-Hurlin panel causality approach and the empirical findings are shown in Table 9. According to the findings, at the 5% statistical significance level, the equity market development plays an important role in clean energy consumption and the clean energy consumption plays an important role in the equity market development. This indicates that there is bidirectional causality between equity market development and clean energy consumption in high-income countries.

Our findings indicate bidirectional causality between ICT and clean energy consumption in high-income countries. This finding of the study is similar to the findings obtained by Huang et al. (2022) for E-7 and G-7 countries and Talan et al. (2023) for G-7 countries, who reveal that ICT is the cause of renewable/clean energy consumption and that renewable/clean energy consumption is also the cause of ICT.

At the same time, Dumitrescu-Hurlin panel causality findings indicate that there is causality between CO2 emissions and economic growth of high-income countries and their clean energy consumption at the 1% significance level. These findings are supported by the finding of Sbia et al. (2014), who found that economic growth and clean energy are the cause of each other, and Cai et al. (2018), who revealed that there is bidirectional causality between CO2 emissions and clean energy consumption.

In conclusion, using panel data of 18 high-income countries from 1990 to 2015, evidence of bidirectional causality is obtained between clean energy consumption and equity market development, ICT, CO2 emissions and economic growth.

Table 9. Dumitrescu-Hurlin panel causality test results

Null Hypothesis	W-Stat.	Zbar-Stat.	Results
lnSMP → lnANE	1.766	2.299**	lnSMP causes lnANE
lnANE → lnSMP	6.551	2.081**	lnANE causes lnSMP
lnICT → lnANE	1.651	1.954*	lnICT causes lnANE
lnANE → lnICT	5.966	14.899***	lnANE causes lnICT
lnCO2 → lnANE	2.156	3.470***	lnCO2 causes lnANE
lnANE → lnCO2	2.283	2.988***	lnANE causes lnCO2
lnGDP → lnANE	3.404	7.213***	lnGDP causes lnANE
lnANE → lnGDP	2.693	5.079***	lnANE causes lnGDP

Note: ***, ** and * show rejection of null hypothesis at 1%, 5% and 10% significance level, respectively.

CONCLUSION

The aim of the current study is to examine the effect of equity market development and ICT on clean energy consumption by including CO2 emissions and economic growth variables as explanatory variables, using panel data from 18 high-income countries for the period 1990-2015. In order to predict the relationship between equity market development-ICT-clean energy consumption, the study models created by Paramati et al. (2016), Alam et al. (2021), Tiwari et al. (2022), Talan et al. (2023) and Yu et al. (2023) are used and the model of this study is developed and analysed. The panel data analysis process consists of certain stages and firstly, the cross-section dependence of the variables chosen in the model and the presence of slope homogeneity in the model are investigated, and then whether the variables contain a unit root is examined by CADF and CIPS unit root tests. In the second stage, the long-term relationship between the variables is tested with the Westerlund, Kao and Pedroni cointegration methods. In the next stage, the long-term elasticities of the variables are estimated by PCSE and FGLS methods, and in the last stage, the causality between the variables is tested with the Dumitrescu-Hurlin panel causality method.

The empirical findings obtained from the study show that the shock in any country among high-income countries spreads to other countries and as a result of the homogeneity of the model, the slope does not diversify among high-income countries. Clean energy consumption, equity market development, ICT, economic growth and CO2 emissions, which are integrated at I(1) level, are cointegrated. At the same time, it is found that the clean energy consumption of high-income countries is positively affected by equity market development, ICT and economic growth, and CO2 emissions negatively affect it. The latest findings of the study prove that there is bidirectional causality between clean energy consumption and equity market development, ICT, economic growth and CO2 emissions.

Finally, these empirical findings of the study offer important advice to policy makers and firms. First, as Paramati et al. (2017b) state, policy makers need to develop sustainable policies to ensure that all listed firms adopt energy-efficient and greener technologies in their production activities. Because stock markets have an important effect in the economy. At the same time, policymakers should offer policies to leverage ever-growing stock markets to finance clean and renewable energy projects by providing lucrative investment incentives and creating a healthy business environment for the investment community. Policy makers should increase inflows of foreign direct investment by providing tax incentives and transform these investments into renewable/clean energy projects and help economies increase the share of renewable/clean energy consumption rather than CO2-intensive traditional energy sources. Also, when the study of Talan et al. (2023) is evaluated, policy makers should encourage ICT infrastructure and deployment and simplification of production from renewable/clean energy sources with ICT to maximize the efficiency of renewable/clean energy sources. Finally, companies should also encourage environmentally friendly activities as these activities can increase clean energy consumption (Alam et al., 2021).

FUTURE RESEARCH DIRECTIONS

The study has some limitations and these limitations offer suggestions for future studies. First of all, in this study, the relationship between equity market development-ICT-clean energy consumption is examined only on the basis of high-income countries, but in future studies, this subject can be investigated based on different country groups such as developed, developing, BRICS, OECD, G-7. Secondly, PCSE and FGLS methods are used as second generation long-term estimators in this study, but new methods such as CS-ARDL, AMG, CCE can be used in future studies. Lastly, in this study, market capitalization of listed domestic companies (% of GDP) is used as an indicator of equity market development and fixed telephone subscriptions (per 100 people) as an indicator of ICT. But, in future studies, different equity market development indicators (market capitalization of listed domestic companies (current US$), stocks traded, total value (% of GDP), stocks traded, total value (current US$) and different ICT indicators (Secure Internet servers (per 1 million people), Mobile cellular subscriptions (per 100 people) can be used.

REFERENCES

Ahmed, Z., Nathaniel, S. P., & Shahbaz, M. (2021). The criticality of information and communication technology and human capital in environmental sustainability: Evidence from Latin American and Caribbean countries. *Journal of Cleaner Production*, *286*, 125529. doi:10.1016/j.jclepro.2020.125529

Al-Mulali, U., Ozturk, I., & Lean, H. H. (2015). The influence of economic growth, urbanization, trade openness, financial development, and renewable energy on pollution in Europe. *Natural Hazards*, *79*(1), 621–644. doi:10.100711069-015-1865-9

Alam, M. S., Apergis, N., Paramati, S. R., & Fang, J. (2021). The impacts of R&D investment and stock markets on clean-energy consumption and CO2 emissions in OECD economies. *International Journal of Finance & Economics*, *26*(4), 4979–4992. doi:10.1002/ijfe.2049

Anton, S. G., & Nucu, A. E. A. (2020). The effect of financial development on renewable energy consumption. A panel data approach. *Renewable Energy*, *147*, 330–338. doi:10.1016/j.renene.2019.09.005

Baltagi, B. H., Feng, Q., & Kao, C. (2012). A Lagrange Multiplier test for cross-sectional dependence in a fixed effects panel data model. *Journal of Econometrics*, *170*(1), 164–177. doi:10.1016/j.jeconom.2012.04.004

Bano, S., Liu, L., & Khan, A. (2022). Dynamic influence of aging, industrial innovations, and ICT on tourism development and renewable energy consumption in BRICS economies. *Renewable Energy*, *192*, 431–442. doi:10.1016/j.renene.2022.04.134

Blomquist, J., & Westerlund, J. (2013). Testing slope homogeneity in large panels with serial correlation. *Economics Letters*, *121*(3), 374–378. doi:10.1016/j.econlet.2013.09.012

Breusch, T. S., & Pagan, A. R. (1980). The lagrange multiplier test and its applications to model specification in econometrics. *The Review of Economic Studies*, *47*(1), 239–253. doi:10.2307/2297111

Burakov, D., & Freidin, M. (2017). Financial development, economic growth and renewable energy consumption in Russia: A vector error correction approach. *International Journal of Energy Economics and Policy*, *7*(6), 39.

Cai, Y., Sam, C. Y., & Chang, T. (2018). Nexus between clean energy consumption, economic growth and CO2 emissions. *Journal of Cleaner Production*, *182*, 1001–1011. doi:10.1016/j.jclepro.2018.02.035

Carp, L. (2012). Can stock market development boost economic growth? Empirical evidence from emerging markets in Central and Eastern Europe. *Procedia Economics and Finance*, *3*, 438–444. doi:10.1016/S2212-5671(12)00177-3

Çeştepe, H., Çetin, M., Avcı, P., & Bahtiyar, B. (2022). The link between technological innovation and financial development: Evidence from selected OECD countries. *International Journal of Finance & Economics*, 1–17.

Chang, C. L., Ilomäki, J., Laurila, H., & McAleer, M. (2020). Causality between CO2 emissions and stock markets. *Energies*, *13*(11), 2893. doi:10.3390/en13112893

Doytch, N., & Narayan, S. (2016). Does FDI influence renewable energy consumption? An analysis of sectoral FDI impact on renewable and non-renewable industrial energy consumption. *Energy Economics*, *54*, 291–301. doi:10.1016/j.eneco.2015.12.010

Gozgor, G., Mahalik, M. K., Demir, E., & Padhan, H. (2020). The impact of economic globalization on renewable energy in the OECD countries. *Energy Policy*, *139*, 111365. doi:10.1016/j.enpol.2020.111365

Habiba, U., Xinbang, C., & Ahmad, R. I. (2021). The influence of stock market and financial institution development on carbon emissions with the importance of renewable energy consumption and foreign direct investment in G20 countries. *Environmental Science and Pollution Research International*, *28*(47), 67677–67688. doi:10.100711356-021-15321-3 PMID:34263400

Hassine, M. B., & Harrathi, N. (2017). The causal links between economic growth, renewable energy, financial development and foreign trade in gulf cooperation council countries. *International Journal of Energy Economics and Policy*, *7*(2), 76–85.

Huang, Y., Haseeb, M., Usman, M., & Ozturk, I. (2022). Dynamic association between ICT, renewable energy, economic complexity and ecological footprint: Is there any difference between E-7 (developing) and G-7 (developed) countries? *Technology in Society*, *68*, 101853. doi:10.1016/j.techsoc.2021.101853

Ji, Q., & Zhang, D. (2019). How much does financial development contribute to renewable energy growth and upgrading of energy structure in China? *Energy Policy*, *128*, 114–124. doi:10.1016/j.enpol.2018.12.047

Jiang, C., & Chang, Y. (2022). Clean energy projects in Asian economies: Does FDI and stock market matter for sustainable development? *Ekonomska Istrazivanja*, *35*(1), 5843–5857. doi:10.1080/1331677X.2022.2038650

Kao, C. (1999). Spurious regression and residual-based tests for co-integration in panel data. *Journal of Econometrics*, *90*(1), 1–44. doi:10.1016/S0304-4076(98)00023-2

Khan, A., Sicen, L., Khan, B., & Salman, N. (2021). On the influence of demographic structure and industrial growth on environmental quality. *Journal of Environmental Management*, *288*, 112453. doi:10.1016/j.jenvman.2021.112453 PMID:33827026

Khan, H., Weili, L., & Khan, I. (2022). Examining the effect of information and communication technology, innovations, and renewable energy consumption on CO2 emission: Evidence from BRICS countries. *Environmental Science and Pollution Research International*, *29*(31), 47696–47712. doi:10.100711356-022-19283-y PMID:35184242

Kim, J., & Park, K. (2016). Financial development and deployment of renewable energy technologies. *Energy Economics*, *59*, 238–250. doi:10.1016/j.eneco.2016.08.012

Lee, C. C., Chen, M. P., & Yuan, Z. (2023). Is information and communication technology a driver for renewable energy? *Energy Economics*, *124*, 106786. doi:10.1016/j.eneco.2023.106786

Liu, Y., Sohail, M. T., Khan, A., & Majeed, M. T. (2022). Environmental benefit of clean energy consumption: Can BRICS economies achieve environmental sustainability through human capital? *Environmental Science and Pollution Research International*, *29*(5), 6766–6776. doi:10.100711356-021-16167-5 PMID:34460087

Murshed, M. (2020). An empirical analysis of the non-linear impacts of ICT-trade openness on renewable energy transition, energy efficiency, clean cooking fuel access and environmental sustainability in South Asia. *Environmental Science and Pollution Research International*, *27*(29), 36254–36281. doi:10.100711356-020-09497-3 PMID:32556989

Öztürk, L. (2002). Dijital uçurumun küresel boyutları. *Ege Academic Review*, *2*(1), 1–10.

Pao, H. T., & Fu, H. C. (2013). Renewable energy, non-renewable energy and economic growth in Brazil. *Renewable & Sustainable Energy Reviews*, *25*, 381–392. doi:10.1016/j.rser.2013.05.004

Paramati, S. R., Alam, M. S., & Apergis, N. (2018). The role of stock markets on environmental degradation: A comparative study of developed and emerging market economies across the globe. *Emerging Markets Review*, *35*, 19–30. doi:10.1016/j.ememar.2017.12.004

Paramati, S. R., Apergis, N., & Ummalla, M. (2017a). Financing clean energy projects through domestic and foreign capital: The role of political cooperation among the EU, the G20 and OECD countries. *Energy Economics*, *61*, 62–71. doi:10.1016/j.eneco.2016.11.001

Paramati, S. R., Mo, D., & Gupta, R. (2017b). The effects of stock market growth and renewable energy use on CO2 emissions: Evidence from G20 countries. *Energy Economics*, *66*, 360–371. doi:10.1016/j.eneco.2017.06.025

Paramati, S. R., Ummalla, M., & Apergis, N. (2016). The effect of foreign direct investment and stock market growth on clean energy use across a panel of emerging market economies. *Energy Economics*, *56*, 29–41. doi:10.1016/j.eneco.2016.02.008

Pedroni, P. (2004). Panel cointegration: Asymptotic and finite sample properties of pooled time series tests with an application to the ppp hypothesis. *Econometric Theory*, *20*(3), 597–625. doi:10.1017/S0266466604203073

Pesaran, M. H. (2004). *General diagnostic tests for cross section dependence in panels*. University of Cambridge, Faculty of Economics, Cambridge Working Papers in Economics, No.0435.

Pesaran, M. H. (2007). A simple panel unit root test in the presence of cross-section dependence. *Journal of Applied Econometrics*, *22*(2), 265–312. doi:10.1002/jae.951

Pesaran, M. H., Ullah, A., & Yamagata, T. (2008). A bias-adjusted LM test of error cross section independence. *The Econometrics Journal*, *11*(1), 105–127. doi:10.1111/j.1368-423X.2007.00227.x

Pesaran, M. H., & Yamagata, T. (2008). Testing slope homogeneity in large panels. *Journal of Econometrics*, *142*(1), 50–93. doi:10.1016/j.jeconom.2007.05.010

Pradhan, R. P., Arvin, M., Nair, M., Bennett, S., & Bahmani, S. (2017). ICT-finance-growth nexus: Empirical evidence from the Next-11 countries. *Cuadernos de Economía (Santiago, Chile)*, *40*(113), 115–134. doi:10.1016/j.cesjef.2016.02.003

Qamruzzaman, M., & Wei, J. (2018). Financial innovation, stock market development, and economic growth: An application of ARDL model. *International Journal of Financial Studies*, *6*(3), 69. doi:10.3390/ijfs6030069

Qiu, L., Chu, L., Zhou, R., Xu, H., & Yuan, S. (2023). How do carbon, stock, and renewable energy markets interact: Evidence from Europe. *Journal of Cleaner Production*, *407*, 137106. doi:10.1016/j.jclepro.2023.137106

Rahman, S. U., Chen, S., Saud, S., Bano, S., & Haseeb, A. (2019). The nexus between financial development, globalization, and environmental degradation: Fresh evidence from Central and Eastern European Countries. *Environmental Science and Pollution Research International*, *26*(24), 24733–24747. doi:10.100711356-019-05714-w PMID:31240660

Raza, S. A., Qamar, S., & Ahmed, M. (2023). Asymmetric role of non-renewable energy consumption, ICT, and financial development on ecological footprints: Evidence from QARDL approach. *Environmental Science and Pollution Research International*, *30*(8), 20746–20764. doi:10.100711356-022-23549-w PMID:36255586

Razmi, S. F., Bajgiran, B. R., Behname, M., Salari, T. E., & Razmi, S. M. J. (2020). The relationship of renewable energy consumption to stock market development and economic growth in Iran. *Renewable Energy*, *145*, 2019–2024. doi:10.1016/j.renene.2019.06.166

Sadorsky, P. (2010). The impact of financial development on energy consumption in emerging economies. *Energy Policy*, *38*(5), 2528–2535. doi:10.1016/j.enpol.2009.12.048

Sadorsky, P. (2011). Financial development and energy consumption in central and Eastern European frontier economies. *Energy Policy*, *39*(2), 999–1006. doi:10.1016/j.enpol.2010.11.034

Saud, S., Chen, S., Danish, H. A., & Haseeb, A. (2019). Impact of financial development and economic growth on environmental quality: An empirical analysis from belt and road initiative (BRI) countries. *Environmental Science and Pollution Research International*, *26*(3), 2253–2269. doi:10.100711356-018-3688-1 PMID:30456608

Saud, S., Chen, S., Haseeb, A., & Sumayya. (2020). The role of financial development and globalization in the environment: Accounting ecological footprint indicators for selected one-belt-one-road initiative countries. *Journal of Cleaner Production*, *250*, 119518. doi:10.1016/j.jclepro.2019.119518

Sbia, R., Shahbaz, M., & Hamdi, H. (2014). A contribution of foreign direct investment, clean energy, trade openness, carbon emissions and economic growth to energy demand in UAE. *Economic Modelling*, *36*, 191–197. doi:10.1016/j.econmod.2013.09.047

Shahbaz, M., Destek, M. A., & Polemis, M. L. (2018). Do foreign capital and financial development affect clean energy consumption and carbon emissions? Evidence from BRICS and Next-11 countries. *SPOUDAI-Journal of Economics and Business*, *68*(4), 20–50.

Sohail, M. T., Majeed, M. T., Shaikh, P. A., & Andlib, Z. (2022). Environmental costs of political instability in Pakistan: Policy options for clean energy consumption and environment. *Environmental Science and Pollution Research International*, *29*(17), 1–10. doi:10.100711356-021-17646-5 PMID:34837622

Talan, A., Rao, A., Sharma, G. D., Apostu, S. A., & Abbas, S. (2023). Transition towards clean energy consumption in G7: Can financial sector, ICT and democracy help? *Resources Policy*, *82*, 103447. Advance online publication. doi:10.1016/j.resourpol.2023.103447

Taşbaşı, D. (2022). *Araştırma: "Yenilenebilir enerji 2050 yılına kadar küresel enerji karışımının yüzde 50'sini oluşturacak"*. https://temizenerji.org/2022/10/17/arastirma-yenilenebilir-enerji-2050-yilina-kadar-kuresel-enerji-karisiminin-yuzde-50sini-olusturacak/

Tiwari, A. K., Nasreen, S., & Anwar, M. A. (2022). Impact of equity market development on renewable energy consumption: Do the role of FDI, trade openness and economic growth matter in Asian economies? *Journal of Cleaner Production*, *334*, 130244. Advance online publication. doi:10.1016/j.jclepro.2021.130244

Westerlund, J. (2005). New simple tests for panel cointegration. *Econometric Reviews*, *24*(3), 297–316. doi:10.1080/07474930500243019

Westerlund, J. (2007). Testing for error correction in panel data. *Oxford Bulletin of Economics and Statistics*, *69*(6), 709–748. doi:10.1111/j.1468-0084.2007.00477.x

Yilanci, V., Ozgur, O., & Gorus, M. S. (2019). The asymmetric effects of foreign direct investment on clean energy consumption in BRICS countries: A recently introduced hidden cointegration test. *Journal of Cleaner Production*, *237*, 117786. doi:10.1016/j.jclepro.2019.117786

Younis, I., Naz, A., Shah, S. A. A., Nadeem, M., & Longsheng, C. (2021). Impact of stock market, renewable energy consumption and urbanization on environmental degradation: New evidence from BRICS countries. *Environmental Science and Pollution Research International*, *28*(24), 31549–31565. doi:10.100711356-021-12731-1 PMID:33608780

Yu, M., Jin, H., Zhang, H., & Chong, A. Y. L. (2023). ICT, Financial Development and Renewable Energy Consumption. *Journal of Computer Information Systems*, *63*(1), 190–203. doi:10.1080/08874417.2022.2049017

Zhang, Y. J., Fan, J. L., & Chang, H. R. (2011). Impact of China's stock market development on energy consumption: An empirical analysis. *Energy Procedia*, *5*, 1927–1931. doi:10.1016/j.egypro.2011.03.331

KEY TERMS AND DEFINITIONS

Clean-Energy Consumption: It is the consumption of energy obtained from sources that do not pollute the atmosphere when used.

Equity Market Performance: It includes market capitalization of listed domestic companies.

ICT: It is the information, communication, and technology as a fundamental tool that contributes to easing human life.

Panel Data Analysis: It is a method that allows testing the models created by combining time series and cross-section analysis.

Chapter 3
Does Economic Growth Increase CO2 Emissions in BRICS Countries?
Question on Sustainability

Esha Sunil Bharadwaj
Jain University (Deemed), India

Ashish Kumar
Jain University (Deemed), India

Pratik Mukherjee
Jain University (Deemed), India

Susobhan Maiti
https://orcid.org/0000-0002-5725-9844
Jain University (Deemed), India

Suvam Sivam
Jain University (Deemed), India

ABSTRACT

Sustainability is an increasingly important issue in the modern world and an essential concept that is increasingly gaining importance in today's world. It refers to the responsible use of natural resources while ensuring that future generations can meet their needs. The impact of human activities on the environment is becoming more apparent, with climate change and biodiversity loss being some of the major challenges that we face today. Considering the foregoing discussion, the present study comprises two main objectives: firstly, to find out the relationship between economic growth, which is measured by the growth of GDP per capita and CO2 emissions per capita in the BRICS nations, and to find out unidirectional or bidirectional causality between carbon emissions and proxies of economic growth based on the data collected from the world development indicator for the period 1990 to 2022. The results suggest that a bi-directionally causality exists between carbon emissions per capita and economic growth per capita in the BRICS countries.

DOI: 10.4018/978-1-6684-9503-2.ch003

INTRODUCTION

Since the last decade, the need for environmental protection, environment policies and sustainable growth has increased to a very great extent. Numerous researchers across the world have worked on this issue very rigorously. Analysis of the relationship between economic growth and carbon emissions have been conducted for various countries. The parameters mostly being the Gross Domestic Product (GDP) per capita and the most commonly used pollutant for the analysis purpose is the Carbon-di-oxide (CO2) emissions per capita, as it is the highest source of pollution standing at 68%. Other pollutants such as PM10, Sulphur-di-Oxide, Nitrogen are also used in some studies. Apart from GDP per capita, real output per capita, GNP per capita are also used. Some studies have also considered the relationship between the energy consumption growth of a country along with the CO2 emissions per capita and GDP per capita to get even better results. These results are further used by the government and policy makers to make good policies that balances the economic growth and the pollution levels as well. Note that the results do vary from different country to country and from different groups of countries to another group of countries. For example, relationship between CO2 emissions per capita GDP per capita and Energy consumption per capita may different for Turkey and Tunisia, Australia and New Zealand, India and Brazil, Brazil and China, Australia and Turkey, USA and UK, Canada and Sri Lanka, India and Russia, South Africa and Middle East Nations, so on and so forth. The relationship between the variables may also vary among different groups of countries like G20, BRICS, MENA regions, ASEAN, OPEC, G7 and so on. The reason for the different results might be the geographical area of the country, the type of industries in the country, whether it is dominated by service industries or by agriculture-based industries or by heavy industries. Suppose if a country is dominated by the service industry, then it can become a developed market with very less amount of pollution. The level of development the nation is in, whether it is a developing nation or a developed nation or an underdeveloped nation. Adding onto this, it is found that results also vary based on the area of analysis, for example in the national wise analysis the results can be different compared to regional based analysis. This is evident for China, which will be discussed later in the chapter. Since the outcomes are different, different nations have to adopt different policies and control measures to tackle their emission problems and achieve sustainable growth. Thus, controlling the emission problem is not as easy as it seems to be.

The most common methodologies and tools used for analyzing the relationship between the CO2 emissions, economic growth and energy consumption are the long-run cointegration tests, followed by the Granger Causality test which sees the causal relationship between the variables, whether their relationship is unidirectional or bidirectional in nature. Vector Autoregression (VAR), Structural Vector Autoregression (SVAR) to see the effects at the current period, Vector Error Correction Model (VECM), Impulse Response Function (IRF) to see how a shock given to a variable might affect the other variable, Factor Error Variance Decomposition (FEVD) helps to see how a single variable gets affected by multiple other variables, these tools are basically used for multivariate time series analysis. Before using the above tools, the time series data must be made stationary by taking their first differences. The stationary check can be done via ADF test, KPSS test or PP test. The purpose of using such advanced and robust tools is to get a more accurate result that can aid in proper analysis of the actual problem that causes the harmful emissions and help to reach a proper solution to resolve the matter.

REVIEW OF LITERATURE

Numerous papers have been written upon the EKC hypothesis and the inverted U shape hypothesis. Moreover, all most all the literature reviews done for this paper have compared their analysis and results with EKC hypothesis. EKC is nothing but Environmental Kuznets Curve. The hypothesis says that every economy undergoes 3 phases of development, the first phase is where the economy picks up growth along with emissions/pollution picking up, the second phase is when the emission has reached its peak level and the third level is the consolidation level where the emissions decrease as the economy further grows. It assumes that, as and when an economy reaches a certain level economic development the population becomes more environmentally conscious due to which businesses and industries adopt to more sustainable practices which leads to decrease in the emission levels but increases the sustainable growth of the economy. Adding on, there are various ways suggested by Xue (2009) through which a country can surpass the peak emission levels. Xue (2009) talks about the effects of foreign capital "The introduction of foreign capital induces technological transfer to developing countries. The multinational cooperation brings with it new products in which new technologies, including environmentally related technology are embodied." When foreign technologies come in, then along with-it international standards of environmental control also accompany. This introduces the domestic countries to an eco-friendlier production technique which helps the country to skip the phase of high pollution and directly reach the third phase which is the phase of high growth and lesser pollution levels. Xue (2009) shaped the China's EKC curve using carbon dioxide data and sulphur dioxide data. It was found that both sulphur dioxide and carbon dioxide emissions have very high correlations with GDP growth and the EKCs of China appeared to be very much in the second stage and it is yet to arrive at the turning point. It can be inferred that the pollution in China will continue to increase if the government of China does not restrict the pollution levels. Saboori, Sulaiman, and Mohd (2012) performed a cointegration analysis of the EKC curve for Malaysia for the period 1980 to 2009. The results suggested that there is an existence of a long run relationship between per capita CO2 emission and real per capita GDP, when CO2 is a dependent variable. There exists an inverted-U shape relationship between CO2 emissions and GDP in the short term and long term. It showed the absence of causality between carbon emission and economic growth in the short run, but it demonstrated a unidirectional causality from economic growth to CO2 emission in the long run. Analysis done for the MENA (Middle East and North Africa) region shows no evidence of short run causality running from economic growth and C02 emission to energy consumption but shows evidence of short run causality from energy consumption to economic growth and CO2 emission. This reflects in Farhani and Rejeb (2012) that an increase in the energy consumption leads to increase in income and the CO2 emissions. Hence, the policies can be made to increase the usage of more efficient energy which will not have an impact on the economic growth. Another analysis done in the 12 MENA region for the period 1981 to 2005 by Arouri et al. (2012) shows long run energy consumption has a positive significant impact on CO2 emission and the GDP exhibits a quadratic relationship with CO2 emission for the region as a whole. The paper concludes in two divisions one for the regional level and the other for the country level. At the regional level the authors of the study say that most MENA region have built a capacity to manage environment problems, especially air pollution and specific institutions were built for the same. The countries focused on raising the general awareness of the mass, it cut the subsidies on the oil and said that economic growth won't cause environmental degradation. At the country level, the EKC hypothesis was not verified except for Jordan and thus EKC was not valid at the country level. Thus, the econometric relationship derived here suggests that future reductions in the carbon emission per capita can be achieved at the same time as GDP per capita in the MENA region continues to grow. In the case of UK and some extent in Japan, it indicated a relationship between carbon emission and GDP per capita, this was characterised by both emission as an item generated by increased energy consumption but also that after controlling oil prices as a proxy for the price of energy. A unit shock given to the emissions showed a negative impact on the income generation i.e., emissions negatively caused income. The argument for this was that it could be due to the sustainability problem or because of the implementation of stringent environmental regulations in UK and Japan. This shows that emission is a determinant of income in these two countries. For Italy there exists causality from emissions to income with a positive effect. The results show that emissions are declining faster in Italy and US than compared to UK where there is a relatively slower reduction. All the 6 countries, France, USA, UK, Italy, Canada, Japan, had a positive and significant causality effect of per capita growth on per capita CO2 emission. Emissions had a negative and significant impact on GDP growth rates for the four out of the six countries. Out of the six countries Italy seemed to be a bit different as the growth of emission had positive and significant impact on the GDP. Concluded a paper done by (Barassi&Spagnolo, 2012).When energy consumption impacts the emission levels and economic growth levels it would be equally interesting to see how renewable energy sources for electricity generation affect the growth and carbon emissions. To see this, a study was done using the Impulse Response Function by Silva et al. (2012) who had taken four countries into consideration which are US, Portugal, Spain and Denmark for the period 1960 to 2004, it showed that USA had a negative relationship with GDP and carbon emissions when a positive shock was given to the renewable energy source for electricity generation for the first 4 periods from then on the effect became positive and came near to zero. While Portugal showed the strongest GDP and carbon emission decrease until the 5th period. Spain and Denmark showed a close and negative response to the positive shock on renewable energy sources. After the 6th period Spain had the strongest GDP negative effects. One reason that might be possible for this result is that when renewable energy sources are dispensed then it incurs certain amount of costs which in turn affect the GDP negatively. For Portugal, the GDP variation was explained by renewable energy sources when it reached to a 45 percent level after the sixth period. Except for Denmark the contribution of carbon emission to the variation in GDP was relatively small. The impact of renewable energy source on GDP reaches the peak at the 2nd period and then declines after that as the weight of the carbon emissions increase. Lastly in the USA, 50 percent of the carbon emission variation is explained by the GDP shock at the 1st period itself.

A lot of studies and researches have been performed in the North American regions, South American regions, Middle East regions, South African regions, Asiatic regions and south East Asia regions. We have to also look at the relationship of the above-mentioned variables in countries like Australia. It is unique because it is surrounded by water bodies on its all four sides. So, the chances of influence from other countries are very less.Let's say a country like Pakistan is surrounded by a growing country like India and China, so the emissions created in India and China will some or the other way affect the environmental conditions in Pakistan, may be in the short run or in the long run. For Australia this is not the case as its pollution levels will tend to increase or decrease because of its own economic activity. Thus, it becomes very important to study countries like Australia and New Zealand. A study says that no cointegrating relationship exists among variable like economic growth, energy consumption and CO2 emission in Australia (Salahuddin & Khan, 2013). Further they found that the energy consumption had a persistent positive impact on carbon emission, on the contrary side GDP had no impact on the coo2 emissions. There exists a bidirectional causal link between economic growth and energy consumption but there is no link between economic growth and co2 emission in the long run. Interestingly, an insignificant bidirectional causal link exists between economic growth and energy consumption but not between economic growth and co2 emission in the short run. In short, we can say from this result that for Australia the main source of emission is the level of energy consumption and not the growth of GDP per capita. Further, the relationship between energy consumption and economic growth gets stronger in the long run. This basically means to say that consumption of energy leads to economic growth but on the other side economic growth does not lead to increase in emission levels. A review was done for Pakistan where the authors had included few extra variables like industrial value added, financial development and trade openness. The felt that emission levels are also affected by a countries trade openness and industrial value added as this means how much value have the industries of a country created for a certain period of time, because this will mean that industries have also created some emission in order to have produced those items. The authors included financial development because when a nation's financial institutions are strong and trust worthy they attract more capital both foreign and domestic capital. More the capital flows into the country the higher the economic activity in the country which directly leads to emission. This was the thinking behind including the three new variables for analyzing Pakistan. Munir and Khan (2013) found that energy consumption negatively affects co2 emissions. The inverted-U shape is seen for Pakistan with the three new variables mentioned above. Industrial value added and trade openness positively affect co2 emission whereas financial development reduces co2 emission. Another fascinating thing found was that, manufactured imports negatively affected the energy consumption while manufactured exports positively affected the energy consumption so did income and investments. Meaning when Pakistan exports manufactured goods, it consumes more energy than when it imports manufactured goods. this is true because when goods are exported, it basically means goods are being produced within the country, hence leading to energy consumption, but on the other hand when more manufactured goods are imported, the energy consumed for producing goods within the country is very less. Thus, manufactured imports show a negative effect on the energy consumption levels and exports show a positive effect. Cowan et al. (2013) examined the causal link between electricity consumption, economic growth and carbon emission for the BRICS countries for the period 1990 to 2010. In Brazil, there is no evidence of causality from any direction between electricity consumption and economic growth similarly no causality existed between electricity consumption and carbon emission. The result seemed quite obvious because electricity accounts for only a marginal amount of Brazil's total greenhouse gas emissions because the majority part comes from land usage. For Russia it was found that bidirectional causality existed between electricity consumption and economic growth, but no causal relationship between electricity consumption and co2 emission. Further the feedback causal relationship between co2 emission and economic growth reflects that any energy conservation policy might have an adverse impact on the economic growth of the country. Next for India, no evidence of causality was found running in any direction between electricity consumption and economic growth, economic growth and co2 emission. Instead, a unidirectional causality ran from electricity consumption to co2 emissions. In China, no causality was found running in any direction between electricity consumption and economic growth, electricity consumption and co2 emissions. There was also a lack of causal relationship between economic growth and co2 emission. Lastly, for South Africa, a unidirectional causality was seen from economic growth to electricity consumption. Since in all the countries electricity plays a major role, it is recommended that they increase their investments in electricity infrastructure and improve the electricity efficiency in their respective countries. China and India do not show any causal relationship between economic growth and co2 emission, hence we can say that over the period 1990-2010, economic growth did not cause co2 emissions in India and China.A study for the period 1972 to 2010, for Bangladesh, investigated the economic structure and trends of carbon emission with the GDP per capita. It also examined the relationship between economic growth and carbon emission based on the EKC curve. Bangladesh's economic structure shifted greatly from 1980 as the contribution of industries increased in the GDP growth while share of agriculture decreased from 1980 onwards. As the share of industries and services rose the carbon emissions also rose gradually following an increasing trend. The study was unable to prove the EKC curve for Bangladesh, at least for the period 1972-2010 (Alam, 2014). Fei et al. (2014) studied the energy growth nexus by incorporating the effect of clean energy, co2 emissions and technological innovation for Norway and New Zealand over the period from 1971 to 2010. Strong cointegration nexus of clean energy use, economic growth, co2 emission and technological innovations for Norway at a 1 percent significance level and weak cointegration for New Zealand was found at 10 percent significance level. Bidirectional causality existed between clean energy and economic growth in Norway and unidirectional relationship in New Zealand. The literature also said that advancement in technology positively elevates the degree of economic growth for both the countries because technological innovation granger causes economic growth and clean energy use. As per the result, if New Zealand changes its environment policies to make it more efficient or cleaner than it might hamper the economic growth, which is why New Zealand does not apply adequate support on technological innovations in clean energy generation processes.

OBJECTIVE

Considering the foregoing discussion, the present study comprises two main objectives. Firstly, to find out the relationship between economic growths, which is measured by the growth of GDP per capita, and CO2 emissions per capita in the BRICS nations, and to find out unidirectional or bidirectional causality between carbon emissions and proxies of economic growth, which is the second objective.

METHODOLOGY AND DATA SOURCE

Data and Variables

In the present paper, two variables i.e., Gross Domestic Product Per Capita and Carbon di-oxide emissions Per capita have been used. The study is based on time-series data from the period 1990 to 2022 for BRICS countries (Brazil, Russia, India, China, and South Africa) collected from World Development Indicators (WDI) published by the World Bank. Real GDP per capita is used as the proxy for economic growth.

Gross Domestic Product per capita and carbon di-oxide emissions per capita are represented by GDPPC and CO2EPC. GDP per capita of Brazil, Russia, India, China, and South Africa represented by GDPPCB, GDPPCR, GDPPCI, GDPPCC and GDPPCS respectively. CO2 emissions per capita represented by CO2EPCB, CO2EPCR,CO2EPCI, CO2EPCC and CO2EPCS respectively.

Methodology

To understand the linkage between variables, two methods have been used which are as under:

1. Karl Pearson's correlation coefficient and
2. Granger Causality Test

The Pearson product-moment correlation coefficient, generally denoted by *r* is a measure of the correlation i.e., linear dependence between two variables *X* and *Y*.

The formula for *r* is:

$$r = \frac{\sum_{i=1}^{n}(X_i - \bar{X})(Y_i - \bar{Y})}{\sqrt{\sum_{i=1}^{n}\left(X_i - \bar{X}\right)^2}\sqrt{\sum_{i=1}^{n}\left(Y_i - \bar{Y}\right)^2}}$$

Where n is the sample size.

A common practice prior to handling time series data and for performing a co-integration test is to determine the stationarity of the series or its degree of integration, *I*(d). For this Augmented-Dickey-Fuller (ADF) Unit Root Test(Dickey & Fuller, 1979)has been used. It is essential to keep in mind that these tests presuppose the absence of any structural breaks (Gupta et al., 2022).

The present study uses ADF unit root test to examine the stationarity of the data series. It consists of running a regression of first difference of the series against the series lagged once, a constant and a time trend. This can be expressed as follows:

$$\Delta Y_t = \alpha_0 + \alpha_1 t + \alpha_2 Y_{t-1} + \sum_{j=1}^{p} \alpha_j \Delta Y_{t-j} + \varepsilon_t$$

The test for a unit root is performed on the coefficient of Yt-1 in the regression in this ADF process. If the coefficient deviates considerably from zero, the hypothesis that Yt has a unit root is rejected. Stationarity is implied by rejecting the null hypothesis(Dickey & Fuller, 1979)

H_0: α_2=0 i.e., Y_t is a non-stationary series

H_1: α_2<0 i.e., Y_t is a stationary series.

The null hypothesis, H0, is accepted and the series is considered nonstationary or not integrated of order zero, I, if the estimated value of the ADF statistic is greater than McKinnon's critical values (Engle & Granger, 1991). If this fails, more differencing must be carried out until stationarity is obtained and the null hypothesis is rejected (Azolibe, 2020).

To determine whether or not one time period can be helpful in forecasting another, a statistical hypothesis test that can be called the Granger causality test is performed (Gupta et al., 2022). It is claimed that a time series X is a Granger cause of another time series Y if it can be shown that the values of the time series X convey statistically significant information about the values of Y in the future (Chakraborty and Jana, 2021; Maiti and Chakraborty, 2023). The examination is carried out in the following way:

Let *y* and *x* be stationary time series.

The null hypothesis is H_0: X does not Granger-cause Y,

The following regression should be used to determine the correct value to use for Y after it has been lagged:

This is the limited model that must be purchased in order to receive SSRR.

The next step is to strengthen the autoregression by including the lag values of x, as shown in the following equation:

$$y_t = a_0 + a_1 y_{t-1} + a_2 y_{t-2} + \ldots + a_m y_{t-m} + b_p x_{t-p} + \ldots + b_q x_{t-q} + residual_t$$

This is the model that does not have any restrictions for obtaining Sum square residual unrestricted (SSRUR).

The following serves as the test statistic in order to put the hypothesis to the test:

$F = ((SSR_R - SSR_{UR})/c) / SSR_{UR}/(n-l)$

Where c represents the number of lagging terms that were taken, n represents the sample size, and l is the number of parameters that were estimated using the unconstrained equation.

If there are no lagged values of x that are kept in the regression, and then one may accept the null hypothesis that x does not Granger-cause y. This is the only condition under which this hypothesis can be accepted (Chakraborty & Jana, 2021; Gupta et al., 2022)

Johansen Cointegration Test:

The Johansen test tests cointegrating relationships between several non-stationary time series data. Compared to the Engle-Granger test, the Johansen test allows for more than one cointegrating relationship. However, it is subject to asymptotic properties (large sample size) since a small sample size would produce unreliable results. Using the test to find the cointegration of several time series avoids the issues created when errors are carried forward to the next step. Johansen's test comes in two main forms, i.e., Trace tests and Maximum Eigen value tests.

ANALYSIS OF RESULT

A significant, high degree of association exists between the economic growth (GDPPC) and CO2 emissions (CO2EPC) in all BRICS nations as depicted by the Pearson correlation coefficients. The BRICS nations have all experienced significant economic growth in the time period under consideration. Changes in consumption patterns, increased production, population growth and upward economic mobility have contributed to increased CO2 Emissions in the time frame under consideration. The BRICS nations have become important participants in the international trade markets and China alone accounts for 14.3% of global merchandise exports in 2020, up from 2,3% in 1990 (World Trade Organisation). The catalysts for growth and the established reasons for increased emissions have a significant overlap, which is seen in the correlation analysis.

Table 1. Pearson correlation coefficient for variables

Variables	CO2EPCB	CO2EPCR	CO2EPCI	CO2EPCC	CO2EPCS
GDPPCB	.938**	-	-	-	-
GDPPCR	-	.759**	-	-	-
GDPPCI	-	-	.962**	-	-
GDPPCC	-	-	-	.991**	-
GDPPCS	-	-	-	-	.923**

Source: Author's calculation
**Correlation is significant at the 0.01 level (2-tailed).

Table 2. ADF unit root test of GDPPC for BRICS

Variables	In Levels	Result	In First Difference	Result	Conclusion
GDPPCB	-1.020	Do not reject H_0	-4.236***	Reject H_0	I (1)
GDPPCR	0.146	Do not reject H_0	-2.933**	Reject H_0	I (1)
GDPPCI	1.145	Do not reject H_0	-8.09854***	Reject H_0	I (1)
GDPPCC	1.48204	Do not reject H_0	-5.064***	Reject H_0	I (1)
GDPPCS	-0.700	Do not reject H_0	-4.107***	Reject H_0	I (1)

Source: Author's calculation
***, **,* represents statistical significance of 1%, 5% and 10% respectively

The Augmented Dickey-Fuller test shows the presence of unit roots in all 10 time series under consideration of studies in levels. No unit roots are present in all time series in the first difference making all of them integrated of order 1.

The results of the Granger causality test demonstrate the existence of a significant (at most 0.1 level of significance) unidirectional causality between CO2EPC and GDPPC in the BRICS nations. CO2EPC Granger causes GDPPC in all the nations. Bi-directional causality exists in only Brazil and Russia (at most 0.1 level of significance). The nature of the causality may be influenced by a variety of factors across nations such as level of enforcement of environmental standards, penetration levels of green technology, growth policies etc. In countries with strict environmental regulation, economic growth is less likely to

Table 3. ADF unit root test of CO2EPC for BRICS

Variables	In Levels	Result	In First Difference	Result	Conclusion
CO2EPCB	-1.590	Do not reject H_0	-4.363***	Reject H_0	I (1)
CO2EPCR	-1.243	Do not reject H_0	-3.179**	Reject H_0	I (1)
CO2EPCI	-0.449	Do not reject H_0	-6.438***	Reject H_0	I (1)
CO2EPCC	1.261	Do not reject H_0	-4.851***	Reject H_0	I (1)
CO2EPCS	-1.180	Do not reject H_0	-5.820***	Reject H_0	I (1)

Source: Author's calculation
***,**,* represents statistical significance of 1%, 5% and 10% respectively

Table 4. Result of Granger causality test

	Brazil		Russia		India		China		South Africa	
Null Hypothesis	**F-Statistic**	**Prob.**	**F-Statistic**	**Prob.**	**F-Statistic**	**Prob.**	**F-Statistic**	**Prob.**	**F-Statistic**	**Prob.**
GDPPC does not Granger Cause CO2EPC	3.589	0.04	2.718	0.09	0.456	0.64	0.256	0.78	1.748	0.19
CO2EPC does not Granger Cause GDPPC	2.620	0.09	7.701	0.00	9.610	0.00	4.610	0.02	3.446	0.05

Source: Author's calculation

Granger cause CO2 emissions. Similarly, in countries where production methods are not technologically advanced, we can expect CO2 emissions to Granger cause economic growth.

The link between economic growth and carbon dioxide (CO2) emissions in China and other countries has been examined in a number of studies using various econometric models and hypothesis testing techniques. A vector autoregression (VAR) model was used by Wang et al. (2021) to investigate the connection between China's economic development, energy use, and CO2 emissions. The findings demonstrated a two-way causal relationship between economic growth and CO2 emissions as well as between energy use and CO2 emissions. Li et al. (2020) used a time-series analysis using the Environmental Kuznets Curve (EKC) hypothesis to investigate the connection between economic growth and carbon emissions in China. Using a regional panel data model, the results in China showed an inverted U-shaped link between economic growth and carbon emissions, supporting the EKC hypothesis. Using a regional panel data model, Zhang et al. (2019) examined the connections between China's economic development, energy use, population, and CO2 emissions. The authors discovered evidence of both bidirectional causation between population and CO2 emissions as well as evidence of unidirectional causality between economic development and energy consumption and CO2 emissions. Similar to this, Nguyen et al. (2019) used a VAR model to examine the connection between economic development and CO2 emissions in Vietnam. The study discovered evidence of a one-way causal relationship between economic development and CO2 emissions in Vietnam, showing that economic expansion is a major contributor to emissions. Wang et al. (2018) used panel data analysis to investigate the connection between economic development, energy use, and CO2 emissions in China.

The Johansen Cointegration test shows that there is a presence of at most one long term relationship between the two variables in each of the countries. Countries must be considerate of this fact when forming policies for growth, sustainability or conservation.

Table 5. Result of cointegration test of variables for BRICS

	Brazil		Russia		India		China		South Africa	
Hypothesized No. of CE(s)	**Trace Statistic**	**Prob.**	**Trace Statistic**	**Prob.**	**Trace Statistic**	**Prob.**	**Trace Statistic**	**Prob.**	**Trace Statistic**	**Prob.**
None*	24.131	0	22.506	0	35.012	0	16.308	0.04	28.469	0
At most 1*	10.440	0	7.113	0	12.074	0	5.631	0.02	9.633	0

Source: Author's calculation

CONCLUSION

A significant, high degree of association exists between the economic growth (GDPPC) and CO2 emissions (CO2EPC) in all BRICS nations as depicted by the Pearson correlation coefficients. The results of the Granger causality test demonstrate the existence of a significant (at most 0.1 level of significance) unidirectional causality between CO2EPC and GDPPC in the BRICS nations. CO2EPC Granger causes GDPPC in all the nations. Significant Bi-directional causality exists in only Brazil and Russia. We can say that initially when the government tries to control pollution it will have an impact on the GDP growth. The Johansen Cointegration test shows that there is a presence of at most one long term relationship between the two variables in each of the countries. Countries must be considerate of this fact when forming policies for growth, sustainability or conservation.

PRACTICAL IMPLICATION AND POLICY SUGGESTION

The analysis of the chapter is practically helpful for the policy maker to make an economy pollution free as we as to make a green economy to achieve sustainable development goals. This chapter also indicated a major concern and its impact on growing and developing economy. The best possible policy suggestion here can be that the government can initiate renewable energy sources in its production techniques by sacrificing its growth in the short run. On the other hand, in the long run the sustainable techniques adopted will yield to economic growth, thus the economy will see a sluggish growth in the short run but a healthy growth in the long run because by then the economy would have completely shifted to renewable energy sources. The modus operandi should be that the government policies should focus on a slow but continuous shift towards green energy till the majority of the energy sources used are sustainable.

REFERENCES

Alam, J. (2014). On the Relationship between Economic. *IOSR Journal of Economics and Finance*.

Arouri, M. E., Youssef, A. B., M'henni, A., & Rault, C. (2012). Energy consumption,economic growth and CO2 emissions in Middle East and North African countries. *The International Journal of the Political, Economic, Planning, Environmental and Social Aspects of Energy*.

Atil, A., Ben Bouheni, F., Lahiani, A., & Shahbaz, M. (2019). Factors influencing CO_2 emissions in China. *Journal of the Energy Division, 44*(1/2), 95–128.

Barassi, M. R., & Spagnolo, N. (2012). Linear and Non-linear Causality between CO2 Emissions and Economic Growth. International Association for Energy Economics, 16.

Chakraborty, C., & Jana, A. (2021). Economic Growth and Trade-related Variables: An Empirical Study Using Indian Data. In Global Tariff War: Economic, Political and Social Implications (pp. 141–152). Emerald Publishing Limited. doi:10.1108/978-1-80071-314-720211011

Cowan, W. N., Chang, T., Inglesi-Lotz, R., & Gupta, R. (2013). The nexus of electricity consumption, economic growth and CO2 emissions in the BRICS countries. *The International Journal of the Political, Economic, Planning, Environmental and Social Aspects of Energy.*

Dada, M. A. (n.d.). Energy consumption and economic growth dynamics in Nigeria (1981-2018). *The Journal of Energy and Development*, *44*(1/2), 237-258.

Farhani, S., & Rejeb, J. B. (2012). Energy Consumption, Economic. *International Journal of Energy Economics and Policy.*

Fei, Q., Rasiah, R., & Shen, L. J. (2014). The clean energy-growth nexus with CO_2 emissions and technological innovation in Norway and New Zealand. *Energy & Environment.*

Green, F., & Stern, N. (2015). Managing Economic Change and Mitigating Climate Change: China's strategies, policies and trends. In L. Song, R. Garnaut, C. Fang, & L. Johnston (Eds.), China's New Sources of Economic Growth: Vol. 1. Reform, Resources and Climate Change. ANU Press.

Gupta, T., Sharma, D., & Maiti, S. (2022). *Global Trade and Economic Crisis.* doi:10.4018/978-1-6684-5950-8.ch005

Issaoui, F., Toumi, H., & Touili, W. (2016). The Effects of Carbon Dioxide Emissions on Economic Growth, Urbanization, and Welfare: Application to Countries in the Middle East and North Africa. *Journal of the Energy Division*, *41*(1/2), 223–252.

JamaïMouhtadi, M., &SadefoKamdem, J. (2018). Economic growth, energy consumption, and transition in Morocco. *Journal of the Energy Division*, *44*(1/2), 283–298.

Kasperowicz, R. (2015). Economic growth and CO2 emissions: The ECM analysis. *Journal of International Students.*

Li, X., Wang, W., & Li, L. (2020). Research on the dynamic relationship between economic growth and carbon emissions: An empirical study based on the EKC hypothesis. *Journal of Cleaner Production*, *266*, 121956.

Lin, F.-L., Inglesi-Lotz, R., & Chang, T. (2018). Revisit coal consumption, CO_2 emissions and economic growth nexus in China and India using a newly developed bootstrap ARDL bound test. *Energy Exploration & Exploitation*, *36*(3), 450–463. doi:10.1177/0144598717741031

Maâlej, A., & Cabagnols, A. (2020). CO2 emissions and growth. *Journal of the Energy Division*, *46*(1/2), 1–24.

Maiti, S., & Chakraborty, C. (2023). Does Air Pollution Affect Labour Productivity in Indian Manufacturing? Evidence from State-level Data. In The Impact of Environmental Emissions and Aggregate Economic Activity on Industry: Theoretical and Empirical Perspectives (pp. 183-194). Emerald Publishing Limited. doi:10.1108/978-1-80382-577-920231014

Munir, S., & Khan, A. (2013). Impact of Fossil Fuel Energy Consumption on CO 2 Emissions: Evidence from Pakistan (1980-2010). *The Pakistan Development Review.*

Nguyen, V. H., Tran, T. M., & Nguyen, T. K. O. (2019). The relationship between economic growth and CO2 emissions in Vietnam: Evidence from a VAR model. *Environmental Science and Pollution Research International, 26*(24), 24846–24858.

Paramati, S. R., Alam, M. S., & Chen, C.-F. (2016). The Effects of Tourism on Economic Growth and CO2 Emissions: A Comparison between Developed and Developing Economies. *Journal of Travel Research*, 1–13. doi:10.1177/0047287516667848

Rastegaripour, F., Karbasi, A., & Pirmalek, F. (2019). Relationship between CO^2 emissions, energy consumption, and economic growth in Iran. *Journal of the Energy Division, 45*(1/2), 197–212.

Raza, S. S., & Hussain, A. (2016). The Nexus of Foreign Direct Investment, Economic Growth and Environment in Pakistan. *Pakistan Development Review, 55*(2), 95–111. doi:10.30541/v55i2pp.95-111

Runde, D. F. (2017). *Quality Infrastructure: Ensuring Sustainable Economic Growth. Center for Strategic and International Studies*. CSIS.

Saboori, B., Sulaiman, J., & Mohd, S. (2012). Economic growthandCO2 emissions in Malaysia: A cointegration analysis of the Environmental Kuznets Curve. *The International Journal of the Political, Economic, Planning, Environmental and Social Aspects of Energy*.

Salahuddin, M., & Khan, S. (2013). Empirical link between economic growth, energy consumption and co 2 emission in australia. *Journal of Developing Areas*. Advance online publication. doi:10.1353/jda.2013.0038

Shapiro, J. S. (2016). Trade Costs, CO2, and the Environment. *American Economic Journal. Economic Policy, 8*(4), 220–254. doi:10.1257/pol.20150168

Sheehan, P., & Sun, F. (2008). Emissions and economic development: Must China choose? In *China's Dilemma* (p. 23). ANU Press.

Silva, S., Soares, I., & Pinho, C. (2012). The Impact of Renewable Energy Sources on Economic Growth and CO2 Emissions - a SVAR approach. *European Research Studies*.

Tiwari, A. K., Mensi, W., & Menegaki, A. N. (2019). CO_2 emissions, energy consumption, and economic growth in Saudi Arabia. *Journal of the Energy Division, 44*(1/2), 61–94.

Union of Concerned Scientists. (2018). *Climate Accountability Scorecard: Insufficient Progress from Fossil Fuel Companies*. Author.

van der Ree, K. (2019). Promoting Green Jobs: Decent Work in the Transition to Low-carbon, Green Economies. In C. Gironde & G. Carbonnier (Eds.), The ILO @ 100: Addressing the Past and Future of Work and Social Protection. Brill.

Wang, H., Zhu, C., Zhang, Z., & Li, S. (2021). The relationships between economic growth, energy consumption, and CO2 emissions: Empirical evidence from China. *Energy, 226*, 120357.

Wang, J., Zhang, Y., Huang, J., & Huang, X. (2018). Economic growth, energy consumption, and CO2 emissions in China: A panel data analysis. *Journal of Cleaner Production, 195*, 154–164.

Xue, J. (2009). Moving towards low-carbon economic growth. In *China's New Place in a World in Crisis* (p. 24). ANU Press.

Zhang, B., Mu, H., & Zhang, L. (2019). The impact of economic growth, energy consumption, and population on CO2 emissions in China: Evidence from a regional panel data model. *Energy Policy*, *128*, 284–295.

Chapter 4
Empirical Evidence on the Relation Between Progress on Sustainable Development Goals and Adaptation of Islamic Finance

Dogus Emin
Social Sciences University of Ankara, Turkey

Aysegul Aytac Emin
https://orcid.org/0000-0001-8086-0091
Social Sciences University of Ankara, Turkey

ABSTRACT

In this study, the authors aim to investigate the dynamic correlation and lead/lag relation between the progress of the countries on the Sustainable Development Goals (SDGs) and adaptation of Islamic finance. As the proxy of progress level on SDGs, SDG Index of the Bertelsmann Stiftung and the Sustainable Development Solutions Network (SDSN) are used. To be able to determine the adaptation level of Islamic finance, Islamic Finance Adaptation (IFA) Index is introduced. The data set is formed with top eight countries according to Islamic finance assets and top eight countries according to the SDG Index for the period of 2012 and 2021. To assess the relation between progress level on Sustainable Development Goals (SDG Index) and Islamic finance adaptation level (IFA Index), the Dynamic Conditional Correlation Model (DCC Model) of Engle and Sheppard and Geweke Feedback Measure of Geweke are used.

1. INTRODUCTION

Sustainable Development Goals (SDGs) were adopted by the United Nations in 2015 by addressing global challenges such as poverty, inequality, environmental pollution and climate change in order to

DOI: 10.4018/978-1-6684-9503-2.ch004

provide peace and prosperity for both humanity and the planet. The SDGs consist of 17 specific goals and 169 targets to be achieved by 2030 such as ending poverty and hunger, promoting well-being, reducing inequalities, ensuring sustainable consumption and production, combating climate change, accessing justice and promoting peace and sustainable economic growth. It is estimated that to mitigate and adopt SDGs, a substantial amount of investment is required. For instance, it is estimated that by 2030 between 5 and 7 trillion U.S. Dollar per year is required in order to achieve the SDGs. Furthermore, developing countries and low-income countries encounter extensive funding gaps and increasing public expenditures.

Islamic finance is not a system based entirely on commercial interests, but includes religious, moral and social aspects. Islamic finance is a system where all kinds of financing activities are conformed to Islamic values and tenets. Namely, all of the financial transactions and commercial activities must be permissible by the rules of Islamic law (Shariah). Accordingly, Islamic financial system includes financial and commercial activities by ensuring that they are implemented considering religion, economy, morality and social issues. In other words, on the basis of Islamic law, there are aspects of transformation for humanity that is based on justice, equality, security and knowledge (Ayub, 2007). Furthermore, establishing socio-economic justice, poverty alleviation, equitable allocation of resources, equitable distribution of wealth, have a prominent place in Islamic law. Accordingly, welfare and wealth maximization have to be realized through fair and equitable resource allocation, efficient use of resources, maintaining peace and security, helping the poor and disadvantaged, avoidance of wasteful consumption, and creating an informed society. In addition, environmental conservation does not only a social responsibility but also is an essential obligation according to Islamic religion. As Chapra (2000) states, it is essential for Islamic law to achieve its objectives by avoiding unnecessary consumption of non-renewable resources and without damaging the ecosystem.

One of the main objectives of the Islamic law is to ensure the provision of inclusive social justice, equal opportunities and well-being by increasing the level of welfare in the society. In this respect, the Islamic financial system provides a comprehensive framework to enhance the financial inclusion, which is declared as the promoter for 7 out of 17 SDGs, through the principle of risk sharing. The system provides financial instruments for all classes of people that allow them to reduce their unique risks and moderate their consumption (Habib, 2013). Furthermore, Islamic law includes the promotion of all required actions for the protection of the environment, the efficient use of resources and the observance of environmental benefits such as avoidance of wasteful consumption. As it be seen, Islamic finance provide environmental, economic and social benefits for humanity and society in line with the SDGs. For this reason, the Islamic financial instruments can be extremely suitable and crucial for achieving SDGs. In other words, since the basic principles and objectives of Islamic finance are overlap with the majority of the SDGs, Islamic finance is a crucial resource besides conventional finance in reaching the SDGs and has the potential to act as a catalyst. In this regard, innovative Islamic financial instruments such as green sukuk have step up as a new asset class to attract new sources of capital and to accelerate the transition towards the green economy.

The social, environmental and economic aspects of Islamic law that form the basis of Islamic finance are perfectly aligned with the SDGs. In this respect, Islamic finance may play a key role in reaching the SDGs. In this study, we aim to investigate the relation between the progress of the countries on the SDGs and adaptation of Islamic finance. By using the Sustainable Development Goals (SDG) Index of Bertelsmann Stiftung and the Sustainable Development Solutions Network (SDSN) and constructing an Islamic Finance Adaptation (IFA) Index, this study significantly contributes to the literature. Furthermore, as far as we know, in the literature this is the first study that investigates dynamic correlation and the lead/lag relation between sustainable development goals and Islamic finance.

The remainder of this paper structured as follows. In the following section, the SDG Index of the Bertelsmann Stiftung and the Sustainable Development Solutions Network (SDSN) is introduced and Islamic finance is briefly discussed by reviewing the related literature. In section three, the details of data and methodology are explained. In section four, the findings are revealed and finally in the last section the study is concluded.

2. SDG INDEX AND ISLAMIC FINANCE

The Bertelsmann Stiftung ant the Sustainable Development Solutions Network (SDSN) have been producing an annual SDG Index and Dashboards Report since 2016. The Sustainable Development Goals (SDGs) were established by the international community in September 2015, and the SDG Index and Dashboards Report measures how well each nation is doing in relation to these goals. The SDG Index includes 109 indicators for 17 sustainable development goals. The indicators for each SDG are listed in a table in Appendix.

The SDG Index is determined by a three-step process; (i) filtering extreme values from each indicator's distribution, (ii) rescaling the data, and (iii) aggregating the indicators both within and between SDGs. Lafortune et al. (2018) filter the data at the lowest 2.5th percentile as the minimum value for the normalization to reduce the impact of extreme values, which can skew the outcomes of a composite index. To rescale the data, each variable is rescaled from 0 to 100, with 0 signifying the lowest performance (2.5th percentile) and 100 describing the technical optimum, in order to make the data consistent across indications. Before the aggregating process, the individual components have to be weighted and combined to create a composite SDG Index. The 2030 Agenda considers all SDGs and indicators to be of equal value, hence the SDG Index applies equal weighting at the goal and indicator levels. Finally, two phases are involved in the aggregating process for the SDG Index. Prior to being integrated for each SDG, the rescaled variables are aggregated across targets. The conventional constant-elasticity-of-substitution (CES) function (Arrow et al., 1961) is used to construct the SDG Index in order to provide the greatest flexibility when aggregating data.

The overall score calculates how far a country has come in accomplishing all 17 Goals. The result can be seen as a percentage of SDG accomplishment. If a score is 100, then all SDGs have been accomplished. A country with a score of 50 on a variable is halfway to the ideal value, whereas one with a score of 75 has gone from worst to excellent by 75 percent (Lafortune et al., 2018).

The table shows the SDG scores of several countries for the years 2012 to 2021. The SDG scores range from 0 to 100 and measure the progress of a country towards achieving the United Nations Sustainable Development Goals. According to the table, between 2012 and 2021, the scores of each country generally show a gradual increase over time, indicating progress towards achieving the SDGs. The scores of the countries in the table range from a low of 60.41 (Saudi Arabia in 2012) to a high of 86.48 (Finland in 2021). The Nordic countries (Finland, Denmark, Sweden, and Norway) consistently have high scores, with Finland having the highest score in 2021. The scores of the Gulf countries (Saudi Arabia, Bahrain, United Arab Emirates, and Kuwait) have also increased steadily over time, but are generally lower than those of the Nordic countries. Indonesia and Malaysia have scores that are similar to or slightly higher

Table 1. SDG index scores between 2012 and 2021

	2012	2013	2014	2015	2016	2017	2018	2019	2020	2021
Malaysia	66.59	67.07	67.96	68.91	68.59	69.43	69.51	70.16	70.37	70.33
Indonesia	62.77	63.32	64.16	65.03	65.19	66.94	68.12	68.42	68.48	68.9
Saudi Arabia	60.41	61.14	62.07	61.96	62.51	64.23	64.68	65.54	65.77	66.36
Bahrain	63.17	62.46	63.42	64.58	64.52	64.23	64.48	64.15	64.19	64.16
United Arab Emirates	62.07	63.64	64.61	65.46	65.98	66.85	67.52	68.98	68.85	68.78
Iran	64.16	64.84	65.24	66.07	66.68	68.05	68.3	68.59	69.39	68.57
Qatar	63.32	63.12	64.21	64.74	64.79	65.49	66.02	66.53	66.58	66.68
Kuwait	62.56	62.44	63.04	63	63.48	63.7	64.55	64.27	64.33	64.17
Finland	85.85	86	86.06	86.41	85.6	86.34	85.89	86.18	86.42	86.48
Denmark	84.55	84.59	84.75	85.33	85.38	85.04	85.13	85.39	85.4	85.61
Sweden	85.3	85.16	85.22	85.29	85.15	85.24	85.06	85.12	85.29	85.15
Norway	81.27	81.06	81.53	81.7	81.61	82.2	81.95	82.25	82.42	82.34
Austria	80.18	80.67	80.78	80.92	80.91	81.68	81.74	82.27	82.21	82.18
Germany	79.53	79.92	80.35	81.64	81.94	81.86	81.87	81.94	82.28	82.16
France	78.89	79.08	79.32	79.52	79.61	80.48	80.75	81.12	81.15	81.22
Switzerland	79.43	79.44	79.99	80.45	80.25	80.44	80.27	80.54	80.64	80.67

Source: Lafortune et al. (2018)

than the Gulf countries, but with more fluctuations over time. Germany, France, Switzerland, and Austria have scores that are consistently high, but slightly lower than those of the Nordic countries.

Some of the most promising Islamic social finance concepts and strategies that could hasten the achievement of the Sustainable Development Goals (SDGs) are Zakat, Waqf, Sadaqat, and Islamic microfinance (Lanzara, 2021). These concepts can be integrated and in line with the SDGs, which can complement one another and be used as tools for the shared responsibility and societal solidarity principles (Rehman & Pickup, 2018).

SDGs and Zakat, Sadaqat and Waqf have a significant number of profound similarities. The primary similarity between Zakat recipients and SDG-1 (end all forms of poverty) and SDG-2 is that they are poor, needy, and/or vulnerable (achieve food security and zero hunger in the world). Also, Zakat embodies the spirit of the SDGs by, for example, eliminating inequality (SDG-10) and promoting economic growth by distributing idle riches to the less fortunate in order to empower them and give them access to socioeconomic chances to develop (SDG-8) (Shaikh & Ismail, 2017). Zakat, Waqf and Sadaqat have generally been used for a long time to assist the poor, vulnerable, and destitute in a variety of ways, such as by providing access to affordable housing (SDG-11), by raising living standards and creating better pathways to quality healthcare (SDG-3), education (SDG-4), and clean water and sanitation (SDG-6) (ESCWA, 2021). In addition to that, Islamic Social Finance (ISF) sector has the potential to make a significant financial contribution to the SDGs. The financial resources available to Zakat, Sadaqat and Waqf institutions allow them to at least assist Muslim nations in achieving the most pertinent and urgent SDGs, such as eradicating poverty and improving access to water and sanitation (Rehman & Pickup, 2018).

On the other hand, according to a number of studies, the Islamic banking sector plays either a minimal or non-existent role in meeting environmental and social goals. Kamla and Rammal (2013) examine the social reporting of 19 Islamic banks and find little indication that these institutions are aiding in social progress or implementing "real plans aimed at eradicating poverty or increasing equitable redistribution of wealth in society" (Kamla & Rammal, 2013, pp. 933). They conclude that the Islamic banks have fallen short of achieving their stated goals in part because they have not made social justice a priority in all aspects of their business operations. Haniffa and Hudaib (2007) find that while Islamic banks' commitment to creditors and employees is at satisfying levels, their commitment to community is quite low. Sairally (2007) find in a study of 48 Islamic financial institutions from 19 different countries that, although some corporate generosity, social responsibility is not a core component of their business strategy. Maali et al. (2006) find that the philanthropic endeavours and employee-related issues of 29 Islamic banks are moderately good in their research of social reporting. None of the Islamic banks, though, disclose any environmental activity. Furthermore, interviews with 18 senior executives of Islamic financial institutions in the Gulf Cooperation Council (GCC) reveal that Islamic banks do not place a high priority on corporate social responsibility (Aribi & Arun, 2015). Their operations are centred on adhering to Shari'ah, and none identified any environmental concerns.

3. DATA AND METHODOLOGY

3.1. Data

In this study to be able to reveal the relation between Islamic finance and progress on SDGs, the data set which is formed with top 8 countries in Islamic finance assets and top 8 countries in SDG Index is investigated for the period of 2012 and 2021. In the first step by using below variables, Islamic Finance Adaptation Index is constructed for 16 countries for each year from 2012 to 2021.

Prior to performing a Principal Component Analysis (PCA), the data set should be evaluated to see if it is suitable. The Kaiser-Meyer-Olkin (KMO) Measure of Sampling Adequacy is employed in this regard. As the KMO was found to be higher than 0.5 in the current context, the data set is appropriate for the PCA.

In the second step Sustainable Development Goals Index scores are collected for the same group of countries and the same time period.

"In implementing the Geweke method, the forecast equations must be estimated with stationary time series, otherwise the forecasts may be subject to spurious correlation" (Fuess & Millea, 2002, pp.5). Therefore, to check for the existence of a unit root, we perform both ADF and PP tests. The null of non-

Table 2. Factors Used in Islamic finance adaptation index

Variables/Factors	Description
Value of Islamic Financial Assets	Islamic financial assets under management of Islamic and conventional institutions
IFA to CFA	The ratio of Islamic Financial Assets to Conventional Financial Assets
VS to VB	The ratio of issuance Value of Sukuk to issuance Value of Bond

stationarity is rejected. On the other hand, since the Geweke technique does not impose any such limits on a VAR system, we do not need to be concerned about any potential heteroscedasticity and autocorrelation issues of our dataset (Brooks, 2008).

3.2. Methodology

To be able to investigate the relation between the progress on the SDGs and Islamic finance adaptation we follow two step procedure. In the first step, we construct a composite index to score the adaptation levels of countries to the Islamic finance. In the second step, by using Dynamic Conditional Correlation (DCC) Model of Engle and Shephard (2001) and Geweke Feedback Measure of Geweke (1982), we investigate the dynamic correlation and lead/lag relation between progress on Sustainable Development Goals and Islamic finance adaptation levels.

3.2.1. Index Construction

To be able to determine the adaptation level of the countries to Islamic finance, an Islamic Finance Adaptation (IFA) Index is constructed with the value of Islamic financial assets, Islamic financial assets to conventional financial assets ratio and the ratio of issuance value of sukuk to issuance value of bond.

The z-score approach, which seeks to convert indicators to a single scale with a mean of zero and standard deviations of one, is used to standardize all indicators by following the below procedure;

$$\text{TM}_{qc} = \frac{x^{t}_{qc} - x^{t}_{qc=\bar{c}}}{\sigma^{t}_{qc=\bar{c}}}$$

The individual components had to be weighted and combined to get the composite IFA Index. The literature recognizes equal weights, mathematical weights, expert weights, and subjective/flexible weights as the four basic ways to designing weights. Equal weighting is retained in this study as each factor has equal importance for identifying Islamic finance adaptation level.

Finally, each normalized individual indicator is multiplied by its final weight and the IFA Index is then aggregated using linear aggregation.

$$IFA_i = w_{pca} Y_i^{VIFA} + w_{pca} Y_i^{IFACFA} + w_{pca} Y_i^{VSVB} \quad (1)$$

3.2.2. Dynamic Conditional Correlation Model

To be able to investigate the dynamic correlation between the progress on the SDGs and Islamic finance adaptation, Dynamic Conditional Correlation model of Engle and Sheppard (2001) is used. The estimation of a univariate GARCH model and the estimation of a bivariate Dynamic Conditional Correlation model are the two phases that make up the DCC model.

The model is defined as:

$$r_{i,t} \mid F_{t-1} \sim N\left(0, H_t\right) \tag{2}$$

where, F_{t-1} is the information set accessible at t-1, H_t is a positive definite conditional variance-covariance matrix, and $r_{i,t}$ are normally distributed return series with zero mean.

The components of $\boldsymbol{H}_t$ are as follows:

$$H_t = D_t R_t D_t \tag{3}$$

$\boldsymbol{D}_t$ is a 2×2 diagonal matrix with $\sqrt{\sigma_{it}^2}$ on the ith diagonal that contains time-varying standard deviations from the estimation of a univariate GARCH model in the first stage. D_t can be expressed in more detail as follows:

$$D_t = \begin{bmatrix} \sqrt{\sigma_{11,t}^2} & 0 \\ 0 & \sqrt{\sigma_{22,t}^2} \end{bmatrix} \tag{4}$$

The univariate GARCH model can be used to calculate $\tilde{A}_{it}^2$ as follows:

$$\sigma_{it}^2 = \omega_0 + \alpha\varepsilon_{t-1}^2 + \beta\sigma_{t-1}^2 \tag{5}$$

where σ_t^2 is the conditional variance at time t, $\omega 0$ is the constant term, ε_{t-1} represents information regarding volatility from the prior period, σ_{t-1}^2 is the estimated conditional variance from the prior period, and α and β are the respective ARCH and GARCH effects. (The presence of conditions $\alpha \geq 0$, $\beta \geq 0$, and $\alpha + \beta < 1$ is required to guarantee the conditional variance σ_t^2's non-negativity.)

Rt is the time-varying correlation matrix of the standardized residuals

$$\varepsilon_t \left(\varepsilon_t = D_t^{-1} r_t \sim N(0, R_t)\right):$$

$$R_t = \begin{bmatrix} 1 & \rho_{12,t} \\ \rho_{21.t} & 1 \end{bmatrix} \tag{6}$$

where, the correlation estimator is $\rho_{12,t} = \dfrac{q_{12,t}}{\sqrt{q_{1,t} q_{2,t}}}$.

The elements in R_t can be calculated by using the following framework:

$$R_t = \left(diag\left(Q_t\right)\right)^{-1} Q_t \left(diag\left(Q_t\right)\right)^{-1} \tag{7}$$

where, Q_t is the proposed dynamic correlation structure.

$\boldsymbol{Q_t}$ can be expressed as:

$$Q_t = (1-\alpha-\beta)\bar{Q} + \alpha\varepsilon_{t-1}\varepsilon'_{t-1} + \beta Q_{t-1} \tag{8}$$

where, εt is the standardized residual matrix from the first stage estimation, and $\bar{Q}$ is the 2×2 unconditional correlation matrix of the standardized residuals ($\bar{Q} = E\left[\varepsilon_t\varepsilon'_t\right]$). The parameters α and β capture the effects of previous shocks and previous dynamic conditional correlations on current dynamic conditional correlations.

Estimation of the Model

The log-likelihood function of the model as follows:

$$\begin{aligned} L(\theta) &= -\frac{1}{2}\sum_{t=1}^{T}(nlog(2\pi) + \log|H_t| + r'_t H_t^{-1} r_t) \\ &= -\frac{1}{2}\sum_{t=1}^{T}(nlog(2\pi) + \log|D_t R_t D_t| + r'_t D_t^{-1} R_t^{-1} D_t^{-1} r_t) \\ &= -\frac{1}{2}\sum_{t=1}^{T}(nlog(2\pi) + 2\log|D_t| + \log|R_t| + \varepsilon'_t R_t^{-1}\varepsilon_t) \end{aligned} \tag{9}$$

where, εt N(0,Rt) are the residuals standardized on the basis of their conditional standard deviations.

The first step parameters can be calculated by maximizing the following function using the quasi-likelihood function:

$$L_1(\varnothing|r_t) = -\frac{1}{2}\sum_{t=1}^{T}\left(nlog(2\pi) + \sum_{t=1}^{n}\left(log(\sigma_{it}^2) + \frac{r_{it}^2}{\sigma_{it}^2}\right)\right) \tag{10}$$

Then the parameters for the second step can be calculated by maximizing the following function:

$$L_2\left(\varphi|\varnothing, r_t\right) = -\frac{1}{2}\sum_{t=1}^{T}\left(\log|R_t| + \varepsilon'_t R_t^{-1}\varepsilon_t - \varepsilon'_t\varepsilon_t\right) \tag{11}$$

3.2.3. Geweke Feedback Measure

To be able to investigate the lead/lag relation between the progress on the SDGs and Islamic finance adaptation, the Geweke Feedback Measure technique is used. As Geweke technique provides cardinal measures of the degree of co-movements and feedback (causality) between the series, it is a quite solid method while investigating whether adaptation of Islamic finance leads to progress on the SDGs.

As Granger causality, Geweke (1982) considers linear projections of two time series. While for the restricted form variable depends on only its own past return -Eq. 12 and Eq. 13-, for the unrestricted form where variable depends on both its own past return and past return of the variable other variable –Eq. 14 and Eq. 15-.

$$x_t = \sum_{i=1}^{\infty} a_{1i} x_{t-i} + u_{1t}, \quad E(u_{1t}) = 0 \ and \ Var(u_{1t}) = \sigma_{u_1}^2 \tag{12}$$

$$y_t = \sum_{i=1}^{\infty} c_{1i} y_{t-i} + v_{1t}, \quad E(v_{1t}) = 0 \ and \ Var(v_{1t}) = \sigma_{v_1}^2 \tag{13}$$

$$x_t = \sum_{i=1}^{\infty} a_{2i} x_{t-i} + \sum_{i=1}^{\infty} b_{2i} y_{t-i} + u_{2t}, \quad E(u_{2t}) = 0 \ and \ Var(u_{2t}) = \sigma_{u_2}^2 \tag{14}$$

$$y_t = \sum_{i=1}^{\infty} c_{2i} y_{t-i} + \sum_{i=1}^{\infty} d_{2i} x_{t-i} + v_{2t}, \quad E(v_{2t}) = 0 \ and \ Var(v_{2t}) = \sigma_{v_2}^2 \tag{15}$$

"The value of $\sigma_{u_1}^2$ measures the accuracy of the autoregressive prediction of x_t based on its previous values, whereas the value of $\sigma_{u_2}^2$ represents the accuracy of predicting the present value of x_t based on the previous values of both x_t and y_t. Furthermore, the value of $\sigma_{v_1}^2$ measures the accuracy of the autoregressive prediction of y_t based on its previous values, whereas the value of $\sigma_{v_2}^2$ represents the accuracy of predicting the present value of y_t based on the previous values of both x_t and y_t" (Schelter et al., 2006, pp.214). Therefore, jointly considering the Equations 12 and 13 and the Equations 14 and 15, Geweke (1982) obtains the systems of Seemingly Unrelated Regressions (SUR) to get the Geweke measures as following:

$$F_{y \to x} = \ln\left[\sigma_{u_1}^2 / \sigma_{u_2}^2\right]$$

$$F_{x \to y} = \ln\left[\sigma_{v_1}^2 / \sigma_{v_2}^2\right] \tag{16}$$

While, the first measure gives the magnitude of the causality from *y* to *x*, the second measure gives the magnitude of the causality from *x* to *y*.

4. FINDINGS

In this section the findings are presented in three sub-sections. According to that, in the first section the results for the Islamic Finance Adaptation Index are revealed. In the second and third sub-sections, the results for dynamic correlation and lead/lag relation between the SDG Index and IFA Index are discussed, respectively.

4.1. IFA Index

The below table shows the Islamic Finance Adaptation scores from 2012 to 2021. With higher scores indicating better adaptation, the results show how much each nation has adapted to Islamic banking. Malaysia has consistently scored the highest, with a score of 0.670 in 2021. Saudi Arabia has shown significant improvement over the years and has the highest score of 0.735 in 2021. Other countries, such as Indonesia, Bahrain, and the United Arab Emirates, have also shown improvement but with varying degrees. The scores of countries that have highest SDG Index scores, such as Finland, Denmark, Sweden, Norway, Austria, Germany, France, and Switzerland, have remained relatively low throughout the years. In other words, these countries have generally scored low in their adaptation to Islamic finance principles. This is not surprising as Islamic finance is not as widely practiced in these countries compared to the countries in the Middle East and Southeast Asia.

Table 3. Islamic finance adaptation index scores between 2012 and 2021

	2012	2013	2014	2015	2016	2017	2018	2019	2020	2021
Malaysia	0.332	0.415	0.483	0.590	0.631	0.632	0.642	0.666	0,662	0.670
Indonesia	0.145	0.310	0.383	0.414	0.434	0.436	0.462	0.492	0.483	0.492
Saudi Arabia	0.289	0.399	0.435	0.564	0.561	0.542	0.581	0.603	0.696	0.735
Bahrain	0.196	0.200	0.231	0.269	0.245	0.243	0.275	0.355	0.387	0.411
United Arab Emirates	0.227	0.241	0.278	0.337	0.389	0.409	0.442	0.485	0.577	0.601
Iran	0.215	0.305	0.335	0.389	0.418	0.466	0.453	0.501	0.557	0.624
Qatar	0.231	0.261	0.320	0.364	0.405	0.469	0.515	0.546	0.566	0.593
Kuwait	0.208	0.225	0.251	0.295	0.337	0.366	0.387	0.456	0.499	0.511
Finland	0.035	0.036	0.034	0.034	0.033	0.035	0.036	0.038	0.039	0.039
Denmark	0.045	0.048	0.045	0.048	0.048	0.048	0.047	0.049	0.049	0.048
Sweden	0.035	0.037	0.037	0.035	0.038	0.039	0.036	0.039	0.040	0.043
Norway	0.046	0.046	0.048	0.047	0.049	0.051	0.051	0.053	0.054	0.053
Austria	0.038	0.039	0.043	0.042	0.043	0.040	0.040	0.039	0.042	0.044
Germany	0.057	0.059	0.061	0.058	0.062	0.066	0.071	0.075	0.080	0.083
France	0.052	0.053	0.052	0.061	0.063	0.064	0.067	0.069	0.070	0.074
Switzerland	0.050	0.061	0.065	0.073	0.087	0.091	0.100	0.112	0.121	0.122

4.2. Dynamic Correlation Between SDG Index and IFA Index

To be able to determine the dynamic conditional correlations between progress on Sustainable Development Goals and adaptation of Islamic finance, in the first step univariate GARCH model is estimated to get the parameter estimations. By using the standardized residuals from the first step, Dynamic Conditional Correlation model of Engle and Sheppard (2001) is estimated. After calculating the annual dynamic correlations between SDG Index and IFA Index for each country, to be able to analyze the dynamic correlation structure between progress on Sustainable Development Goals and adaptation of Islamic finance, average correlations for each country and group of countries are calculated and presented in below table.

Table 4 shows the dynamic correlations between progress on Sustainable Development Goals (SDG) and the adaptation of Islamic finance for top 8 countries in Islamic assets and top 8 countries in the SDG Index. Among the top countries in Islamic finance assets, the top countries with high dynamic correlations between progress on SDGs and adaptation of Islamic finance are United Arab Emirates (0.846), Iran (0.868), and Qatar (0.848). On average, the dynamic correlation for this group of countries is 0.756. Among the top countries according to the SDG Index, the top countries with high dynamic correlations between progress on SDGs and adaptation of Islamic finance are Norway (0.836), Switzerland (0.861), and France (0.807). On average, the dynamic correlation for this group of countries is 0.604. It is important to note that Sweden has a negative dynamic correlation (-0.219) between progress on SDGs and adaptation of Islamic financial instruments.

The first group, which consists of Malaysia, Indonesia, Saudi Arabia, Bahrain, United Arab Emirates, Iran, Qatar, and Kuwait, has a higher average dynamic correlation between progress on Sustainable Development Goals (SDGs) and adaptation of Islamic finance at 0.756. This suggests that these countries have a stronger relationship between the use of Islamic finance and progress towards achieving the SDGs. In contrast, the second group, which consists of Finland, Denmark, Sweden, Norway, Austria, Germany, France, and Switzerland, has a lower average dynamic correlation at 0.604. This indicates a weaker relationship between the use of Islamic finance and progress towards achieving the SDGs in these countries.

The first group consists of countries ranked in the top 8 in Islamic finance while the second group consists of countries ranked in the top 8 according to the SDG Index, which measures progress towards achieving the United Nations' Sustainable Development Goals (SDGs). According to that, there is a weaker relationship between the use of Islamic finance and progress towards achieving the SDGs in the second group of countries. In other words, the use of Islamic finance appears to have less impact on the progress towards achieving the SDGs in these countries compared to the first group. Furthermore, the

Table 4. Country and group averages of dynamic correlations between SDG Index and IFA Index

Top 8 Countries in Islamic Finance Assets								
Malaysia	Indonesia	Saudi Arabia	Bahrain	United Arab Emirates	Iran	Qatar	Kuwait	Average
0.821	0.765	0.780	0.301	0.846	0.868	0.848	0.822	0.756
Top 8 Countries in SDG Index								
Finland	Denmark	Sweden	Norway	Austria	Germany	France	Switzerland	Average
0.567	0.647	-0.219	0.836	0.253	0.642	0.807	0.861	0.604

first group of countries shows a stronger connection between the use of Islamic finance and progress towards achieving the SDGs. This implies that the use of Islamic finance is more effective in contributing to progress towards the SDGs in the first group of countries compared to the second group.

4.3. Lead/Lag Relationship Between SDG Index and IFA Index

In the second step of the analysis, lead/lag relationship between SDG Index and IFA Index is investigated. The below table presents the Geweke Feedback Measures between those two variables at both country level and group level.

Each statistic has an approximate χ^2 distribution with three degrees of freedom under the null hypothesis that adaptation of Islamic finance does not lead progress on Sustainable Development Goals over a period of 1-3 years.

The critical value χ_3^2, is 6.251 for 10% level, 7.815 for 5% level, and 11.345 for 1% level.

Rejection of the null hypothesis is represented by: *, for the 10; **, for the 5; and ***, for the 1% level.

Table 5 presents the Geweke Feedback Measures between two variables, namely the Sustainable Development Goals Index (SDGI) and the Islamic Finance Adaptation Index (IFAI), at both the country and group levels. The Geweke Feedback Measure is a statistical measure used to assess the causal rela-

Table 5. Geweke feedback measures

	Geweke Unidirectional Feedback Measure From SDGI to IFAI ($GCFM_{SDGI \to IFAI}$)	Geweke Unidirectional Feedback Measure From IFAI to SDGI ($GCFM_{IFAI \to SDGI}$)
Malaysia	3.211	6.449**
Indonesia	3.458	11.359***
Saudi Arabia	5.173	12.047**
Bahrain	1.214	5.355
United Arab Emirates	4.453	10.305***
Iran	2.075	9.611**
Qatar	0.998	8.677**
Kuwait	3.039	14.258***
Average of Top Islamic Finance Countries	2.953	9.758**
Finland	2.237	3.356
Denmark	2.856	4.819
Sweden	3.555	2.924
Norway	1.987	5.053
Austria	0.993	1.917
Germany	4.254	6.075
France	3.035	5.026
Switzerland	2.908	6.211
Average of Top SDG Index Countries	2.728	4.423

tionship between two variables. The first column shows the Geweke Unidirectional Feedback Measure from SDGI to IFAI, indicating the degree to which progress on Sustainable Development Goals leads to the adaptation of Islamic financial instruments. The second column shows the Geweke Unidirectional Feedback Measure from IFAI to SDGI, indicating the degree to which the adaptation of Islamic financial instruments leads to progress on Sustainable Development Goals. The table includes data for individual countries, as well as the averages for the top Islamic finance countries and the top SDG Index countries. The statistics have an approximate chi-squared distribution with three degrees of freedom under the null hypothesis that adaptation of Islamic finance does not lead to progress on Sustainable Development Goals over a period of 1-3 years. The critical values for rejection of the null hypothesis are also provided for the 10%, 5%, and 1% levels. The rejection of the null hypothesis is represented by asterisks, with one, two, or three asterisks indicating rejection at the 10%, 5%, or 1% level, respectively.

The table shows that the top Islamic finance countries (Malaysia, Indonesia, Saudi Arabia, Bahrain, UAE, Iran, Qatar, and Kuwait) have a significant positive Geweke Feedback Measure from IFAI to SDGI, indicating that adaptation of Islamic finance instruments in these countries lead to the progress on Sustainable Development Goals. On the other hand, Geweke Feedback Measures from SDGI to IFAI, suggesting that progress on Sustainable Development Goals the does not lead the adaptation of Islamic finance. The top SDG Index countries (Finland, Denmark, Sweden, Norway, Austria, Germany, France, and Switzerland) also show a positive Geweke Feedback Measure from IFAI to SDGI, indicating that the adaptation of Islamic finance instruments tends to lead to progress on Sustainable Development Goals in these countries too. However, as the results are not statistically significant, we conclude that in the countries that have highest SDG Index scores, there is no significant lead/lag relationship between progress on Sustainable Development Goals and Islamic finance adaptation.

The average Geweke Unidirectional Feedback Measure from the SDG Index to the IFA Index for the top Islamic finance countries is 2.953. On the other hand, the average Geweke Unidirectional Feedback Measure from the IFA Index to the SDG Index for the top Islamic finance countries is 9.758, which is statistically significant at the 5% level. For the top SDG Index countries, the results suggest a weaker lead/lag relationship between the two variables. The average Geweke Unidirectional Feedback Measure from the SDG Index to the IFA Index for the top SDG Index countries is 2.728, which is not statistically significant. The average Geweke Unidirectional Feedback Measure from the IFA Index to the SDG Index for the top SDG Index countries is 4.423, which is also not statistically significant. Overall, the results suggest a positive relationship between the adaptation of Islamic finance and progress on Sustainable Development Goals in the top Islamic finance countries and the top SDG Index countries. The results suggest that the adaptation of Islamic financial instruments can lead progress on Sustainable Development Goals, particularly in the top Islamic finance countries.

5. CONCLUSION

The Sustainable Development Goals (SDGs) were established by the United Nations in 2015 to address global challenges such as poverty, inequality, environmental pollution, and climate change by 2030. Islamic finance, which incorporates religious, moral, and social aspects, aligns with the SDGs by promoting justice, equal opportunities, well-being, and environmental conservation. However, the industry's actual contribution to these goals has fallen short of its potential. This study aims to investigate the relationship between progress on the SDGs and the adaptation of Islamic financial instruments by constructing an Islamic Finance Adaptation (IFA) Index and analysing the dynamic correlation and lead/lag relation between the SDGs and Islamic finance.

The Sustainable Development Solutions Network (SDSN) and the Bertelsmann Stiftung have been publishing an annual SDG Index and Dashboards Report since 2016 to measure each nation's progress towards achieving the Sustainable Development Goals (SDGs) established by the international community in 2015. This study investigates the relationship between Islamic finance and progress on SDGs using data from the top 8 countries for Islamic finance and the top 8 countries according to the SDG Index from 2012 to 2021. The study involves developing a composite index to score the adaptation levels of the countries to Islamic finance and investigating the dynamic correlation and lead/lag relation between progress on SDGs and Islamic finance adaptation levels using a Dynamic Conditional Correlation (DCC) model and Geweke Feedback Measure. An Islamic Finance Adaptation (IFA) Index is constructed using the value of Islamic financial assets, Islamic financial assets to financial assets ratio, and the ratio of sukuk issuance value to bond issuance value to determine the adaptation level of countries to Islamic finance.

The Islamic Finance Adaptation scores rank countries by their level of adaptation to Islamic finance, with Malaysia consistently scoring the highest. Saudi Arabia has also improved significantly, while countries that have highest SDG Index scores have remained relatively low. The top countries with high dynamic correlations between progress on SDGs and adaptation of Islamic finance are United Arab Emirates, Iran, and Qatar, with an average dynamic correlation of 0.756. Norway, Switzerland, and France have the highest dynamic correlations between progress on SDGs and adaptation of Islamic financial instruments among the top 8 countries according to the SDG Index, with an average dynamic correlation of 0.604. The first group -top Islamic finance countries- has a stronger relationship between the use of Islamic finance and progress towards achieving the SDGs, while the second group –top countries according to SDG Index- has a weaker relationship. Finally, the adaptation of Islamic finance led to progress on Sustainable Development Goals in the top Islamic finance countries.

REFERENCES

Aribi, Z. A., & Arun, G. T. (2015). Corporate Social Responsibility and Islamic Financial Institutions (IFIs): Management Perceptions from IFIs in Bahrain. *Journal of Business Ethics, 129*(4), 785–794. doi:10.100710551-014-2132-9

Arrow, K. J., Chenery, B. H., Minhas, S. B., & Solow, M. R. (1961). Capital-Labor Substitution and Economic Efficiency. *The Review of Economics and Statistics, 43*(3), 225–250. doi:10.2307/1927286

Ayub, M. (2007). *Understanding Islamic Finance* (1st ed.). John Wiley & Son's Ltd.

Brooks, C. (2008). *Introductory Econometrics for Finance* (2nd ed.). Cambridge University Press. doi:10.1017/CBO9780511841644

Chapra, M. U. (2000). *The Future of Economics: An Islamic Perspective*. The Islamic Foundation.

Engle, R. F., & Sheppard, K. (2001). *Theoretical and Empirical Properties of Dynamic Conditional Correlation MVGARCH*. Economic Working Paper Series 8554, University of California.

ESCWA. (2021). *International Dialogue on the Role of Islamic Social Financing in Achieving the SDGs.* Available online: https: //www.unescwa.org/news/international-dialogue-role-islamic-social-financing-achieving-sdgs

Fuess, S., & Millea, M. (2002). Do Employers Pay Efficiency Wage? Evidence from Japan. *Journal of Labor Research, 23*(2), 279–292. doi:10.100712122-002-1007-5

Geweke, J. (1982). Measurement of Linear Dependence and Feedback between Multiple Time Series. *Journal of the American Statistical Association, 77*(378), 304–313. doi:10.1080/01621459.1982.10477803

Habib, A. (2013). Financial Inclusion and Islamic Finance: Organizational Formats, Products, Outreach and Sustainability. In Economic Development and Islamic Finance. World Bank Publications.

Habib, A., Mohieldin, M., Verbeek, J., & Aloulmagd, F. (2015). *On Sustainable Development Goals and the Role of Islamic Finance.* Policy Research Working Paper 7266, World Bank Group.

Haniffa, R., & Hudaib, M. (2007). Exploring the Ethical Identity of Islamic Banks via Communication in Annual Reports. *Journal of Business Ethics, 76*(1), 97–116. doi:10.100710551-006-9272-5

Kamla, R., & Rammal, G. H. (2013). Social Reporting by Islamic banks: Does Social Justice Matter? *Accounting, Auditing & Accountability Journal, 26*(6), 911–945. doi:10.1108/AAAJ-03-2013-1268

Lafortune, G., Fuller, G., Moreno, J., Schmidt-Traub, G., & Kroll, V. (2018). *SDG Index and dashboards. Detailed Methodological Paper.* Bertelsmann Stiftung and Sustainable Development Solutions Network Sustainable Development Solutions Network.

Lanzara, F. (2021). Islamic Finance and Sustainable Development Goals. A Bibliometric Analysis from 2000 to 2021. *European Journal of Islamic Finance, 18*, 1–19.

Maali, B., Casson, P., & Napier, C. (2006). Social Reporting by Islamic Banks. *Abacus, 42*(2), 266–289. doi:10.1111/j.1467-6281.2006.00200.x

Rehman, A. A., & Pickup, F. (2018). Zakat for the SDGs. *Islamic Finance News, 15*(34), 1–14.

Sairally, S. (2007). Evaluating the 'Social Responsibility' of Islamic Finance: Learning from the Experience of Socially Responsible Investment Funds. In M. Iqbal, S.S. Ali, & D. Muljawan (Eds.), Advances in Islamic Economics and Finance. Islamic Research & Training Institute, Islamic Development Bank Group.

Schelter, B., Winterhalder, M., Eichler, M., Peifer, M., Hellwig, B., Guschlbauer, B., Lucking, C., Dahlhaus, R., & Timmer, J. (2006). Testing for Directed Influences among Neural Signals using Partial Directed Coherence. *Journal of Neuroscience Methods, 152*(2), 210–219. doi:10.1016/j.jneumeth.2005.09.001 PMID:16269188

Shaikh, S. A., & Ismail, A. G. (2017). Role of Zakat in Sustainable Development Goals. *International Journal of Zakat, 17*(2), 1–9. doi:10.37706/ijaz.v2i2.21

APPENDIX

Table 6. Sustainable development goals and indicators included in SDG Index

Sustainable Development Goals	Indicator/Proxy
No Poverty	Poverty headcount ratio at $1.90/day (%) Poverty headcount ratio at $3.20/day (%) Poverty rate after taxes and transfers, Poverty line 50% (% population)
Zero Hunger	Prevalence of obesity, BMI ≥ 30 (% adult population) Cereal yield (t/ha) Prevalence of stunting (low height-for-age) in children under 5 years of age (%) Prevalence of wasting in children under 5 years of age (%)
Good Health and Well-being	Maternal mortality rate (per 100,000 live births) Neonatal mortality rate (per 1,000 live births) Mortality rate, under-5 (per 1,000 live births) Incidence of tuberculosis (per 100,000 population) HIV prevalence (per 1,000) Age-standardised death rate due to cardiovascular disease, cancer, diabetes, and chronic respiratory disease in populations age 30–70 years (per 100,000 population) Traffic deaths rate (per 100,000 population) Adolescent fertility rate (births per 1,000 women ages 15-19) Percentage of surviving infants who received 2 WHO-recommended vaccines (%) Universal Health Coverage Tracer Index (0-100 Subjective Wellbeing (average ladder score, 0-10) Life Expectancy at birth (years) Daily smokers (% population age 15+)
Quality Education	Net primary enrolment rate (%) Mean years of schooling (years) Population age 25-64 with tertiary education (%) Students performing below level 2 in science (%)
Gender Equality	Estimated demand for contraception that is unmet (% women married or in union, ages 15-49) Ratio of female to male labour force participation rate Seats held by women in national parliaments (%) Gender wage gap (Total, % male median wage)
Clean Water and Sanitation	For all other countries: Population using at least basic drinking water services (%) For all other countries: Population using at least basic sanitation services (%) For high-income & OECD countries: population using safely managed water services (%) For high-income & OECD countries: population using safely managed sanitation services (%)
Affordable and Clean Energy	Access to electricity (% population) Access to clean fuels & technology for cooking (% population) CO2 emissions from fuel combustion / electricity output (MtCO2/TWh) Share of renewable energy in total final energy consumption (%)
Decent Work and Economic Growth	Adults (15 years and older) with an account at a bank or other financial institution or with a mobile-money-service provider (%) Unemployment rate (% total labor force) Employment-to-Population ratio (%) Youth not in employment, education or training (NEET) (%)
Industry, Innovation and Infrastructure	Population using the internet (%) Mobile broadband subscriptions (per 100 inhabitants Quality of overall infrastructure (1= extremely underdeveloped; 7= extensive and efficient by international standards) Research and development researchers (per 1,000 employed) Triadic Patent Families filed (per million population)
Reduced Inequalities	Gini Coefficient adjusted for top income (1-100) Palma ratio
Sustainable Cities and Communities	Improved water source, piped (% urban population with access) Satisfaction with public transport (%) Levels of particulate matter smaller than 2.5 microns (PM2.5)
Responsible Consumption and Production	Municipal Solid Waste (kg/year/capita) E-waste generated (kg/capita) Percentage of anthropogenic wastewater that receives treatment (%) Production-based SO2 emissions (kg/capita) Net imported SO2 emissions (kg/capita) Nitrogen production footprint (kg/capita) Net imported emissions of reactive nitrogen (kg/capita) Non-Recycled Municipal Solid Waste (MSW in kg/person/year times recycling rate)

Sustainable Development Goals	Indicator/Proxy
Climate Action	Energy-related CO2 emissions per capita (tCO2/capita)
Life Below Water	Ocean Health Index Goal - Biodiversity (0-100) Ocean Health Index Goal - Clean Waters (0-100) Ocean Health Index Goal - Fisheries (0-100) Fish caught by trawling (%)
Life on Land	Mean area that is protected in terrestrial sites important to biodiversity (%) Mean area that is protected in freshwater sites important to biodiversity (%) Red List Index of species survival (0-1)
Peace, Justice and Strong Institutions	Proportion of the population who feel safe walking alone at night in the city or area where they live (%) Government Efficiency (1-7) Property Rights (1-7) Corruption Perception Index (0-100)
Partnerships for the Goals	Government Health and Education spending (% GDP) For all other countries: Tax revenue (% GDP) For high-income and all OECD DAC countries: International concessional public finance, including official development assistance (% GNI)

Source: Lafortune et al. (2018)

Chapter 5
The Effects of Climate Change on Human Health and the Healthcare Sector

Ranjit Barua
https://orcid.org/0000-0003-2236-3876
Omdayal Group of Institutions, India

Sumit Bhowmik
Omdayal Group of Institutions, India

Debasish Banerjee
Omdayal Group of Institution, India

Sudipto Datta
https://orcid.org/0000-0003-2161-6878
Indian Institute of Science, Bangalore, India

ABSTRACT

Just a few of the ways that climate change has influenced human lives and health over the past several decades include changes in extreme weather and climatic events, changes in the labor force, changes in food security, and changes in the prevalence and geographic distribution of infectious diseases. Indicators for climate change and health were developed as useful tools to track the complex network of interdependent relationships that connect climate change and human health. While the weather has an impact on the timing and severity of outbreaks, climate primarily influences the range of infectious diseases. This chapter aims to offer an outline of the impacts of climate change on human health in analysis, research potential, and applications in the healthcare sector.

DOI: 10.4018/978-1-6684-9503-2.ch005

1. INTRODUCTION

Recent years have seen significant scientific and popular discussions over climate change and its direct and indirect effects on human health. According to (Prüss-Üstün et al., 2006), some 2.5 million people each year pass away from non-infectious diseases that can be directly linked to environmental variables like lead exposure, air pollution, harsh weather, stressful work environments, and exposure to cigarette smoke. More specifically, lead exposure is thought to be responsible for 3% of the burden of cerebrovascular illness and 2% of the burden of ischemic heart disease (McMichael et al., 2003). About 2% of all cardiovascular diseases worldwide are caused by outdoor air pollution (McMichael et al., 2003). Climate variability and change are other elements that may impact human health either directly or indirectly by altering the biological and ecological mechanisms that control the spread of many infectious diseases (Bezirtzoglou et al., 2011). Thermal stresses brought on by more frequent and intense heat waves, such as heart and respiratory conditions and heat exhaustion, as well as fatalities and injuries from extreme weather events, are only a few examples of direct effects on human health (Mouchet et al., 1997) (Githeko et al., 2000). Malnutrition, food- and water-borne illnesses, vector-borne illnesses, and increased morbidity as a result of exposure to high temperatures and air pollution are examples of indirect consequences (Githeko et al., 2000; Ma et al., 2022). The relationship between human health and climate change is supported by science and has existed for a long time. But until recently, there wasn't much attention paid to leveraging the science of health informatics to address this aspect of healthcare and wellness. Figure 1 shows the impact of the climate change on human health. This chapter provides a brief overview of current worldwide efforts on two fronts: how health informatics may lower the carbon footprint of healthcare, and how it can incorporate new types of data to provide insights into how climate change will affect people's health. The two guiding principles of health informatics are connectivity, or the linking and integrating of health data from many sources, and sustainability, or improving the efficiency and effectiveness of healthcare as a whole. By strengthening its adherence to these principles, health informatics will be better positioned as a subject and a profession to support and direct technological developments that address global health concerns related to climate change.

Figure 1. Impact of the climate change in human health

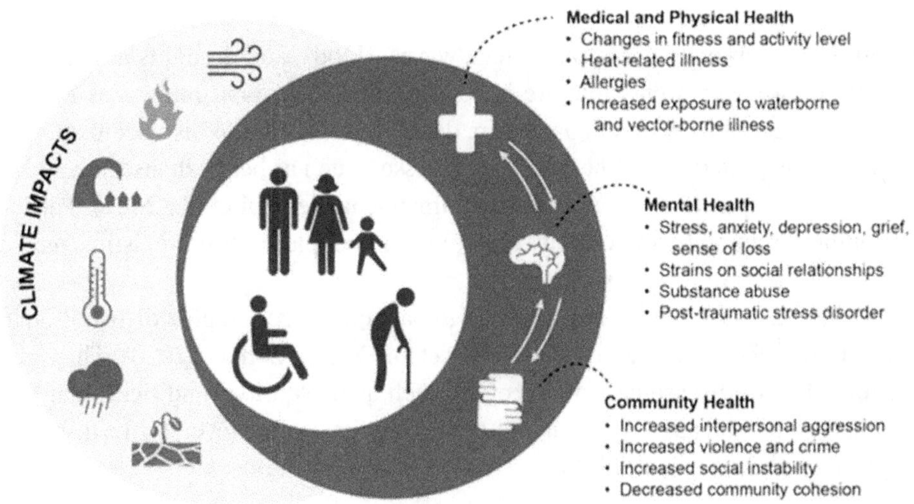

Human health may be significantly impacted by climate change, which may place additional strain on healthcare systems. The effects of climate change on human health may put more burden on the healthcare infrastructure (Patz et al., 1996). More people may need emergency medical care or treatment for injuries or diseases when the frequency and severity of extreme weather events rise (Stark et al., 2009). Additionally, deteriorating air quality may prompt more people to seek medical attention for respiratory conditions like asthma, chronic bronchitis, and COPD (Hansel et al., 2016). The expansion of aquatic illnesses and disease-carrying insects may potentially result in a rise in the demand for medical care (Vezzulli et al., 2013; Tarek et al., 2019). Through its energy use and waste output, the healthcare industry itself also contributes to climate change. In order to run equipment and maintain acceptable temperatures, hospitals and healthcare institutions use a lot of energy, which increases greenhouse gas emissions (D'Amato et al, 2020; Barrett et al., 2015). With warming and variations in outbreaks linked to weather extremes, climate change models also involve changes in the distribution of infectious illnesses (Nikendei, et al., 2020; Lu et al., 2022). Several vector-borne diseases or their distributions are already shifting in altitude as a result of global warming (Rossati et al., 2017; Ogden et al., 2019). Additionally, more severe weather events foster the development of infectious disease outbreaks by eradicating insect breeding grounds, driving rodents from their burrows, and contaminating clean water supplies.

2. EFFECT ON HUMAN HEALTH

Climate change has a wide range of impacts on human health, both direct and indirect (El-Sayed et al., 2020). Toxins in the environment can aggravate ailments like cancer, heart disease, and respiratory problems. People with lower incomes are more likely to live in polluted areas and have access to contaminated water (Joshi et al., 2020). Additionally, children and pregnant women are more at risk for pollution-related health problems. Some of these issues include chemical pollution, air pollution, climate change, harmful microbes, lack of access to healthcare, poor infrastructure, and poor water quality (Ray et al., 2020). Increased exposure to heat stress, changing patterns of infectious disease, and worsened air quality are just a few simple implications of climate change on health (Figure 2). Changes in the hydrologic cycle, higher ground-level ozone, and enhanced pollen production are some other consequences. Here are some ways in which climate change can affect human health:

Extreme weather occurrences: Hurricanes, heat waves, floods, and droughts are just a few examples of the extreme weather events that are becoming more often and intense as a result of climate change. These occurrences may cause harm, death, and population displacement, posing an acute health danger. Increased risk of heat-related illnesses, such as heat exhaustion and heatstroke, is a result of rising temperatures brought on by climate change (Ebi et al., 2021). Particularly at risk are vulnerable groups like the elderly, children, and people with pre-existing medical disorders (Pacheco et al., 2021).

Infectious diseases: Climate change may have an impact on the geographic distribution and transmission patterns of infectious diseases (Ellwanger et al., 2020; Thomas., 2020). The reproduction and movement habits of disease-carrying vectors, such as mosquitoes and ticks, can be impacted by rising temperatures and altered rainfall patterns. Malaria, dengue fever, Lyme disease, and the Zika virus can all spread as a result of this.

Figure 2. Social effects of climate change on human society, livelihood, and infectious diseases (Khan et al., 2019)

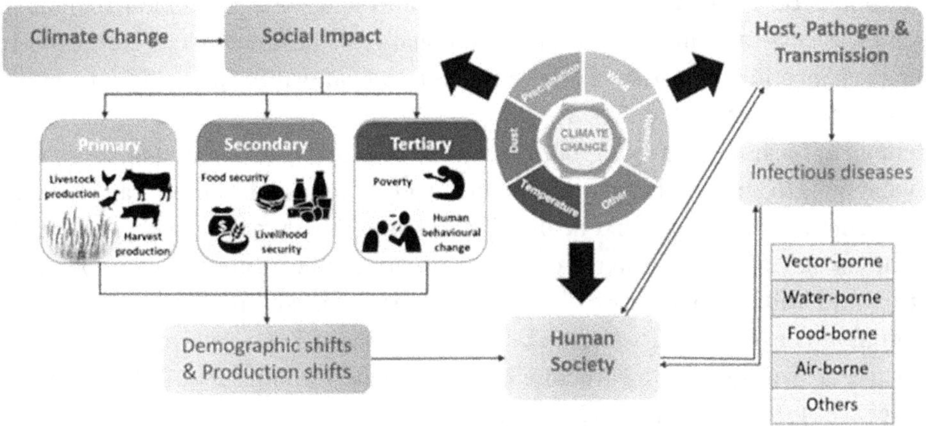

Air pollution: By raising the concentration of pollutants and encouraging the production of ground-level ozone, climate change can deteriorate air quality. Chronic obstructive pulmonary disorders (COPD), including asthma, bronchitis, and other respiratory illnesses, are linked to air pollution exposure (Janson et al., 2020).

Malnutrition and food security: Climate change has an impact on agricultural output, causing crop failures, alterations in the availability of food, and a rise in food prices (Boliko., 2019). Malnutrition and nutritional deficits may result from these changes, especially in communities that are already at risk (Al-Jawaldeh et al., 2022).

Waterborne infections: Changes in temperature and precipitation patterns can have an impact on the quality of the water and raise the risk of waterborne infections (Levy et al., 2018). Floods and periods of heavy rain can contaminate water supplies with bacteria, causing cholera and other waterborne illness outbreaks (Abbass et al., 2022).

3. VULNERABILITY INDICATORS OF CLIMATE CHANGE

Vulnerability indicators of climate change refer to the factors that determine the susceptibility of a population or system to the impacts of climate change (Workman et al., 2021). These indicators help identify areas or groups that are more likely to be affected and provide insights into the specific risks they face. Here are some commonly used vulnerability indicators:

Socioeconomic factors: Socioeconomic indicators include poverty rates, income levels, education levels, access to basic services, and employment opportunities (Johnson et al., 2023). Populations with low socioeconomic status often have limited resources and capacity to cope with and adapt to climate change impacts.

Geographic location: The physical location of a population plays a significant role in vulnerability (Romanello et al., 2021). Coastal areas, for example, are susceptible to sea-level rise, storm surges, and increased flood risk. Similarly, populations in arid or semi-arid regions may be vulnerable to drought and water scarcity.

Health status and access to healthcare: The overall health status of a population, including prevalence of diseases, access to healthcare facilities, and quality of healthcare, can influence vulnerability to climate change impacts, particularly in relation to heat waves, air pollution, and the spread of infectious diseases (Pega et al., 2015).

Age and demographics: Certain age groups, such as the elderly, children, and pregnant women, are more vulnerable to climate change impacts due to physiological differences and increased sensitivity to extreme weather events and health risks. Demographic factors such as population density, urbanization, and migration patterns can also influence vulnerability (Geng et al., 2022).

Infrastructure and built environment: The quality and resilience of infrastructure, including housing, transportation networks, and water and sanitation systems, can determine vulnerability. Poorly constructed or inadequate infrastructure can exacerbate the impacts of climate-related disasters and affect access to essential services (Huang et al., 2022).

Ecosystem dependence: Communities that rely heavily on natural resources and ecosystem services for their livelihoods, such as agriculture, forestry, and fisheries, are vulnerable to climate change-induced disruptions in these systems. Changes in precipitation patterns, temperature, and ecosystem functioning can directly affect their livelihoods and food security (Malhi et al., 2020).

Governance and institutional capacity: The effectiveness of governance structures, policies, and institutions in addressing climate change and implementing adaptation measures plays a crucial role in vulnerability. Strong governance and institutional capacity are essential for coordinating response efforts, providing early warnings, and ensuring equitable distribution of resources and support.

These susceptibility indicators are frequently combined to determine how vulnerable an area or population is overall to the effects of climate change. Understanding these variables aids in the identification and prioritisation of adaptation plans and actions to lessen vulnerability and increase resilience by policymakers and stakeholders. Disasters brought on by climate change, the loss of livelihoods, and forced migrations can all have a negative influence on mental health. In areas affected by climate change-related disasters, higher rates of anxiety, depression, post-traumatic stress disorder (PTSD), and other mental health issues have been noted. It is significant to highlight that different regions can experience different effects of climate change on human health (Zinsstag et al., 2018). These health concerns frequently have an outsized impact on vulnerable people, including low-income areas, marginalized groups, and developing nations. For the sake of preserving human health and well-being, it is essential to take proactive steps to mitigate and adapt to climate change.

4. THE EFFECTS OF CLIMATE CHANGE ON INFECTIOUS DISEASES

Infectious illness occurrence, distribution, and transmission patterns are all impacted by climate change in different ways (Gray., 2022; Setoguchi et al., 2022). Here are a few ways that infectious diseases have been impacted by climate change:

Water-borne diseases: Modifications in precipitation patterns, such as more frequent and intense downpours, can cause flooding and contaminate water supplies. This raises the possibility of developing watery illnesses like cholera, dysentery, and other diarrheal illnesses. Events of significant rainfall can overload sewage systems, resulting in sewage mixing with drinking water sources and outbreaks (Caminade et al., 2019).

Vector-borne diseases: Disease-carrying vectors like mosquitoes, ticks, and sandflies can modify their geographic range and population density as a result of climate change. The transmission season can be extended and the optimal habitat for these vectors can increase due to warmer temperatures and changing precipitation patterns, which will let them to flourish in new places. This may cause vector-borne illnesses including Lyme disease, malaria, dengue fever, chikungunya and the Zika virus, to spread to areas where they had not previously been common (Semenza et al., 2012).

Respiratory illnesses: As air quality deteriorates due to climate change, respiratory illnesses rise. Increased concentrations of air pollutants and the production of ground-level ozone can be caused by rising temperatures and changes in the atmosphere. Long-term exposure to air pollution raises the incidence of respiratory infections and aggravates respiratory disorders like asthma (Barne et al., 2013).

Food-borne illnesses: Climate change can have an effect on food safety and help spread these infections (Lake et al., 2018). The production, storage, and transportation of food can be impacted by rising temperatures and changes in precipitation patterns, increasing the danger of pathogen contamination with organisms like Salmonella and E. coli. The distribution of diseases found in seafood can also be impacted by changes in water temperature.

NTDs or neglected tropical diseases: NTDs are endemic in some regions, and climate change has an impact on both their prevalence and distribution (Booth., 2018; Molyneux et al., 2021). Variations in temperature and humidity can have an impact on the life cycle and transmission of diseases, such as leishmaniasis, schistosomiasis, and onchocerciasis, which can affect the people at risk (Hotez., 2018).

Only two environmental factors, temperature and climatic conditions, can have an impact on skin problems. The following is a discussion of skin ailments that can occur in both hot and cold weather:

Skin conditions brought on by a hot climate: a. Heat Rash (Prickly Heat): This condition develops when sweat glands become clogged, leaving the skin itchy and covered with tiny red bumps. It frequently occurs in hot, humid weather. a. Sunburn: Excessive sun exposure results in red, painful skin and can have long-term effects like early aging and a higher risk of developing skin cancer (Huang et al., 2022). Despite not directly being a skin disorder, heatstroke is a serious condition brought on by extended exposure to high temperatures. It may cause heated, dry skin, a quick heartbeat, and confusion in the mind (Wang et al., 2021).

Skin Disorders Resulting from the Cool Climate (Cold Weather)

Dry skin: The skin can become dry, flaky, and itchy due to cold temperatures, low humidity levels, and interior heating (Esperland et al., 2022). Chilblains, which can appear on the fingers, toes, ears, and nose as a result of exposure to the cold, are unpleasant, red, and itchy pimples.

Frostbite: Skin and underlying tissue damage from prolonged exposure to subfreezing conditions can result in frostbite. It results in the affected parts turning pallid, numb, and ultimately blistering.

Related Skin Diseases

Eczema (dermatitis): Eczema is a persistent skin disorder marked by red, swollen, and itchy spots. Extreme temperatures, both hot and cold, have the potential to cause it or make it worse (Spiewak., 2023.

Psoriasis: This autoimmune skin disorder results as thick, flaming, and scaly patches of skin. Changes in the weather, particularly in the cold and dry conditions, might make symptoms worse (Bu et al., 2022).

Rosacea: Rosacea is a chronic inflammatory skin disorder that occasionally results in pimples that resemble acne, facial redness, and small, visible blood vessels (Barakji et al., 2022). Flare-ups can be caused by extreme temperatures, both hot and cold.

Cold urticaria: This uncommon condition causes hives and itching when exposed to cold air or water because the skin reacts to the cold.

In extreme weather, it's crucial to take the right precautions to preserve your skin. Use sunscreen, put on protective clothing, and drink plenty of water while it's hot outside. Always moisturize your skin in chilly temperatures, and layer your clothing to stay warm. For correct diagnosis and treatment, it is recommended to see a dermatologist if you have any recurring skin problems. While climate change can affect the dynamics of infectious diseases, it is crucial to remember that other factors also play a role in their incidence. Other important aspects include socioeconomic conditions, population density, hygienic practises, access to healthcare, and others. In order to reduce the effects of climate change on infectious illnesses and safeguard the population's health, appropriate adaptation plans must be put into place alongside comprehensive public health interventions.

5. CONCLUSION

Infectious illness epidemiology is affected by climate change. When climate variables alter dramatically, VBDs may first appear or then reappear within populations. Changes in climate, soils, forest cover, and land use have improved the environmental appropriateness for infectious illnesses. Humanity will certainly face new opportunities to respond to novel or emerging infectious diseases as the climate catastrophe develops. Systems that use climate data to pinpoint areas that are particularly vulnerable to disease outbreaks are currently being developed. Areas with developing diseases may benefit from targeted vector-control initiatives, which have already been utilized to successfully combat some outbreaks. Good management of currently available antibiotics and ongoing research to produce new antibiotics are crucial given the concern over antibiotic resistance and the diminishing availability of next-generation antibiotics. Dermatologists should be aware that the burden and geographic distribution of infectious illnesses, many of which have cutaneous symptoms and may be encountered during routine practice, will be impacted by climate change.

CONFLICTS OF INTEREST

The authors declare that there are no conflicts of interest regarding the publication of this chapter.

ACKNOWLEDGMENT

The authors would like to thank Dept. of Mechanical Engineering, Omdayal Group of Institutions, and thanks to Mrs. Nibedita Bardhan for language proof reading.

REFERENCES

Abbass, K., Qasim, M. Z., Song, H., Murshed, M., Mahmood, H., & Younis, I. (2022). A review of the global climate change impacts, adaptation, and sustainable mitigation measures. *Environmental Science and Pollution Research International, 29*(28), 42539–42559. doi:10.100711356-022-19718-6 PMID:35378646

Al-Jawaldeh, A., Nabhani, M., Taktouk, M., & Nasreddine, L. (2022). Climate Change and Nutrition: Implications for the Eastern Mediterranean Region. *International Journal of Environmental Research and Public Health, 19*(24), 17086. doi:10.3390/ijerph192417086 PMID:36554966

Barakji, Y. A., Rønnstad, A. T. M., Christensen, M. O., Zachariae, C., Wienholtz, N. K. F., Halling, A. S., Maul, J. T., Thomsen, S. F., Egeberg, A., & Thyssen, J. P. (2022). Assessment of Frequency of Rosacea Subtypes in Patients With Rosacea: A Systematic Review and Meta-analysis. *JAMA Dermatology, 158*(6), 617–625. doi:10.1001/jamadermatol.2022.0526 PMID:35385049

Barne, C., Alexis, N. E., Bernstein, J. A., Cohn, J. R., Demain, J. G., Horner, E., Levetin, E., Nei, A., & Phipatanakul, W. (2013). Climate change and our environment: The effect on respiratory and allergic disease. *The Journal of Allergy and Clinical Immunology. In Practice, 1*(2), 137–141. doi:10.1016/j.jaip.2012.07.002 PMID:23687635

Barrett, B., Charles, J. W., & Temte, J. L. (2015). Climate change, human health, and epidemiological transition. *Preventive Medicine, 70*, 69–75. doi:10.1016/j.ypmed.2014.11.013 PMID:25434735

Bezirtzoglou, C., Dekas, K., & Charvalos, E. (2011). Climate changes, environment and infection: Facts, scenarios and growing awareness from the public health community within Europe. *Anaerobe, 17*(6), 337–340. doi:10.1016/j.anaerobe.2011.05.016 PMID:21664978

Boliko, M. C. (2019). FAO and the Situation of Food Security and Nutrition in the World. *Journal of Nutritional Science and Vitaminology, 65*(Supplement), S4–S8. doi:10.3177/jnsv.65.S4 PMID:31619643

Booth, M. (2018). Climate Change and the Neglected Tropical Diseases. *Advances in Parasitology, 100*, 39–126. doi:10.1016/bs.apar.2018.02.001 PMID:29753342

Bu, J., Ding, R., Zhou, L., Chen, X., & Shen, E. (2022). Epidemiology of Psoriasis and Comorbid Diseases: A Narrative Review. *Frontiers in Immunology, 13*, 880201. doi:10.3389/fimmu.2022.880201 PMID:35757712

Caminade, C., McIntyre, K. M., & Jones, A. E. (2019). Impact of recent and future climate change on vector-borne diseases. *Annals of the New York Academy of Sciences, 1436*(1), 157–173. doi:10.1111/nyas.13950 PMID:30120891

D'Amato, G., Chong-Neto, H. J., Monge Ortega, O. P., Vitale, C., Ansotegui, I., Rosario, N., Haahtela, T., Galan, C., Pawankar, R., Murrieta-Aguttes, M., Cecchi, L., Bergmann, C., Ridolo, E., Ramon, G., Gonzalez Diaz, S., D'Amato, M., & Annesi-Maesano, I. (2020). The effects of climate change on respiratory allergy and asthma induced by pollen and mold allergens. *Allergy*, *75*(9), 2219–2228. doi:10.1111/all.14476 PMID:32589303

Ebi, K. L., Capon, A., Berry, P., Broderick, C., de Dear, R., Havenith, G., Honda, Y., Kovats, R. S., Ma, W., Malik, A., Morris, N. B., Nybo, L., Seneviratne, S. I., Vanos, J., & Jay, O. (2021). Hot weather and heat extremes: Health risks. *Lancet*, *398*(10301), 698–708. doi:10.1016/S0140-6736(21)01208-3 PMID:34419205

El-Sayed, A., & Kamel, M. (2020). Climatic changes and their role in emergence and re-emergence of diseases. *Environmental Science and Pollution Research International*, *27*(18), 22336–22352. doi:10.100711356-020-08896-w PMID:32347486

Ellwanger, J. H., Kulmann-Leal, B., Kaminski, V. L., Valverde-Villegas, J. M., Veiga, A. B. G. D., Spilki, F. R., Fearnside, P. M., Caesar, L., Giatti, L. L., Wallau, G. L., Almeida, S. E. M., Borba, M. R., Hora, V. P. D., & Chies, J. A. B. (2020). Beyond diversity loss and climate change: Impacts of Amazon deforestation on infectious diseases and public health. *Anais da Academia Brasileira de Ciências*, *92*(1), e20191375. doi:10.1590/0001-3765202020191375 PMID:32321030

Esperland, D., de Weerd, L., & Mercer, J. B. (2022, December). Health effects of voluntary exposure to cold water - a continuing subject of debate. *International Journal of Circumpolar Health*, *81*(1), 2111789. doi:10.1080/22423982.2022.2111789 PMID:36137565

Geng, X., Yu, Z., Zhang, D., Li, C., Yuan, Y., & Wang, X. (2022). The influence of local background climate on the dominant factors and threshold-size of the cooling effect of urban parks. *The Science of the Total Environment*, *823*, 153806. doi:10.1016/j.scitotenv.2022.153806 PMID:35150695

Githeko, A. K., Lindsay, S. W., Confalonieri, U. E., & Patz, J. A. (2000). Climate change and vector-borne diseases: A regional analysis. *Bulletin of the World Health Organization*, *78*(9), 1136–1147. PMID:11019462

Gray, K. (2022). Climate change, human health, and health informatics: A new view of connected and sustainable digital health. *Frontiers in Digital Health*, *4*, 869721. doi:10.3389/fdgth.2022.869721 PMID:35373178

Hansel, N. N., McCormack, M. C., & Kim, V. (2016). The Effects of Air Pollution and Temperature on COPD. *COPD*, *13*(3), 372–379. doi:10.3109/15412555.2015.1089846 PMID:26683097

Hotez, P. J. (2018). Human Parasitology and Parasitic Diseases: Heading Towards 2050. *Advances in Parasitology*, *100*, 29–38. doi:10.1016/bs.apar.2018.03.002 PMID:29753341

Huang, Y., Li, B., Biswas, A., & Li, Z. (2022). Factors dominating the horizontal and vertical variability of soil water vary with climate and plant type in loess deposits. *The Science of the Total Environment*, *811*, 152172. doi:10.1016/j.scitotenv.2021.152172 PMID:34883182

Huang, Y., Song, H., Wang, Z., Cheng, Y., Liu, Y., Hao, S., Li, N., Wang, Y., Wang, Y., Zhang, X., Sun, B., Li, Y., & Yao, X. (2022, October 22). Heat and outpatient visits of skin diseases - A multisite analysis in China, 2014-2018. *Heliyon*, *8*(10), e11203. doi:10.1016/j.heliyon.2022.e11203 PMID:36339999

Janson, C., Henderson, R., Löfdahl, M., Hedberg, M., Sharma, R., & Wilkinson, A. J. K. (2020). Carbon footprint impact of the choice of inhalers for asthma and COPD. *Thorax*, *75*(1), 82–84. doi:10.1136/thoraxjnl-2019-213744 PMID:31699805

Johnson, T. F., Isaac, N. J. B., Paviolo, A., & González-Suárez, M. (2023). Socioeconomic factors predict population changes of large carnivores better than climate change or habitat loss. *Nature Communications*, *14*(1), 74. doi:10.103841467-022-35665-9 PMID:36693827

Joshi, M., Goraya, H., Joshi, A., & Bartter, T. (2020). Climate change and respiratory diseases: A 2020 perspective. *Current Opinion in Pulmonary Medicine*, *26*(2), 119–127. doi:10.1097/MCP.0000000000000656 PMID:31851023

Khan, M. D., Thi Vu, H. H., Lai, Q. T., & Ahn, J. W. (2019). Aggravation of Human Diseases and Climate Change Nexus. *International Journal of Environmental Research and Public Health*, *16*(15), 2799. doi:10.3390/ijerph16152799 PMID:31390751

Lake, I. R., & Barker, G. C. (2018). Climate Change, Foodborne Pathogens and Illness in Higher-Income Countries. *Current Environmental Health Reports*, *5*(1), 187–196. doi:10.100740572-018-0189-9 PMID:29446033

Levy, K., Smith, S. M., & Carlton, E. J. (2018). Climate Change Impacts on Waterborne Diseases: Moving Toward Designing Interventions. *Current Environmental Health Reports*, *5*(2), 272–282. doi:10.100740572-018-0199-7 PMID:29721700

Lu, L. C., Chiu, S. Y., Chiu, Y. H., & Chang, T. H. (2022). Sustainability efficiency of climate change and global disasters based on greenhouse gas emissions from the parallel production sectors - A modified dynamic parallel three-stage network DEA model. *Journal of Environmental Management*, *317*, 115401. doi:10.1016/j.jenvman.2022.115401 PMID:35660833

Ma, J., Guo, Y., Gao, J., Tang, H., Xu, K., Liu, Q., & Xu, L. (2022). Climate Change Drives the Transmission and Spread of Vector-Borne Diseases: An Ecological Perspective. *Biology (Basel)*, *11*(11), 1628. doi:10.3390/biology11111628 PMID:36358329

Malhi, Y., Franklin, J., Seddon, N., Solan, M., Turner, M. G., Field, C. B., & Knowlton, N. (2020). Climate change and ecosystems: threats, opportunities and solutions. *Philosophical transactions of the Royal Society of London. Series B, Biological sciences, 375*(1794), 20190104. https://doi.org/ doi:10.1098/rstb.2019.0104

McMichael, A. J., Campbell-Lendrum, D. H., Corvalán, C. F., Ebi, K. L., Githeko, A., Scheraga, J. D., & Woodward, A. (2003). *Climate change and human health: risks and responses*. World Health Organization.

Molyneux, D., Bush, S., Bannerman, R., Downs, P., Shu'aibu, J., Boko-Collins, P., Radvan, I., Wohlgemuth, L., & Boyton, C. (2021). Neglected tropical diseases activities in Africa in the COVID-19 era: The need for a "hybrid" approach in COVID-endemic times. *Infectious Diseases of Poverty*, *10*(1), 1. doi:10.118640249-020-00791-3 PMID:33397494

Mouchet, J., & Carnevale, P. (1997). Impact des transformations de l'environnement sur les maladies à transmission vectorielle [Impact of changes in the environment on vector-transmitted diseases]. *Sante (Montrouge, France)*, *7*(4), 263–269. PMID:9410453

Nikendei, C., Bugaj, T. J., Nikendei, F., Kühl, S. J., & Kühl, M. (2020). Klimawandel: Ursachen, Folgen, Lösungsansätze und Implikationen für das Gesundheitswesen [Climate change: Causes, consequences, solutions and public health care implications]. *Zeitschrift für Evidenz, Fortbildung und Qualität im Gesundheitswesen*, *156-157*, 59–67. doi:10.1016/j.zefq.2020.07.008 PMID:32859556

Ogden, N. H., & Gachon, P. (2019). Climate change and infectious diseases: What can we expect? *Canada communicable disease report = Releve des maladies transmissibles au Canada*, *45*(4), 76–80. doi:10.14745/ccdr.v45i04a01

Pacheco, S. E., Guidos-Fogelbach, G., Annesi-Maesano, I., Pawankar, R., D' Amato, G., Latour-Staffeld, P., Urrutia-Pereira, M., Kesic, M. J., & Hernandez, M. L. (2021). Climate change and global issues in allergy and immunology. *The Journal of Allergy and Clinical Immunology*, *148*(6), 1366–1377. doi:10.1016/j.jaci.2021.10.011 PMID:34688774

Patz, J. A., Epstein, P. R., Burke, T. A., & Balbus, J. M. (1996). Global climate change and emerging infectious diseases. *Journal of the American Medical Association*, *275*(3), 217–223. doi:10.1001/jama.1996.03530270057032 PMID:8604175

Pega, F., Shaw, C., Rasanathan, K., Yablonski, J., Kawachi, I., & Hales, S. (2015). Climate change, cash transfers and health. *Bulletin of the World Health Organization*, *93*(8), 559–565. doi:10.2471/BLT.14.150037 PMID:26478613

Prüss-Üstün, A., & Corvalán, C. (2006). *Preventing disease through healthy environments: towards an estimate of the environmental burden of disease*. World Health Organization.

Ray, C., & Ming, X. (2020). Climate Change and Human Health: A Review of Allergies, Autoimmunity and the Microbiome. *International Journal of Environmental Research and Public Health*, *17*(13), 4814. doi:10.3390/ijerph17134814 PMID:32635435

Romanello, M., van Daalen, K., Anto, J. M., Dasandi, N., Drummond, P., Hamilton, I. G., Jankin, S., Kendrovski, V., Lowe, R., Rocklöv, J., Schmoll, O., Semenza, J. C., Tonne, C., & Nilsson, M. (2021). Tracking progress on health and climate change in Europe. *The Lancet. Public Health*, *6*(11), e858–e865. doi:10.1016/S2468-2667(21)00207-3 PMID:34562381

Rossati, A. (2017). Global Warming and Its Health Impact. *The International Journal of Occupational and Environmental Medicine*, *8*(1), 7–20. doi:10.15171/ijoem.2017.963 PMID:28051192

Semenza, J. C., Suk, J. E., Estevez, V., Ebi, K. L., & Lindgren, E. (2012). Mapping climate change vulnerabilities to infectious diseases in Europe. *Environmental Health Perspectives*, *120*(3), 385–392. doi:10.1289/ehp.1103805 PMID:22113877

Setoguchi, S., Leddin, D., Metz, G., & Omary, M. B. (2022). Climate change, health, and health care systems: A global perspective. *Gastroenterology*, *162*(6), 1549–1555. doi:10.1053/j.gastro.2022.02.037 PMID:35247463

Spiewak, R. (2023). Diseases from the Spectrum of Dermatitis and Eczema: Can "Omics" Sciences Help with Better Systematics and More Accurate Differential Diagnosis? *International Journal of Molecular Sciences*, *24*(13), 10468. doi:10.3390/ijms241310468 PMID:37445645

Stark, K., Niedrig, M., Biederbick, W., Merkert, H., & Hacker, J. (2009). Die Auswirkungen des Klimawandels. Welche neuen Infektionskrankheiten und gesundheitlichen Probleme sind zu erwarten? [Climate changes and emerging diseases. What new infectious diseases and health problem can be expected?]. *Bundesgesundheitsblatt, Gesundheitsforschung, Gesundheitsschutz*, *52*(7), 699–714. doi:10.100700103-009-0874-9 PMID:19536444

Tarek, F., Hassou, N., Benchekroun, M. N., Boughribil, S., Hafid, J., & Ennaji, M. M. (2019). Impact of rotavirus and hepatitis A virus by worldwide climatic changes during the period between 2000 and 2013. *Bioinformation*, *15*(3), 194–200. doi:10.6026/97320630015194 PMID:31354195

Thomas, M. B. (2020). Epidemics on the move: Climate change and infectious disease. *PLoS Biology*, *18*(11), e3001013. doi:10.1371/journal.pbio.3001013 PMID:33232329

Vezzulli, L., Colwell, R. R., & Pruzzo, C. (2013). Ocean warming and spread of pathogenic vibrios in the aquatic environment. *Microbial Ecology*, *65*(4), 817–825. doi:10.100700248-012-0163-2 PMID:23280498

Wang, F., Shi, C., Dong, J., & Nie, H. (2021, December). Association between ambient temperature and atopic dermatitis in Lanzhou, China: A time series analysis. *Environmental Science and Pollution Research International*, *28*(47), 67487–67495. doi:10.100711356-021-15198-2 PMID:34254239

Workman, A., Jones, P. J., Wheeler, A. J., Campbell, S. L., Williamson, G. J., Lucani, C., Bowman, D. M. J. S., Cooling, N., & Johnston, F. H. (2021). Environmental Hazards and Behavior Change: User Perspectives on the Usability and Effectiveness of the AirRater Smartphone App. *International Journal of Environmental Research and Public Health*, *18*(7), 3591. doi:10.3390/ijerph18073591 PMID:33808395

Zinsstag, J., Crump, L., Schelling, E., Hattendorf, J., Maidane, Y. O., Ali, K. O., Muhummed, A., Umer, A. A., Aliyi, F., Nooh, F., Abdikadir, M. I., Ali, S. M., Hartinger, S., Mäusezahl, D., de White, M. B. G., Cordon-Rosales, C., Castillo, D. A., McCracken, J., Abakar, F., ... Cissé, G. (2018). Climate change and One Health. *FEMS Microbiology Letters*, *365*(11), fny085. Advance online publication. doi:10.1093/femsle/fny085 PMID:29790983

Chapter 6
Green Marketing:
Sustainable Strategies for Success

Shriya Tripathi
Ganpat University, India

Disha Sharma
Amity University, Raipur, India

ABSTRACT

Consumers in the 21st century are more informed and discerning than ever about the products they buy. This is positively obvious with regards to assessing the ecological effect and manageable acts of the brands they support. That is why many of the world's most significant and effective organizations are seeking green marketing drives. Green marketing has become an essential business strategy as consumers' awareness of environmental issues has grown. Purchasers are progressively worried about the natural effect of their utilization decisions, and they are bound to purchase items or administrations that line up with their ecological qualities. Businesses can draw in these environmentally conscious customers, set themselves apart from their rivals, and enhance their brand image by employing green marketing strategies.

Consumers in the 21st century are more informed and discerning than ever about the products they buy. This is positively obvious with regards to assessing the ecological effect and manageable acts of the brands they support. That is why many of the world's most significant and effective organizations are seeking green marketing drives.

Green marketing has become an essential business strategy as consumers' awareness of environmental issues has grown. Purchasers are progressively worried about the natural effect of their utilization decisions, and they are bound to purchase items or administrations that line up with their ecological qualities. Businesses can draw in these environmentally conscious customers, set themselves apart from their rivals, and enhance their brand image by employing green marketing strategies.

Green marketing is a procedure that organizations use to advance harmless ecosystem items or administrations to purchasers. A critical review of the literature on green marketing, including its definition,

DOI: 10.4018/978-1-6684-9503-2.ch006

significance, obstacles, and potential areas of study, is provided in this paper. According to the literature review, environmental sustainability is attracting more and more attention to green marketing. Be that as it may, it likewise features a few difficulties like the absence of normalization, greenwashing, and high execution costs. The need for standardization, the impact of green marketing on organizational performance, and the role of green marketing in consumer behavior are some of the future research directions.

Green marketing manages an organization's publicizing methodologies to lay out a brand as being harmless to the ecosystem. Obviously, it has likewise turned into a well-known promoting initiative, given the eco-cognizant purchaser today. Throughout recent years, there has been a 71% ascent in online searches for sustainable merchandise worldwide. Up to 55% of customers are prepared to put more in eco-friendly brands. So there's little uncertainty that green showcasing truly does to be sure to sell.

REVIEW OF LITERATURE

According to Polonsky (2011) green marketing is characterized as the work of a company's strategy for its product, price, place, and promotion effectively in a eco-friendly manner. As per him, all promoting strategies ought to minimize unfavorable effects on the environment.

According to Chitra (2007) the green marketing mix is the blend of product, price, place, promotion, process, people, and physical evidence, and further discusses how the product should be green, to be used for healthy consumption for the target audience. The price is the monetary value or the worth paid by the customers. The physical distribution could be involved in the logistics and store keeping with an eco-friendly mindset in the distribution of products to end customers. He further discussed regarding eco-friendly approach to creating awareness and the process of eco-friendly products for final consumption.

Abzari, Shad, Sharbiyani, and Morad (2013) and Mahmoud, Ibrahim, Ali, and Bleady (2017) identified a substantial positive association between green marketing mix and consumer purchase intention. Mahmoud, Ibrahim, Ali, and Bleady (2017) found that environmental information acts as the mediator of this connection.

Leonidou, Katsikeas, and Morgan (2013)'s analysis directed towards the application of green marketing programs by the company had noteworthy performance payoffs. Consequences implied that green products and distribution certainly affected product-market performance, whereas green pricing and promotion boosted its yield on assets. Furthermore, industry-level environmental reputation leads the link between green marketing mechanisms and firms' product-market and financial performance.

Green product helps conserve and recover the environment along with the preservation of energy or resources and lessening or omission of the use of poisonous materials, pollution, and wastes (Ottman, Stafford, & Hartman, 2006). In other words, the green product is the one that sustains less harm to the nature (Dahl & Persson, 2008).

Yaacob and Zakaria (2011) advised that in broad-spectrum, customers tend to buy green products as it works as a one step closer towards environmental improvement. In few cases, direct personal benefits, namely the health benefits of organic products consumed by them or be the energy-saving part in electronic devices are commonly observed.

Suki (2016) conferred that eco-friendly product purchase was wedged by knowledge for green marketing as well as the functional value quality. Functional value price, emotional value, and conditional value were not affected. According to Ranjan and Kushwaha (2017) buying decision is sturdily influenced by views in favor of green products.

According to a study by Rahbar and Wahid (2011), ideally, Malaysian consumers tend to purchase more of green products when it comes to cleaning products or pesticides over the normal conventional ones.

Chang (2011) in his study analyzed that considerable consumers find the price of eco-friendly labeled products more expensive than non-eco-friendly.

According to D'Souza, Lamb, and Taghian (2006) consumers inclined towards green products will continue towards their choice even though the quality served is comparatively inferior but will closely check the green labels for information. Price-sensitive customers check for all relevant information regarding green labels before making a buying decision.

Green prices ought to be economical and competitive in order to keep up with the sales (Soonthonsmai, 2007). It is presumed that in the beginning the cost and price of green products will be on the upper end but over a period of time, it will be less due to awareness, experiential learning, increase in knowledge, and advancement in technology (Fan & Zeng, 2011).

According to Shil (2012) green place refers to the managing of logistics to diminish the emissions by the movement of goods caused by transportation, thereby reducing carbon footprint. Awan (2011) stated that place is not a cost generator factor, Certainly if used effectively can result in an increase in the sources of revenues.

HOW DOES GREEN MARKETING WORK?

Green advertising works by empowering customers to purchase items produced using reused materials or items that don't hurt the climate in any capacity. For instance, an oil company could publicize its kind of engine oil as being produced using reused materials instead of unrefined petroleum separated from the beginning. A vehicle producer could publicize its vehicles as being controlled by power rather than gas or diesel fuel.

With green marketing, the items are normally made in a feasible manner, don't contain harmful materials, are delivered with reused or sustainable materials, don't have over-the-top bundling, and are intended to be reused.

At the point when a company can make its items in a manner that is eco-friendly, it shows that they're focused on maintainability and social obligation. Also, it assists with brand notoriety. At the point when you make your items in a harmless to the ecosystem way, you can zero in on green promoting to tell everybody that you're really trying to help the climate.

Likewise, this could set aside your cash in packaging and energy use over the long haul. While many organizations don't become environmentally friendly on the grounds that the forthright expense is normally costlier, they probably won't understand that it'll create reserve funds over the long haul.

In any case, it's vital to take note that you can't simply advertise something as green in the event that it's not. There are guidelines and lawful principles for green advertising.

Promoting environmentally friendly goods and services to customers is known as green marketing. A system looks to speak to customers who are worried about the effect of their utilization decisions on the climate. In this chapter, we will walk through the idea of green advertising, its significance, and the difficulties that accompany executing it.

WHAT IS GREEN MARKETING MIX?

Figure 1. Green marketing mix

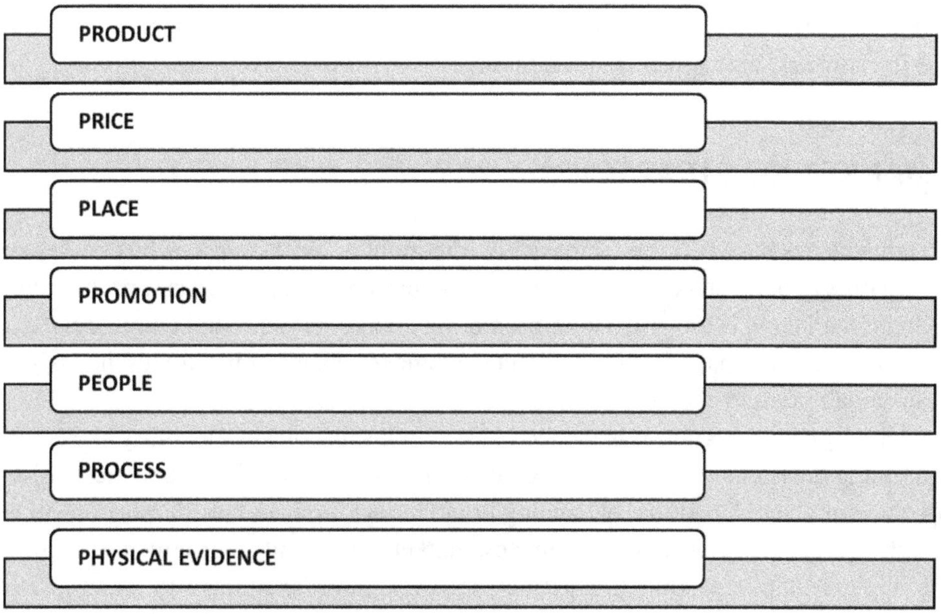

- **Product:** The product ought to be planned and created in such a way that they utilizes fewer resources and are without are eco-friendly, in addition, they contains no poisonous substance, whose utilization can be unsafe. Additionally, the product should be built it keeping in mind the optimum utilization of resources, especially for scarce resources.
- **Price:** In green promoting, cost assumes a major part, as the clients will follow through on the extra price, just when they are of the view that they will get exceptional quality items, regarding plan, execution, allure, taste, or whatever else.
- **Promotion:** Green promoting should be possible in three ways, for example, there can be promotions that show the association in the midst of the product and the climate, advertisements that advance a green and natural way of life, or promotions that grandstand a corporate image of ecological obligation.
- **Place:** Place characterizes the accessibility of the items thus the advertisers ought to select an optimal method for making such items accessible as it will incredibly affect the clients.
- **People:** The Green Marketing Mix's "people" component emphasizes the significance of getting internal stakeholders including partners and workers to share the company's environmental ideals. Companies should highlight employee involvement in environmental activities, highlight staff sustainability efforts, and make sure that the organization's culture and practises reflect its commitment to sustainability.

- **Process:** The term "process" refers to the internal practises and protocols that businesses use to reduce their negative environmental effects. This entails integrating sustainable practises along the entire value chain, such as trash reduction, recycling initiatives, and energy-efficient manufacturing processes.
- **Physical Evidence:** The emphasis of the physical evidence component is on offering verifiable proof of the environmental characteristics of a good or service. Indicators of sustainability, such as eco-friendly packaging, labels, or emblems

GREEN MARKETING STRATEGIES

Green marketing tactics have become increasingly important in recent years as businesses and consumers alike have come to understand the need to address environmental issues and promoting sustainable practices. Green marketing is the practice of promoting goods, services, and commercial activities that have little environmental impact. It seeks to meet customer demands while simultaneously taking environmental sustainability into account.

Environmentally conscious consumers who place a high value on sustainability and eco-friendliness in their purchasing decisions are targeted by green marketing methods. These tactics assist companies in standing out from the competition, enhancing brand recognition, and promoting a greener future.

A thorough awareness of consumer preferences, market trends, and environmental concerns is necessary for the implementation of green marketing strategies. In order to identify target groups and their particular environmental factors, businesses must perform extensively.

1. Get Certified

There are a few outsider green business confirmations that can assist you with promoting your supportability rehearses.

To get guaranteed, research the various accreditations, including Green America, Green Business Certificate Inc., or Green Seal.

For the most part, you should meet a specific degree of norms and send in an application. These confirmations assist with proving your cases that you're a green business.

2. Advertise Your Green Message

When you have the confirmation and you're the genuine article, now is the ideal time to spread the news. You can compose sites, post about practicing environmental awareness on your web-based entertainment, add confirmation stickers to your site, and so on.

Advancing your maintainability endeavors will assist with building your image and client dependability from eco-cognizant shoppers.

3. Integrate Sustainability Into Your Company Culture

Try not to simply fall into the green showcasing trap without truly focusing on the reason. You ought to integrate green practices into your organization culture. Each representative ought to realize that you need to make things as eco-accommodating as could be expected.

4. Support Eco-Friendly Initiatives

At the point when your organization goes to give to good cause or back to neighborhood and local projects and initiatives, ensure you pick ones that are green. It assists with your reliable image message and it will uphold a reason you care about.

Beyond making eco-friendly products, entrepreneurs can execute different strategies to generate profits from the advantages of green promotion. The underlined can act as an important measure in the footprints of green marketing:

- Employing eco-friendly paper and inks for print showcasing materials
- Using electronic advertising rather than print media.
- Responsibility for taking care of garbage removal rehearses
- Utilizing eco-friendly or reused materials for packaging products.
- Utilizing effective warehousing and supply chain management techniques
- Utilizing environment-friendly power and farming practices
- Taking conscious steps to counterbalance carbon emissions and footprints.

5. Classify Green Products and Exclusively Highlight in Green Marketing

If a brand practices environmentally friendly methods like working with sustainable dealers or using resources from paper and electronic goods that have been recycled. In this case, a green marketer's job is to spread the word about it. A company will be able to attract customers who value sustainability thanks to this.

6. Utilizing Recycled Resources

During production, efforts should be made to use as numerous recycled materials as likely. Reusing includes refining different materials, and it helps with diminishing water and air contamination levels. The effects of greenhouse gases and their emissions will ultimately be lessened as a result of this. For instance, Mush is a sustainable Indian company that makes towels and other products from bamboo.

7. Using Electronic Mail to Reach Targeted Customers

Mediums such as newsletters and email notifications work well to influence both the intended and new customers. Because it does not necessitate any promotional flyers, retargeting previous customers through the mail is more accessible. A brand is able to reduce its use of paper thanks to this. However, email marketing is more effective at obtaining valuable customer feedback.

Sendpulse is one such tool, one can send many messages and emails to the current client base. They add a few new features, like a drag-and-drop editor that curates individual emails.

8. Improve the Delivery and Supply Structure

The automobiles that businesses use to transport and deliver goods have a significant impact on the environment. It is novel to invest in new vehicles and equipment because it can significantly reduce the company's carbon emissions. Electric vehicles are being used by a number of businesses to improve delivery systems. Because it significantly reduces air pollution, several businesses support this. For effective green marketing, these examples must be used and featured on the website to increase traffic. Your organization's carbon footprint can be reduced by upgrading its equipment and vehicles.

9. Green Initiatives

In order to promote the company's environmental consciousness and raise awareness of the significance of protecting the environment, organize a beach clean-up event in conjunction with a local environmental group.

10. Bulk Email Service

Think About Using a Bulk Email Service Use a bulk email service to effectively reach a large audience, promote sustainable products or services, and reduce your carbon footprint.

11. Make Use of Recycled Materials

Make a line of promotional items out of recycled materials and advertise them and the positive effects they have on the environment through a variety of marketing channels.

12. Feature That Your Organization Is Eco-Accommodating

Make a promoting effort that exhibits your organization's eco-accommodating practices and items, featuring their natural advantages and interesting to eco-friendly shoppers.

13. Put Resources Into Web-Based Entertainment Promoting

Utilize web-based entertainment to advance your organization's feasible items and eco-accommodating practices. For a wider audience and a sense of community, choose paid advertising and relevant hashtags.

PRINCIPLES OF GREEN MARKETING

Figure 2. Principles of green marketing

Consumer-Oriented Marketing

This is founded on the notion that a company should discover as much as it can about its clients and exert every effort to meet their requirements and desires. In other words, it means that the business must consider and plan its marketing initiatives from the viewpoint of the client. The business not only promotes its goods but also looks into how and why consumers use them. Nike, for instance, focuses on the customer experience. The group spends a lot of time and money helping those who run or participate in sports. For instance, Nike offers free online preparation programs where you may submit questions about your schooling and get helpful advice in return. According to the notion, a company should approach its marketing initiatives from the perspective of the customer in order to build a long-lasting and profitable relationship with its customers.

Customer Value Marketing

This principle states that rather than focusing the majority of a company's marketing budget on advertising and selling, it should focus on understanding and improving the quality of its products and services over a span of time. In order to cultivate long-term customer engagement, loyalty, and relationships, enlightened marketing emphasizes continuously increasing the value consumers derive from the company's market offering. The business can get value back from customers by adding value for them. Instead of simply altering the packaging of the product or investing a significant amount of money in the advertisement, this idea holds that the business ought to allocate its resources in ways that enhance the value of the service or product it provides. this is on the grounds that, when the worth is added to the item, they will be esteemed by the clients.

Innovative Marketing

For a business to engage in innovative marketing, its products and services must be updated frequently. The organisation should adapt using a variety of techniques, including new product creation, market segmentation, positioning strategies, product standards and specifications, customer satisfaction, customer value, packaging, and delivery of goods and services. Enlightened advertisers trust that the most effective way to accomplish development is to offer purchasers something they need. In addition, for a business to achieve long-term marketing success, it must be able to successfully market new products. . We are all aware that the world and the preferences and tastes of customers are constantly shifting. Therefore, in order to avoid easily losing customers, the business should constantly seek out new and improved methods.

Sense-of-Mission Marketing

According to sense-of-mission marketing, a company must establish a broad social mission that it will follow throughout its years of operation. Well-informed marketers employ this guiding concept to decide on the company's marketing content, brand image, business model, stakeholder connections, and innovation tactics. The sense of mission speaks to an organization's ethos or social responsibility. Every business has some sort of social obligation, whether they are conscious of it or not, and this idea helps to guarantee that these responsibilities are constructive. The organization's primary objective needs to be fully described in social terms. and not in the item. This is because of the way that assuming an organization expresses the mission that has some friendly government assistance concealed in it, the representatives feel glad towards the efforts for a worthy objective and efforts in the correct direction.

For examples:

- Johnson and Johnson's lead Johnson's Baby brand devotes itself to a mission of thoughtful towards infants and the extraordinary sustenance they need, then, at that point, utilizes that information to give guardians protected and compelling child care items.
- Procter & Gamble's Tide, one of the best laundry detergents in the world, has a solid sense of mission and helps charities like Habitat for Humanity.

The company's guiding principle is that it should contribute to people's happiness. They want to give joy and foster the capability of kids by being their primary decision in quite a while of toys, quality, play esteem, plan, security and advantages.

Societal Marketing

Societal marketing is a showcasing idea that goes past the conventional spotlight on consumer loyalty and company benefits. It considers the welfare of society as a whole over the long term. Societal marketing plans to adjust the interests of customers, organizations, and society by thinking about the social, ecological, and moral ramifications of showcasing exercises.

Here are a few critical parts of cultural showcasing:

Consumer Satisfaction: Social marketing, like traditional marketing, recognizes the significance of satisfying customer requirements and providing value. Nevertheless, it also emphasizes the obligation of businesses to take into account the broader societal effects of their goods and services.

Consumer Welfare: By offering products or services that truly benefit customers, social marketing aims to improve consumers' well-being. It encourages businesses to put customer satisfaction, health, and safety first in their marketing efforts.

Society's Welfare: Societal marketing acknowledges that businesses and society are interdependent. It emphasizes the significance of taking into account the effects that marketing activities have on society, the economy, and the environment. Sustainability, resource conservation, social equity, and community development are all part of this.

Ethical Contemplations: Ethical behavior is highly valued in social marketing. It encourages businesses to promote fair competition, be truthful in advertising, avoid deceptive practices, respect consumer privacy, and adhere to ethical standards in marketing.

Environmental Responsibility: Cultural showcasing recognizes the ecological difficulties confronting the world. It encourages businesses to develop eco-friendly products, reduce their carbon footprint, encourage sustainable sourcing and production, and adopt environmentally friendly practices.

Long-term Viewpoint: Societal marketing looks at business success over the long term. It acknowledges that meeting society's needs and addressing societal issues are closely linked to sustainable profitability. Companies can achieve sustainable growth by cultivating consumer trust and loyalty by considering the long-term effects of their marketing decisions.

The goal of social marketing is to link business objectives with the greater good. It perceives that organizations have an obligation to contribute emphatically to society and that doing so can likewise be valuable for their own prosperity. Companies have the potential to strengthen their relationships with customers and have a beneficial effect on society as a whole by taking into account the welfare of society as a whole and incorporating ethical and sustainable practices.

CHALLENGES IN GREEN MARKETING

Businesses may find it challenging to put green marketing strategies into action. The lack of uniformity in the definition of what constitutes "green" is one of the main obstacles. Various associations and guidelines bodies have various standards for what constitutes a harmless to the ecosystem item or administration. This can make it hard for organizations to understand what norms to stick to and what claims to make about their items or administrations.

The possibility of greenwashing presents another obstacle. "Greenwashing" refers to making exaggerated or deceptive claims regarding a good or service's environmental impact. This can prompt purchaser wariness and a loss of confidence in the brand. Businesses must ensure that their claims are accurate, verifiable, and transparent in order to avoid greenwashing.

Lastly, green marketing strategies can be expensive for businesses to implement. Promoting environmentally friendly products or services may necessitate additional marketing costs, and developing them may necessitate significant investments in research and development. However, the potential advantages of attracting environmentally conscious customers and enhancing the brand's image may be able to offset these expenses.

- Need for Standardisation: Only 5% of the showcasing messages from "Green" campaigns are wholly evident, and standardization is required to support these claims. There is no standardization that backs up these claims. The certification of an organic product currently lacks any standards. Unless certain regulatory agencies are involved in issuing the certifications, there will be no way to verify the information. For such branding and permitting, there should be a standard quality control board in place.
- The idea holds that urban, educated Indian consumers are becoming more aware of the benefits of eco-friendly goods. However, it is still a novel idea to most people. The buyer should be informed of the environmental dangers and made aware of them. The new environmental movements will need a lot of time and work to spread their message to the broader population. Indian consumers are quite aware of the importance of using natural and herbal beauty products, thanks to India's Ayurvedic heritage. Indian customers are exposed to healthy habits such as yoga and eating natural foods. In those perspectives, the customer is already aware and will lean over to notice the green goods.
- Resilience and persistence The environment has to be viewed as a significant long-term speculative potential by investors and businesses, and advertising needs to consider the long-term advantages of this new green movement. It will take a lot of patience and won't yield results right away. As a different concept and way of thinking, it will have its own period of acceptability.
- Restricting Green Myopia Focusing on customer benefits, or the primary justification for a customer's purchase, is the first rule of green marketing. If you do this correctly, you can convince customers to switch brands or even pay more for the environmentally friendly choice. It won't be useful to create a product that is entirely green in all respects but falls short in terms of client pleasure. There will be green myopia as a result. Furthermore, if green products are highly expensive, their market acceptance will decline.

TRADITIONAL MARKETING VS. GREEN MARKETING

The terms "traditional marketing" and "conventional marketing" are used interchangeably. It involves tactics like billboards, direct mail, television commercials, print commercials, and telemarketing. Green marketing, also referred to as sustainable marketing or environmental marketing, focuses on promoting environmentally friendly goods and services. It includes procedures that underscore the natural advantages of an item or the eco-accommodating acts of an organization.

Table 1. Traditional v. green marketing

BASIS OF COMPARISON	TRADITIONAL MARKETING	GREEN MARKETING
Parties involved	Business and customer	Business, customer, and environment
Main objectives	Customer Satisfaction and meeting the company's objectives	Customer Satisfaction, meeting the company's objectives and sustainable marketing
Company's Responsibility	Economical responsibility	Social responsibility
Ecological demands	Legal obligations	Beyond the law- Environmental culture
Green pressure groups	Passive behavior	Public relations and cooperation
Decision making	Short Term Orientation	Long Term Orientation
Main focus	Tangible goods	Products as services
Thinking Pattern	Fragmented thinking	Integrated Thinking
Philosophical Basis	Anthropocentric	Biocentric
Ecological Cost	No/ Underpaid ecological costs	A full accounting of ecology costs.

The focus and strategies of traditional marketing and green marketing are distinct approaches to product or service promotion. Let us contrast the two:

It's critical to take note of that these two advertising approaches are not fundamentally unrelated. In order to promote their eco-friendly practices and reach a larger audience, some businesses may combine elements of traditional and green marketing. Lastly, the decision between customary promoting and green advertising relies upon the main interest group, the idea of the item or administration, and the organization's qualities and objectives.

COMPANIES PRACTICING GREEN MARKETING

Starbucks

Starbucks is regarded as a pioneer in environmental responsibility. Several environmental programs benefit from investments and donations from the company. As a matter of fact, Starbucks committed more than $140 million to the improvement of sustainable power sources. Green advertising models from this brand range these undertakings:

- The sources of espresso, tea, and paper supply
- Recyclable bundling
- Climate cordial outlet tasks and plans

In any case, close by these prescribed procedures, Starbucks lays out client faithfulness and entrusts with their customized encounters via virtual entertainment that features consciousness of the organization's practical decisions, plan of action, and activities and itemized CSR reports accessible on the web. This two-pronged strategy creates a positive cycle that boosts brand confidence and drives sales.

Ben and Jerry's

Ben and Jerry's effectively showcases the social issues that the organization thinks often about through their site on their Qualities page. They frequently discuss the significance of committing to environmentally friendly business practices in blogs.

You can show your target audience that you care about the same issues they do with green marketing. You can further develop your image notoriety, while likewise contributing and giving to a reason that you care about.

Johnson & Johnson

With innovative products, recycling initiatives, and strong sustainability commitments, Johnson & Johnson has established a legacy of environmentally friendly practices. Somewhere between 2013 and 2017, the organization eliminated plastic microbeads from its whole array of corrective items. Following Johnson & Johnson's participation in the 2018 New Plastics Economy Global Movement, it also outperformed the competition in product packaging by controlling the use of plastic. Johnson & Johnson has a website devoted to DIY recycling projects as part of its marketing efforts. All of this demonstrates how a comprehensive green (pro-environment) impression on customers can result from a collection of intelligent marketing strategies and sustainable practices.

Patagonia

Patagonia laid down a good foundation for itself as a considerable representative of natural worries, mixing benefit with reason through its entryway, "The Footprint Chronicles", featuring the whole assembly cycle of every item. It was a critical green-promoting effort that displayed the brand's straightforwardness to purchasers, empowering them to make more educated purchasing choices. Patagonia discussed not only their contributions but also the shortcomings of their products. Besides, the American open-air clothing retailer is the first organization in quite a while to succeed as a confirmed B Corp, a certification given exclusively to those associations with solid ecological practices.

World Wildlife Fund (WWF)

The World Wildlife Fund (WWF) is a major player in environmental protection. Each of their initiatives provides a number of inspiring examples of green marketing. Additionally, their contact with worldwide organizations permits them to produce assets to keep their tradition of environment intercession rehearses alive. WWF accomplice advertising is one of the most productive ways for different organizations to join the trend of nature safeguarding and diminishing the ills of environmental change. With 65% of funding coming from individuals and bequests, WWF has become the most highly funded conservation organization in the world thanks to the symbiotic relationship it has developed through numerous partnerships.

Apple

Apple's environmentally friendly iPhone model, the iPhone 14 series, contributed to the global movement for sustainability. During the send-off, a great showcasing effort featured the organization's supportability endeavors by widely utilizing reused metals to bind circuit sheets and work out the Taptic Motor. The entire gold used in the camera wires and a number of printed circuit boards was recycled, resulting in a significant reduction in the new models' carbon footprint. Apple's plans to become carbon-neutral in full by 2030 further support this green marketing campaign.

Unilever

Unilever is another business that has adopted green marketing strategies and has put these practises into practise. Recycled materials are used by Unilever. Nearly 75% of industrial waste is delivered to recycling facilities. By the year 2039, they hope to have zero trash emissions worldwide. Additionally, they have demonstrated an interest in lowering GHG emission levels.

Unilever is enhancing both the global environmental consciousness and the image of its brand by employing such eco-friendly marketing strategies. Such green marketing techniques can enhance sales and draw in new clients.

H&M

H&M has rolled out numerous improvements in environmental strategies to make a "Conscious collection'. This made them unmistakable as one of the most naturally capable organizations in the world. They are popular for making practically half of their attire items from eco-friendly materials. H&M likewise presented a reusing program that permits shoppers to return old garments. Some of H&M's garments are made with recycled polyester and natural cotton.

IKEA

One of the most convincing green promotion models can be tracked down in IKEA's business and showcasing models. The Swedish worldwide aggregate likewise backs its aggressive sustainability claims with real work and offers proof. For example, the 2021 IKEA Supportability Report expressed that the organization's complete environment impression diminished by 1.6 million tons of CO2 same — which is more than a 5% decrease in five years. In light of the rising issues over animal brutality, IKEA reported that by 2025, half of its food items would be vegan. The deliberate circling back to the brand's commitments and their committed detailing engraves trust in the shoppers' brains.

NIKE

In 2019, Nike advanced its most memorable ecological mindfulness crusade, which has turned into a top green-marketing model. The company launched the "Move to Zero" marketing campaign, featuring seven critical stages to advance sustainability, like lessening fossil fuel by-products, utilizing environmentally friendly resources, and so on. Through everything, the predominant message of the athletic equipment goliath remained reliable: Help protect the future of the sport." Global warming affects athletes' exhibitions, and Nike's convincing sustainable endeavors reverberated through its customer base when it started integrating athletic ethics into the promotional campaign

CONCLUSION

Green marketing is a basic technique for organizations that need to engage environmentally conscious buyers, lessen their natural effects, and further develop their brand image. In any case, it additionally accompanies difficulties like the absence of normalization, greenwashing, and high execution costs. The need for standardization, the impact of green marketing on organizational performance, and the role of green marketing in consumer behavior are some of the future research directions. Businesses can successfully implement green marketing strategies and contribute to a more sustainable future by addressing these obstacles and following best practices.

In conclusion, green marketing is essential for advancing ethical business practises and solving environmental issues. It entails the creation and marketing of environmentally friendly products as well as corporate adoption of eco-aware business practises.

Businesses may use green marketing to demonstrate their dedication to sustainability, which is appealing to a rising number of environmentally conscious customers. It aids businesses in standing out from the competition, luring in a devoted clientele, and enhancing the reputation of their brands.

Businesses can increase consumer knowledge of the value of sustainable choices by highlighting the environmental advantages of their goods and services. By promoting responsible consumption, waste reduction, and the use of renewable resources, green marketing strategies can help create a more sustainable future.

However, it is imperative for businesses to put their green marketing claims into practise and make sure they are supported by real-world deeds. Customers are growing more discriminating, and they demand authenticity and transparency from brands. The practise known as "greenwashing," which involves making false or inflated environmental claims, can damage a company's brand and lose customer confidence.

Businesses should incorporate sustainability into their whole business strategies and operations if they want to excel at green marketing. This entails implementing eco-friendly production techniques, lowering carbon emissions, and promoting moral supplier chains. Building trust and credibility with customers requires openness, concise communication, and adherence to accepted environmental standards.

Green marketing, sometimes referred to as sustainable marketing or environmental marketing, is a strategy for advertising goods and services that places an emphasis on their environmental friendliness and sustainability. It encourages businesses to adopt sustainable and socially conscious business practices and promotes environmentally friendly goods and methods, communicating the advantages of doing so to consumers who care about the environment. It tries to strike a balance between commercial goals and environmental considerations, helping to create a society that is greener and more sustainable.

Overall, businesses have possibilities and obligations related to green marketing. Companies can make a difference by coordinating their marketing initiatives with sustainable practices.

Pick green marketing for the best outcomes in the future!

REFERENCES

Abzari, M., Shad, F. S., Sharbiyani, A. A., & Morad, A. P. (2013). Studying the effect of green marketing mix on market share increase. *European Online Journal of Natural and Social Sciences*, *2*(3), 641–653.

Ansar, N. (2013). Impact of Green Marketing on Consumer Purchase Intention. *Mediterranean Journal of Social Sciences*, *4*(11), 650–655. doi:10.5901/mjss.2013.v4n11p650

Awan, U. (2011). Green Marketing: Marketing Strategies for the Swedish Energy Companies. *International Journal of Industrial Marketing*, *1*(2), 1–19. doi:10.5296/ijim.v1i2.1008

Chang, C. (2011). Feeling ambivalent about going green – Implication For Green. *Journal of Advertising*, *40*(4), 19–31. doi:10.2753/JOA0091-3367400402

Chitra, K. (2007). In search of the Green Consumers: A perceptual Study. *Journal of Services Research*, *7*(1), 173–191.

D'Souza, C., Lamb, P., & Taghian, M. (2006). An empirical study on the influence of environmental labels on consumers. *Corporate Communications*, *11*(2), 162–173. doi:10.1108/13563280610661697

Dahl, U., & Persson, S. (2008). *Purchasing of environmental friendly computers*. Jonkoping: Bachelor Thesis within Business Administration

Fan, H., & Zeng, L. (2011). *Implementation of Green Marketing Strategy in China: A Study of the Green Food Industry* [MSc thesis]. Dept. Bus. Administration Econ. Study, Univ. Gavle.

Gault, R. (1907). A history of the questionnaire method of research in psychology. *Research in Psychology*, *14*(3), 366–383. doi:10.1080/08919402.1907.10532551

Hashem, T., & Al-Rifai, N. (2011). The Influence of Applying Green Marketing Mix by Chemical Industries Companies in Three Arab States in West Asia on Consumer's Mental Image. *The International Journal of Social Sciences (Islamabad)*, *2*(3), 92–101.

Leonidou, C. N., Katsikeas, C. S., & Morgan, N. A. (2013). "Greening" the marketing mix: Do firms do it and does it pay off? *Journal of the Academy of Marketing Science*, *41*(2), 151–170. doi:10.100711747-012-0317-2

Maheshwari, A., & Malhotra, D. G. (2011). Green marketing: A study on Indian youth. *International Journal of Management and Strategy*, *2*(3).

Mahmoud, T. O., Ibrahim, S. B., Ali, A. H., & Bleady, A. (2017). The Influence of Green Marketing Mix on Purchase Intention: The Mediation Role of Environmental Knowledge. *International Journal of Scientific and Engineering Research*, *8*(9), 1040–1048. doi:10.14299/ijser.2017.09.006

Ottman, J., Stafford, E. R., & Hartman, C. L. (2006). Avoiding Green Marketing Myopia: Ways to Improve Consumer Appeal for Environmentally Preferable Products. *Environment*, *48*(5), 22–36. doi:10.3200/ENVT.48.5.22-36

Polonsky, M. J. (2011). Transformative green marketing: Impediments and opportunities. *The Journal of Business*, *64*(12), 1311–1319.

Rahbar, E., & Wahid, N. A. (2011). Investigation of green marketing tools' effect on consumers' purchase behavior. *Business Strategy Series*, *12*(2), 73–83. doi:10.1108/17515631111114877

Ranjan, R. K., & Kushwaha, R. (2017). Impact of Green Marketing Strategies on Consumer Purchase Behaviour. *Review of Management*, *7*(3), 9–22.

Secondi, L., & Pirani, E. (2011). Eco-Friendly Attitudes: What European Citizens Say and What they do. *International Journal of Environmental Research*, *5*(1), 67–84.

Sharma, Y. (2011). Changing Consumer Behaviour with Respect to Green Marketing: A Case Study of Consumer Durables and Retailing. *International Journal of Multidisciplinary Research*, *1*(4), 152–162.

Shil, P. (2012). Evolution and Future of Environmental Marketing. *Asia Pacific J. Marketing Manage. Rev*, *1*(3), 74–81.

Shiu, E., Hair, J., Bush, R., & Ortinau, D. (2009). *Marketing Research* (European Edition). McGraw-Hill Higher Edition.

Singh, G. (2013). Green: The New Colour of Marketing in India. *ASCI Journal of Management*, *42*(2), 57–72.

Soonthonsmai, V. (2007). *Environmental or Green Marketing As Global Global Competitive Edge: Concept, Synthesis And Implications*. EABR (Business) &ETLC (Teaching) *Conference Proceedings*, Venice, Italy.

Suki, N. (2016). Consumer environmental concern and green product purchase in Malaysia: Structural effects of consumption values. *Journal of Cleaner Production*, *132*, 204–214. doi:10.1016/j.jclepro.2015.09.087

Verru, S. (2013). The impact of consumer behavior on Green Marketing: An Analysis. *International Journal of Advanced Trends in Computer Science and Engineering*, *2*(1), 569–575.

Wannimayake, W., & Randiwela, P. (2008). Consumer attractiveness towards Green Products of FMCG sector: An empirical study. Oxford Business and Economics Conference Program, 1-19.

Yaacob, M., & Zakaria, A. (2011). Customers' awareness, perception and future prospects of green products. *The Journal of Commerce*, *3*(2), 1–10.

Chapter 7
Development of Quaternary Concrete

Sagar Dhengare
https://orcid.org/0000-0001-6285-6064
Yeshwantrao Chavan College of Engineering, Nagpur, India

R. L. Sharma
Lovely Professional University, India

ABSTRACT

The amount of unregulated trash produced by industry, particularly in developing countries, has resulted in an increase in environmental challenges. The cultivation of this type of quaternary concrete material is both a sustainable and cost-effective solution to pollution. From the standpoint of producing quaternary cement concrete products from waste materials, the chapter provides a quick overview of which materials are suitable for use in construction as cement replacements. A number of writers and academics conduct a variety of physic-mechanical tests. Various approaches are employed in various studies. As a result, resource-efficient cements with appropriate qualities can be produced to strike the necessary balance between the industry's objective for high-performance concrete and increasingly stringent environmental laws. On the other side, the increase in construction activity in the country caused a scarcity of most supplies used in making concrete, and consequently, cement costs will rise as a result.

INTRODUCTION

Worldwide, over ten billion tons of concrete are being produced each year. Approximately 8% of the world's carbon dioxide (CO2) emissions are attributable to Portland cement. Carbon dioxide belongs to the so-called greenhouse gases, which contribute to global warming. In addition to CO2 emissions, the burning of Portland-cement residue at temperatures around 1400 °C is costly in terms of vestige fuel usage. The disposal of agro and industrial waste materials is causing environmental problems in and around the society. Cements with mineral admixtures remain largely underutilized. This is to a certain extent due to lack of technology transfer and pilot projects that demonstrate the economy, durability, and

DOI: 10.4018/978-1-6684-9503-2.ch007

safety of structures built with such cements. Countless cement plants still do not have provisions for the production, storage, and handling of blended cements. Obviously, there is growing need replacement cements, to develop trustworthy analytical models for their behavior, and to set into the world guidelines for their mix design and field implementation. Normal Concrete is the extremely castoff man-made product in construction manufacturing, but just like a penny it has two looks, no uncertainty it is the most valuable thing in construction industry, but it has a depressing side likewise. Cement and concrete industry are well counsel to be part of a resolute, globally supported approach to meet sustainable development and environmental goals. There is growing credence that quaternary blended cements can be optimized with a synergistic effect, allowing component ingredients to recompense for any mutual shortcomings. Therefore, resource-competent cements with adapted properties can be developed to achieve the needed balance between the industry's mission for high-performance concrete and increasingly limiting environmental regulations. Principles governing the actions of such composite cements can be examined. In other words we can say that this waste materials can be a possible solution to replace cement in optimum ratio.

Cement utilized in concrete creates carbon dioxide that is damaging the environment, so now the scientists and researchers in the pasture of Civil Engineering are pushing their interest in the direction of the use of materials that can trigger tiniest harm to the earth. Globally, over ten billion tons of concrete are being manufactured every year. Almost, eighth part out of hundred of the ecosphere's carbon dioxide secretions are attributable to Portland product. Carbon dioxide be appropriate to the called greenhouse fumes, which promote to global warming. In addition to carbon dioxide emanations, the burning of OPC scum at temperatures around 1400 ^{0}C is costly in terms of vestige fuel usage. The disposal of farming and industrial waste materials is causing environmental problems in and around the society. Cements with mineral admixtures remain largely underutilized. This is to a certain extent due to lack of technology transfer and pilot projects that demonstrate the economy, durability, and safety of structures built with such cements. Countless cement plants still do not have provisions for the production, storage, and handling of blended cements. Obviously, there is growing need replacement cements, to develop trustworthy analytical models for their behavior, and to set into the world guidelines for their mix design and field implementation. A measure undertaken in this path is the use of waste manufactured goods in conjunction with or in substitute of cement. Several of these materials are previously in use, like silica fume, fly ash, etc.

Climate transform is considered as the major environmental challenge for the world. Technological encroachment has resulted in Cement making companies being able to produce higher volumes compared to the past. However, the higher production levels have also been largely labeled as the leading cause of pollution. CO2 Emissions from Cement manufacturing are one of the major contributors in global warming and climate change. Cement belongs to the most often used building materials and its production is increasing over the world. The quaternary concrete will make environment ecofriendly, solves disposal by problems by utilizing agro and industrial waste reduces pollution, provide low house costing.

1. DEVELOPMENT OF QUATERNARY CONCRETE

Alena Vimmrová et al. (2014) performed the work on Calcined gypsum-lime-metakaolin files as a model of best arrangement and results were obtained that reveal the use of these quaternary products are very crucial when replace. Axel Schöler et al. (2015) deliberated hydration of quaternary melds encompassing explosion-furnace slag, silicious fly ash, and limestone powder. Following the application as a sequence in separate relationships, the calculations indicated gigantic force disparity which is extraordinarily considerable. Aashay Arora et al. (2019) verified fundamental insights into the compressive test and flexural test results on ultra-high-performance concrete which shows greater results than the controlled concrete. Erhan Güneyisi et al. (2010) investigated the potency and soak up drop homes of person-compressing concretes combining multi-structure mix together aggregate alloys for different combinations. And results are quite good as compared to normal Portland concrete. Kasım Mermerdas et al. (2013) presented tentative calculation along with demonstrating of wiping decline comportment of metakaolin as well as calcined kaolin unified concretes after that results obtained are quite good for the betterment of the society as it is using up to 20% of waste materials. Eva Kuzielová et al. (2017) analyzed connection stuck between sponginess in addition to automated estates of multicomponent organisms involving of Portland cement–dross–silica emission–metakaolin and results are greater than the conventional one. Kamel Mohamed Rahla et al. (2019) examined the virtual sustainability evaluation of second mix together specific applying accompanying Cementitious Items coupled with OPC. The conclusions proved that as soon as at hand is the substitution of food production and industrial byproducts used as alternative purposes the strength is at the higher side. Reza Saleh Ahari et al. (2015) considered Thixotropy in addition to fundamental summary assets of identity-securing existing covering countless ancillary cementitious items action as quaternary combinations. Results give positive parameters in all tests performed in this paper. Aashay Arora et al. (2018) described Microstructural sorting-then rheology-built ring folder high-quality next to including submission created for Particularly-Great ranking-Running Concrete and results showed higher compressive strength properties than the other combination of normal concrete. Niragi Dave et al. (2016) (2017) created an tentative study of intensity and stability estates of quaternary adhesive ring binder then mortar and results showed that the combination of quaternary replacements is useful for sustainability and global warming effect. Findings going on quaternary specific microstructure, depth, toughness pondering the shape of multifactor are taken into consideration and got the possible alternative. Chao Fan et al. (2018) examined that in this project hot house rubbish radiations stood lowered in support of recommended physical strength than the traditional concrete making. Aysegul Petek Gursel et al. (2016) studied the life expectancy-phase methodology to environmentally friendly, automated, and endurance goods of green real blend through rice stubble powder as comparison with the simple one or plane concrete formation. M.M. Hossain et al. (2016) delivered the results for the sturdiness of grout in addition to solid got upward of pozzolans being as a preferential spare of stick as similarity through the opening performances of the conventional methods for binder, mortar, and concrete. Nihat Kabay et al. (2015) studied the estates of certain including pumice stone powder and fly residue when adhesive extra resources as well as the findings are better than the zero replacement. P. Kathirvel et al. (2013) derived the properties of asset plus strength of quaternary strengthen actual ended thru soar powder, rice peapod dust and granite residue what is more results showed that the replacement level up to 25%. Z. Makhloufi et al. (2012a) (2012b) (2014) (2016) (2015) recommended battle of sandstone plasters thru quaternary files to sulfuric caustic resolution in addition the results obtained are better than the initial one. Various trials were done for the optimum combinations. The power of granite putties per quaternary loose-leaf folder as ooze consequence by demineralize aquatic remained considered in addition results are forwarded for the various other tests so that optimum percentage replacement ratio is to be calculated. Paper designated the properties and outcome of quaternary cementitious arrangements encompassing sandstone, flash kiln slag, then ordinary pozzolan on the power-driven comportment of marble cements with various combinations. Paper observed that the upshot of atomic number 12 sulfate on the robustness of limestone putties based on quaternary unified categories of glue showed the best combinations by compressive strength tests. Mohammed Seddik Meddah et al. (2014) premeditated the probable custom of double and merged sand stone lumps of cement in concrete creation as an substitute riposte for the discharge problem. Liwu Mo et al. (2015) developed bend plus power-driven goods of quaternary combined strengthen comprehending milled gritty flash kiln slag, fly cinders, what is more magnesium oxide, results occurred were good for the strength. Ahmed Omran et al. (2016) examined the performance of glass-powder as replaceable materials to concrete in field applications and results are quite better in strength parameter. Chuslip et al. (2009) studied the effects of LOI of ground bagasse ash on the compressive strength and sulfate resistance of mortars. Weight of binder could produce higher compressive strengths than a control mortar after 28 and 90 days, respectively. Morales et al. (2009) examined the study of calcining conditions on the microstructural features of sugar cane waste ash (SCWA). It is observed that iron oxide (magnetite Fe3O4) is located in the calcium-rich particles. The results show that the sugar cane wastes ash calcined at 800 and 1000 ^{0}C have properties indicative of high pozzolanic activity. The X-ray diffraction patterns, the TEM images and the pozzolanic activity tests show the influence of different factors on the activation of these ashes. Calligaris et al. (2015) evaluated the pozzolanicity of Sugar Cane Straw Ashes (SCSA), thermal treated, at different curing times. It concludes that this addition of SCSA in OPC has caused a delay in the cement setting time. Moreover, the behaviour of the C3S and Calcite contents in both OPC and OPC/SCSA samples were determined by refinement of the XRD pattern using the Rietveld method. Schettino et al. (2015) determined the chemical, physical and mineralogical characteristics of sugarcane bagasse ash to investigate its use in the production of ceramic floor tile. Technological properties such that flexural strength, apparent density, linear shrinkage, and water absorption were determined using standard procedures. The result indicates that the sugarcane bagasse ash waste is rich in quartz particles, and has potential to be used as an alternative raw material for the production of ceramic floor tile. Soares et al. (2016) explained the pozzolanic behavior of sugar cane bagasse ash-SCBA by comparing it to amorphous and crystalline silica. It is shown that calcium silicate hydrate is formed in lime solution with SCBA but the reaction is slow and does not consume all the material. Comparing its results with the obtained in tests with silica fume and with crushed quartz show a better agreement with the latter. Characterization of cement pastes shows 20% of cement replacement by sugar cane bagasse ash leads to only a minor reduction in the amount of calcium hydroxide formed. Analysis of the microstructure of the cement pastes revealed the presence of calcium hydroxide in samples prepared with partial replacement by silica fume, quartz and sugar cane bagasse ash. Deepika K P et al. (2016) investigated the effect of using copper slag as a replacement of fine aggregate on the strength properties. In this report, M25 grade concrete was used and tests were conducted for various proportions of copper slag replacement with sand of 40%, 80% and 100% in concrete. The obtained results were compared with those of nominal concrete made with ordinary Portland cement and sand. Further split tensile strength and flexural strength studies were also performed to check its durability aspects. Jagmeet Singh et al. (2019) conducted the study on feasibility of use of Copper Slag (CS) as aluminosilicate material for developing Alkali-activated Cementitious Material (ACM). Further, the addition of mineral admixtures in alkali-activated CS cement improves its performance; however, the addition of MK was found to be more effective than FA in terms of compressive strength and microstructure development. Nikita Gupta et al. (2019) intended to design self-compacting concrete mixes incorporating copper slag as fine aggregate replacement. Compressive strength and split tensile strength increased as the content of copper slag escalated. The results indicate that Compressive strength and flexural Strength is increased due to high toughness of copper slag. B. Janakiramaiah1 et al. (2016) focused on the characteristics of copper slag and its effects on the engineering properties of cement, mortars and concrete. The mechanical properties of copper slag and reinforced concrete under dynamic compression. The results showed that the dynamic compressive strength of copper slag reinforced concrete generally improved with the increase in amounts of copper slag used as a sand replacement up to 20%, compared with the control concrete, beyond which the strength was reduced. Pavan Hiremath et al. (2016) The production and worldwide use of chicken eggs on an industrial or domestic level leads to a considerable quantity of shell residue, which is considered as a waste or is used as a complement in agriculture. The use of eggshell powder as fillers for the plastic industry will not only provide a renewable source of filler for the plastic industry but also generate a non – food source of economic development for the famers in the rural areas. Pavan Hiremath et al. (2018) studied various physical and mechanical properties of egg shell powder filled hybrid composite shave been found. Novelty/Improvement: The cost of egg shell powder fillers is less than that of the polyester matrix. The use of eggshell powder as fillers for the plastic industry will not only provide a renewable source of filler for the plastic industry but also generate a non – food source of economic development for the famers in the rural areas. Pradeep Sharma et al. (2018) carried out research to determine the optimum percentage of eggshell powder as partial cement Replacement and

Figure 1. Findings of literature review

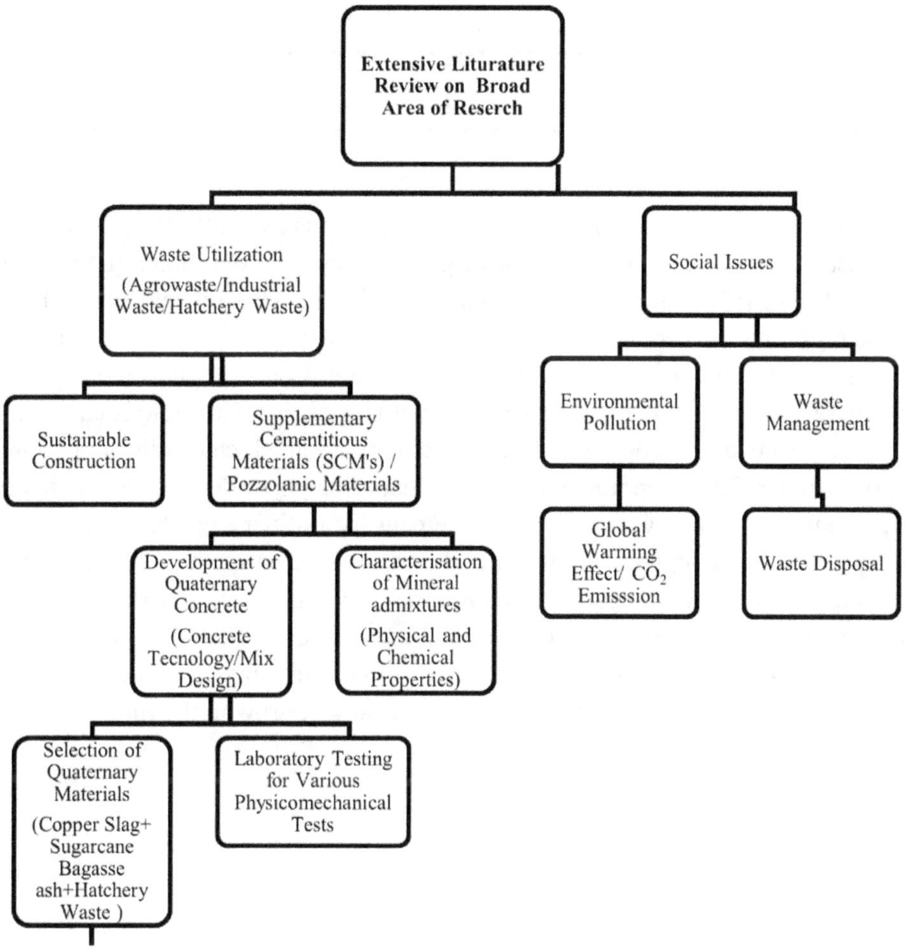

Figure 2. Methodology followed for manufacturing of quaternary concrete

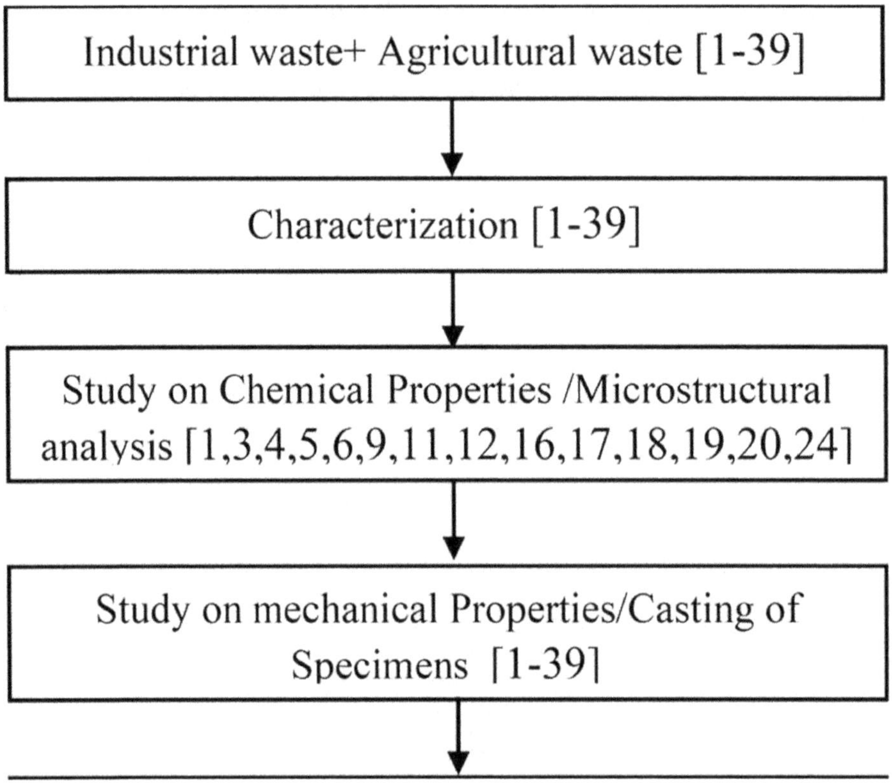

Copper Slag as partial replacement of Fine aggregate. This slag is currently used for many purposes like land filling, construction of abrasive tools, roofing granules, cutting tools and rail road ballast material. Bandhavya G.B et al. (2017) currently India has taken a major initiative on developing the infrastructure such as express highways, power projects and industrial structure etc., to meet the requirements of globalization, in the construction of building and other structure. Concrete plays the key role and a large quantum of concrete is being utilized in every construction practice. Concrete is cast and compressive test and tensile tests were carried out to find the best combination which results in optimum percentage of strength. Atul Kumar (2018) reported the results of experiments evaluating the use of egg shell powder from egg production industry as partial replacement for ordinary Portland cement concrete. The chemical composition of the egg shell powder and compressive strength of the cement concrete was determined. an experimental investigation of compressive strength, split tensile strength, and Flexural strength was undertaken to use egg shell powder and admixtures as partial replacement for cement in concretein this developing world, many countries are going to urbanization due to population growth. Since India is one of developing countries, the construction will grow in the future.

Figure 3. Test results of different combinations of quaternary materials (compressive)

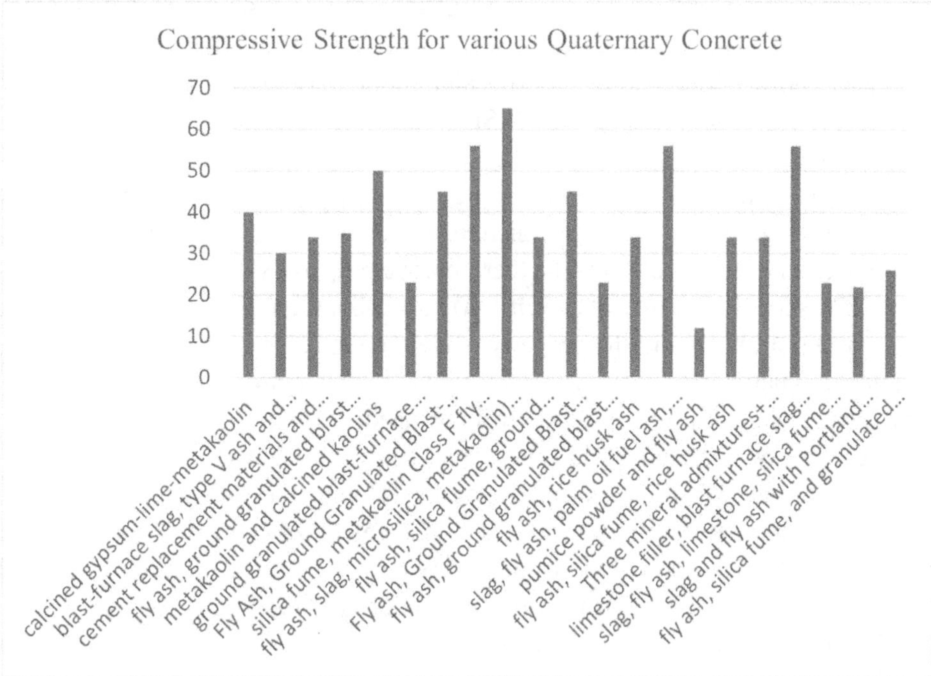

2. PHYSICAL PROPERTIES OF MATERIALS

This search paper studies about pozzolan properties including, specific gravity, particle size, specific surface, chemical and mineral compositions, amorphousness and water demand, bulk density, conductivity, hardness of the materials. This paper addresses the possibility and the effectiveness of using Pozzolans. For this determination, different supplementary materials studied from the literature i.e., fly ash, rice husk ash. Lime, metakaolin, sugarcane bagasse ash, copper slag, egg shell powder, wood saw dust, glass powder etc. The physical and chemical properties of these pozzolanic materials were resolute. The efficiency of each one of these mineral admixtures used as OPC replacement material in monitoring alkali silica reaction are studied and investigated. Results reveal that the performance of Natural Pozzolana as compared with that of other pozzolanic material is very pleasing and can be successfully used in construction.

The test results showed that all the combinations are giving satisfactory result as compared to conventional concrete. Some combinations need attention for recommendations and some need revisions.

3. DISCUSSION ON CHEMICAL COMPOSITIONS AND RESEARCH GAP IDENTIFICATION

1. Metakaolin is a largely amorphous dehydration product of kaolinite, Lime (calcium oxide) is a white solid with strongly basic properties. Lime reacts readily with water to produce slaked lime, which is the chemical compound calcium hydroxide. Eggshell contains of 1% magnesium carbonate, 1%

calcium phosphate, 4% organic matter, and 94% calcium carbonate, The phase of the copper slag is mainly fayalite and magnetite, based on the XRD results. The sugarcane bagasse ash has a high concentration SiO2.. The rice husk ash content is 92 to 95% silica, highly porous, and lightweight. Fly ash consists primarily of oxides of silicon, aluminum iron and calcium. The primary component of silica fume is amorphous silicon dioxide SiO2 in a range of 85% - 98%. GGBS is relatively high in calcium oxide.

2. The existing mixing method of binary blending supplementary materials with cement and ternary combination has higher result oriented concrete. Although there is a possibility of beyond progress by using quaternary blend which can further improve properties of concrete.
3. Different SCM's are, RHA, FA, silica fume, Meta-kaolin, GGBS, SCBA, ESP etc. Concern with all, trials will be study to find the property namely compressive strength of M20 grade cement concrete using quaternary mixes incorporating OPC, sugarcane bagasse Ash, copper slag and hatchery waste along with super plasticizer if any needed.
4. After doing extensive literature review on various keywords such as quaternary concrete, mineral admixture, waste material utilization and disposal, it is found that Slag of Copper material is mainly used as sand as alternative and never used as replacement for cement. Sugarcane bagasse ash is agriculture waste and hatchery waste from poultry farm never used in quaternary concrete.
5. Proposed research work is based on new quaternary blend (OPC+ Sugarcane bagasse ash+ hatchery waste + Copper slag), such combination never used earlier.
6. The development of quaternary concrete will be the best alternative for concretes. To make people aware about use of these materials will be a great challenge for researchers.
7. All these materials mentioned above are pozzolanic in behaviors and having supplementary cementitious properties after characterizing the materials as per literature surveys.
8. Combinations of various different materials having the greater strength properties than the traditional concrete making methods.

4. DRAWBACKS/LIMITATIONS OF USING AVAILABLE COMBINATIONS

With reference to variable combinations however we are in need of replacement of cement but not inn full form as research is going on there on various physical chemical parameters. As per literature whatever combinations tried for best, some get better results and some fails. Reason behind is addition of excess percentages of waste materials which are not suited to other materials properties. Therefore, suitable materials is suggested to do such practices. Many materials can not be use more than 30% as replacement of OPC. As we use PCC then with addition of waste materials changes its properties. Hence, it is suggested that whenever researchers tries to make combination they should use OPC only to get precise outcomes.

Many times the materials what we are procuring is about not in the proper form what we required as per Indian standard. The factors responsible for the limitations.

1. To get materials in with respect to loss on ignition factor
2. Materials with rich in chemical composition as per IS
3. Formation of byproduct on desired temperature
4. Skilled labor

Whatever the combinations available for quaternary mixes should match the IS requirements of chemical and physical properties. If not so there will be loss of strength and structure gets collapse and the combination is of no use. In some cases one of materials is having one chemical component is more in percentage and another is missing then in sime cases there is chances of failure. So you will not get proper basic test results such as slump, compaction factor, setting time, consistency, surface area etc.

5. RECOMMENDATION/SUGGESTIONS, CONCLUSION, AND FUTURE SCOPE

Some conclusions from the literature survey are listed are as follows,

The quaternary unified bolster completes glowing in power and robustness features. The calculation of countless forms of yarns in the quaternary blend is probable for surge the strength limitations of the growth. Project on this topic wide scope for future investigation. Further investigations must be conducted regarding cracking, creep, temperature development and deformation. As the topic is unique and concept is very good, we are doing it experimentally and numerically, so the chance of acceptance of paper in very good journal is expected.

- Ø After doing extensive literature review on various keywords such as quaternary concrete, mineral admixture, waste material utilization and disposal, it is found that Copper slag material is mainly used as partial replacement of fine aggregate and never used as replacement for cement. Sugarcane bagasse ash is agro waste and hatchery waste from poultry farm never used in quaternary concrete.
- Ø Proposed research work is based on new quaternary blend (OPC+ Sugarcane bagasse ash+ hatchery waste + Copper slag), such combination never used earlier. As per Indian standard code if the such materials having the % chemical composition in addition of SiO2, Fe2O3, Al2O3 is about more than 70% then and then only you can proceed to add it as pozzolanic materials. All these materials satisfies the condition hence we can add it as pozzolanic materials as replacement of cement.
- Ø The developed quaternary concrete could be the best substitute of OPC concretes. It's utilization in construction may help in reducing the burden on natural raw materials used in OPC and will reduce our dependence on OPC and promote the utilization of waste materials in construction.
- Ø The following suggestions are made for exploring effective utilization of quaternary blended cement in construction industry:
    - The quaternary blended cement performs well in strength and durability factors, which is evidenced in the microstructure also.
    - The addition of various types of fibers in the quaternary mix is expected to increase the tensile and flexural strength of the specimens. The study has a wide scope for future investigation.
    - Further investigations have to be carried out regarding cracking, creep, temperature development and deformation.

REFERENCES

Ahari, R. S. (2015). Thixotropy and organizational breakdown belongings of Self-consolidating concrete comprising several supplementary cementitious materials. *Cement and Concrete Composites*, *59*, 26–37. doi:10.1016/j.cemconcomp.2015.03.009

Aparecida, M., Schettino, S., Nilson, J., & Holanda, F. (2015). Characterization of sugarcane bagasse ash waste for Its Use in Ceramic Floor Tile. *Procedia Materials Science*, *8*, 190–196. doi:10.1016/j.mspro.2015.04.063

Arora, A. (2019). *Essential insights into the compressive and flexural response of binder- and aggregate-optimized ultra-high-performance concrete (UHPC)* (Vol. 98). Elsevier Ltd.

Arora, A., Aguayo, M., Hansen, H., Castro, C., Federspiel, E., Mobasher, B., & Neithalath, N. (2018). Microstructural padding- and rheology-based binder collection and depiction for Ultra-high-Performance Concrete (UHPC). *Cement and Concrete Research*, *103*(February), 179–190. doi:10.1016/j.cemconres.2017.10.013

Bandhavya, G. B., Sandeep, K., & Bindhushree, G. B. (2017). An experimental study on partial replacement of cement with egg shell powder in concrete. Academic Press.

Calligaris, G. A. (2015). Assessing the pozzolanic activity of cements with added sugar cane straw ash by synchrotron X-ray diffraction and Rietveld analysis. *Constr. Build. Mater.*, *98*, 44–50.

Chusilp, N., Jaturapitakkul, C., & Kiattikomol, K. (2009). Effects of LOI of ground bagasse ash on the compressive strength and sulfate resistance of mortars. *Construction & Building Materials*, *23*(12), 3523–3531. doi:10.1016/j.conbuildmat.2009.06.046

Dave, N. (2016). Investigational analysis of métier and robustness properties of quaternary cement binder and mortar. *Construction & Building Materials*, *107*, 117–124. doi:10.1016/j.conbuildmat.2015.12.195

Dave, N. (2017). Study on quaternary concrete micro-structure, strength, durability since the inspiration. *Construction & Building Materials*, *109*, 217–224.

De Soares, M. M. N. S., Garcia, D. C. S., Figueiredo, R. B., Teresa, M., Aguilar, P., & Cetlin, P. R. (2016). Comparing the pozzolanic behavior of sugar cane bagasse ash to amorphous and crystalline SiO 2. *Cement and Concrete Composites*, *71*, 20–25. doi:10.1016/j.cemconcomp.2016.04.005

Deepika, K. P. (2016). Utilization of copper slag as a partial replacement of fine aggregate in concrete. Academic Press.

Fan, C. (2018). Reducing greenhouse gas productions for prearranged concrete compressive gift. *Construction & Building Materials*, *167*, 918–928. doi:10.1016/j.conbuildmat.2018.02.092

Güneyisi, E. (2010). Strength and drying shrinkage properties of self-compacting concretes integrating multi-system unified inanimate admixtures. *Construction & Building Materials*, *24*(10), 1878–1887. doi:10.1016/j.conbuildmat.2010.04.015

Gupta, N., & Siddique, R. (2019). Strength and micro-structural properties of self- Compacting concrete incorporating copper slag. *Construction & Building Materials*, *224*, 894–908. doi:10.1016/j.conbuildmat.2019.07.105

Gursel, A. P. (2016). A life-cycle slant to environmental, power-driven, and durability belongings of 'green' concrete assortments with rice husk ash. *Journal of Cleaner Production*, *112*, 823–836. doi:10.1016/j.jclepro.2015.06.029

Hiremath & Shettar. (2016). *Investigation on Mechanical and Physical Properties of GFRP-Egg Shell Powder Investigation on Mechanical and Physical Properties of GFRP-Egg Shell Powder Hybrid Composites*. Academic Press.

Hiremath, P., Shettar, M., Shankar, M. C. G., & Mohan, N. S. (2018). Investigation on Effect of Egg Shell Powder on Mechanical Properties of ScienceDirect Investigation on Effect of Egg Shell Powder on Mechanical Properties of GFRP Composites. *Materials Today: Proceedings*, *5*(1), 3014–3018. doi:10.1016/j.matpr.2018.01.101

Hossain, M. M. (2016). Durability of mortar and tangible made up of pozzolans as a limited auxiliary of cement: A review. *Construction & Building Materials*, *116*, 128–140. doi:10.1016/j.conbuildmat.2016.04.147

Kabay, N. (2015). Chattels of concrete with pumice stone powder and fly ash as cement auxiliary materials. *Construction & Building Materials*, *85*, 1–8. doi:10.1016/j.conbuildmat.2015.03.026

Kathirvel, P., Saraswathy, V., Karthik, S. P., & Sekar, A. S. S. (2013). Strength and Durability Possessions of Quaternary Cement Concrete thru with Fly Ash, Rice Husk Ash and Limestone Powder. *Arabian Journal for Science and Engineering*, *38*(3), 589–598. doi:10.100713369-012-0331-1

Kumar, A. (2018). Egg Shell Powder Partial Replaced with Cement Concrete. *International Journal of Scientific Research*, *7*(9), 2016–2019.

Kuzielová, E., Žemlička, M., Bartoničková, E., & Palou, M. T. (2017). The correlation between permeability and mechanical properties of multicomponent systems comprising of Portland cement–slag–silica fume–metakaolin. *Construction & Building Materials*, *135*, 306–314. doi:10.1016/j.conbuildmat.2016.12.105

Makhloufi. (2012a). Confrontation of limestone mortars through quaternary binders to sulfuric acid explanation. *Construction & Building Materials*, *26*(1), 497–504. doi:10.1016/j.conbuildmat.2011.06.050

Makhloufi, Z. (2012b). The strength of limestone mortars with quaternary files: Discharge consequence by demineralized marine. *Construction & Building Materials*, *36*, 171–181. doi:10.1016/j.conbuildmat.2012.04.112

Makhloufi, Z. (2014). Durability of limestone mortars created on quaternary binders imperiled to sulfuric acid using drying-immersion cycles. *Computers & Chemical Engineering*, *71*, 579–588.

Makhloufi, Z. (2016). Upshot of magnesium sulfate on the durability of limestone mortars constructed on quaternary blended epoxy resin. *Cement and Concrete Composites*, *65*, 186–199. doi:10.1016/j.cemconcomp.2015.10.020

Makhloufi, Z., Chettih, M., Bederina, M., Kadri, E. H., & Bouhicha, M. (2015). Effect of quaternary cementitious systems containing limestone, blast furnace slag and natural pozzolan on mechanical behavior of limestone mortars. *Construction & Building Materials*, *95*, 647–657. doi:10.1016/j.conbuildmat.2015.07.050

Meddah, M. S. (2014). Probable use of binary and amalgamated limestone adhesives in concrete creation. *Construction & Building Materials*, *58*, 193–205. doi:10.1016/j.conbuildmat.2013.12.012

Mermerdaş, K. (2013). Experimental estimation and modeling of drying contraction performance of metakaolin and calcined kaolin blended concretes. *Construction & Building Materials*, *43*, 337–347. doi:10.1016/j.conbuildmat.2013.02.047

Mo, L., Liu, M., Al-Tabbaa, A., Deng, M., & Lau, W. Y. (2015). Deformation and mechanical goods of quaternary blended cements holding ground granulated blast furnace slag, fly ash and magnesia. *Cement and Concrete Research*, *71*, 7–13. doi:10.1016/j.cemconres.2015.01.018

Morales, E. V., Villar-cociña, E., Frías, M., Santos, S. F., & Jr, H. S. (2009). Cement & Concrete Composites Effects of calcining conditions on the microstructure of sugar cane waste ashes (SCWA): Influence in the pozzolanic activation. *Cement and Concrete Composites*, *31*(1), 22–28. doi:10.1016/j.cemconcomp.2008.10.004

Omran, A., & Tagnit-Hamou, A. (2016). Recital of glass-powder concrete in pitch applications. *Construction & Building Materials*, *109*, 84–95. doi:10.1016/j.conbuildmat.2016.02.006

Patil. (2015). Properties and effects of copper slag in concrete. *International Journal of Science and Research, 2*, 46–50.

Rahla, K. M., Mateus, R., & Bragança, L. (2019). Comparative sustainability assessment of binary blended concretes using Supplementary Cementitious Materials (SCMs) and Ordinary Portland Cement (OPC). *Journal of Cleaner Production*, *220*, 445–459. doi:10.1016/j.jclepro.2019.02.010

Ramaiah, Babu, & Yadav. (2016). *A Study of Concrete Using Copper Slag as a Partial Replacement of Fine Aggregate*. Academic Press.

Schöler, A. (2015). Hydration of quaternary Portland cement combinations containing blast-furnace slag, silicious fly ash and limestone residue. *Cement and Concrete Composites*, *55*, 374–382. doi:10.1016/j.cemconcomp.2014.10.001

Sharma, P., & Kumar, M. (2018). An experimental investigation on partial replacement of cement with egg shell powder and fine aggregate with copper slag in concrete. Academic Press.

Singh, J., & Singh, S. P. (2019). Development of Alkali-activated Cementitious Material using Copper Slag. *Construction & Building Materials*, *211*, 73–79. doi:10.1016/j.conbuildmat.2019.03.233

Vimmrová, A., Keppert, M., Michalko, O., & Černý, R. (2014). Calcined gypsum-lime-metakaolin binders: Project of optimal composition. *Cement and Concrete Composites*, *52*, 91–96. doi:10.1016/j.cemconcomp.2014.05.011

Chapter 8
The Anthropology and Anthropocene Urban Socio-Ecology Resilience Approach to Resilience Planning and Green Innovations

José G. Vargas-Hernandez
Tecnológico Superior oe Jalisco Mario Molina, Zapopan, Mexico

Carlos A. Rodriguez-Maillard
https://orcid.org/0000-0003-2406-196X
Universidad Cristòbal Colòn, Mexico

Omar C. Vargas-González
https://orcid.org/0000-0002-6089-956X
Tecnológico Nacional de México, Ciudad Guzmán, Mexico

ABSTRACT

This study aims to analyze some of the existing implications between urban anthropology and Anthropocene urban socio-ecology planning resilience. Beginning from the assumption that urban anthropology gives support to create and develop any urban planning based on the Anthropocene urban socio-ecology resilience, the methods employed are the analytical-descriptive based on an ethnographic interpretation and reflection of the theoretical and empirical literature review. The analysis concludes that urban anthropology fundamentals give support to strengthen the Anthropocene orientation of the urban socioecological planning resilience.

DOI: 10.4018/978-1-6684-9503-2.ch008

INTRODUCTION

The term resilience was introduced by Holling (1973) in the field of socioecological research to describe the capacity of the socioecosystem to return to its original form after being deformed by the impact of external pressures from the external environment, in other words defined as the capacity to absorb shocks and perturbations of social-ecological systems and show their adaptability and transformation capacity often related to a natural hazard and climate change (Land, 2019). Resilience influences urban planning and innovation, urban infrastructure construction and development leading to urban and regional economic resilience as one of the contents of resilience research in urban planning and governance. Improving urban socio-ecological resilience contributes to disaster prevention, management and reduction, which strengthens urban resilience processes (Yang, Lili & Hongchi, 2022). Resilience in anthropology has been used to explain different types of socio-environmental phenomena: generalist or specialized resource socio-environmental phenomena: generalist or specialized strategies in the attainment of resources; uncertainty and surprise in the management of resources; adaptive capacity and degree of centralization in the use and management of resources and degree of centralization in the use and management of resources (Escalera & Ruiz, 2011). Some approaches to urban anthropology do not consider it a science and there are those who prefer theories, particular or universal, and those who choose to interpret rather than explain.

Second, the term urban anthropology is a scientific discipline epistemologically closely linked and interrelated with other scientific theories. Anthropology, like any social science, predicts diverse phenomena (Alvarez, 2016; Brady, 2019; Grimmer et al., 2021; Hindman, 2015; Hofman et al., 2017; Hofman et al., 2021). The scientific status of anthropology goes beyond the conception of a soft, interpretive, humanistic phenomenon to interpret meanings to use quantitative techniques, apply formal methods in experimental designs to postulate scientific theories predicting phenomena and formulating scientific laws linked and interrelated to other scientific theories. The phenomena of sociology, anthropology and political science differ from individual behavior, according to Bunge (1999), who argues that predicting the behavior of an unknown individual is different from predicting the behavior of a system. However, the social sciences have transferred the prediction of small groups to complex social systems by means of sophisticated computational methods (Bunge, 1999). A scientific law is a logical-mathematical representation of the relationships between variables (Pfeifer, 2006).

Scientific predictions arise from the verification and application of a theory capable of explaining specific phenomena. Research in urban anthropology and socio-ecology resilience in urban planning and green innovation areas using interdisciplinary methodologies is changing peoples' societies relevant to global research more now in the face of the climate change phenomenon the world is facing.

Thirdly, since 1945, the post-war period accelerated changes in human civilizations with negative impacts on the biosphere that have further increased since the end of the cold war (Eriksen, 2016; Snow, 1998). This displacement originating in the large urban centers of industrialized countries marks the limiting period that leads to validate the ethnographic method. Alternatives to the ideologies and practices of global capitalism are human lives not dominated by the state and market forces (Escobar, 2020). Other natural phenomena are also causing urban ecological and resilience dysfunctionalities such as, for example, heavy storms that flood urban communities, immobilize and disrupt urban transportation, energy, water and waste networks. Cities and any urban area are practically marinated in fossil fuels largely controlled by corporations. Stakeholder and actor interactions contribute haphazardly to identifying the inherent complexity of urban socio-ecological meta-problems and contribute to intensifying the complexity required to address the process.

Anthropomorphic representations in research have not been convincing in favor of the functional hypothesis. Science solves testable hypotheses (Hirst, 2019). Socioeconomic and organizational science research focuses on sympathy for the population of subjects who move from the realm of an observer to becoming an active participant (Bilharz, 1998; Kolb, 1998). Interdisciplinarity and transdisciplinarity of urban anthropology applied to urban socio-ecological planning and green innovation areas are limited in space to carry out the development of disciplines in multidimensional crisis including ethical values and moral issues leading to an ontological crisis of homo sapiens (Tarantino-Curseri, 2018). Ethnographic writing, interpretivism and perspectives in urban anthropology must be replaced by experimentation, theorization, causality, prediction, mathematical modeling and epistemology. Participant observation is an ethnographic methodology and one of the most popularized and important techniques developed by Malinowski (1961) in anthropological research that can be applied to urban anthropology and its impact on urban social-ecological resilience planning and development (DeWalt, DeWalt & Wayland, 1998; Kawulich, 2005; Jorgenson, 2015).

The conceptual framework of the institutional analysis model is proposed by René Lourau (1979) and Lapassade (1986) that can be applied to analyze urban anthropology in the Anthropocene of socio-ecological urban planning and green innovation areas. The ethnomethodological urban anthropology perspective of urban socioecological resilience planning and green innovation areas points to the inherent transformation of the sociohistorical process that affects the production of non-contemplative knowledge that claims meaning and challenges to decipher without historically shaped typologies. Unstructured techniques without a specific direction leads to collect deeper and broader responses and the original terms to know the native categories and perspectives of urban anthropology in socio-ecological urban planning (Pizarro, 2014). Interviews through defined questions are recommended as an ethnographic research tool at the lowest level of structuring to extract more information linking urban anthropology and urban social-ecological resilience planning.

This study aims to analyze the Anthropology and Anthropocene urban socio-ecology resilience approach to resilience planning and green innovations, to raise awareness on the impact of climate change and green innovations. To achieve this goal, first organizational and urban anthropology is analyzed, followed by urban socioecological resilience planning and urban green innovation areas. Finally, some conclusions are offered.

BACKGROUND

Qualitative-quantitative division of methods and techniques in urban anthropology and urban socio-ecology resilience planning and innovation is objectionable (Aldrich, 2014). Recent research addressing the quantitative and qualitative approaches to reveal the consequences of the COVID-19 pandemic crisis on the theoretical framework of urban, society and institutional governance resilience, was conducted by (Olimid, Georgescu & Gherghe, 2022). Quantitative methods and techniques in urban anthropology and urban socio-ecology resilience planning and innovation aim to collect and analyze numerical data (Roni et al., 2020).

Formal methods must meet requirements in operationalization of terms, data analysis, logical consistency, justification of statements and replication of results, although neither formal theories were dominant nor were the dominant theories ever dominant (Reynoso, 1995). The discipline of urban anthropology is capable of formalizing produced knowledge. Formal computational methods used for the ethnographic analysis in urban anthropology and archeological data based on agent models, cellular automata, diffusion and percolation models, network analysis, etc., proposed for a new research form open to theoretical development and experimentation (Díaz et al. 2007).

Urban anthropological experiments, predictions and theories can be compared with natural and physical sciences to determine the scientific status (Llanos, 2009) but are not sufficient scientific criteria for good science if it is conducted without laws and predictions. The most common experiments in urban anthropology and social sciences are natural experiments (Diamond & Robinson, 2010; Dunning, 2012), field experiments (Baldassarri & Abascal, 2017) and naturalistic experiments (Bernard, 2011). The new anthropology of the Anthropocene needs to retain and refine some of the defining traits of early- to mid-20th century anthropology, although research priorities have changed dramatically in the last century (Hylland, 2022).

To carry out this bibliometric review, electronic, literary and scientific sources were used as the basis for recent research on relationship between Anthropology and Anthropocene urban socio-ecology planning resilience. The study was conducted by searching for the keywords "urban anthropology", Urban Socio-Ecology resilience planning", "Anthropocene urban socio-ecology planning resilience" and " urban green innovations areas " in the main scientific search engines, previous works were reviewed in detail to define those that would be included in this study.

ORGANIZATIONAL AND URBAN ANTHROPOLOGY

Organizational anthropology has close relationships with urban anthropology and that includes the demands for participatory actions in practical anthropology of urban imaginaries and organizational facilitation (Gravano, 1992). This situation integrates the organizational and urban anthropological perspectives and the ethnological methodology tools to the analysis of organizational studies, transversal to research financing demands. Social sciences can make possible scientific predictions on specific events and problems in urban socio-ecological planning and development despite the arbitrariness of human behavior and the assumption that is not useful to predict behavior (Nießen et al., 2020).

The debates in the analysis of organizational and urban anthropology bridging the studies between organizations and urban areas from an anthropological perspective to analyze the interrelationships between urban anthropology and business and work organizations, from a sociohistorical orientation to develop reflections situating the analysis of urban areas and neighborhoods in post-industrial and global societies. Urban anthropology can be analyzed from a socio-historical through its interrelationships with organizations and business.

The work of Warner ratified the analysis of Durkheim from a sociohistorical perspective has an influence on anthropology and sociology related to the social structures of tribal and industrialized complex societies, advancing the analysis of interrelationships between urban anthropology and organizations. Industrial technology and social research become relevant and, consequently, may improve the understanding of technology in organizational anthropology and urban socio- ecological planning and innovation (Roca, 1998).

Organizational and urban anthropology has conceptual, methodological, and theoretical tools aimed to develop the interrelationships between organizations and urban settlements, leading to create the meanings that both workers in organizations and citizens have of their industrial heritage. The studies of Mayo (1945) and collaborators Roethlisberger and Dickson (1939) at the industrial plant of Hawthorne Wester Electric Co. pioneering the investigations of organizational anthropology applied to work productivity. Job design in urban work settlements and organizations is characterized by the development of multiple skills based on redundancy of a variety of functions instead of redundancy of parts being capable to perform the job.

A sociohistorical perspective analysis of the interrelationships between urban anthropology and work organizations is conducted on the anthropological practices in organizations and urban settings that have challenges affecting research and academic training and development. A multidisciplinary perspective on urban anthropology and green and socio-ecological innovation attempts to understand traditional technology in production process. Technological training of firms reinforces the converging of scientific research on green innovation at universities. Firms can share the green technological innovation training fostered by public green investment. Technological and social capital determine the level of socio-ecological and urban anthropology in green innovation of companies. Accumulation of social, economic, and symbolic capital becomes relevant to undertake activities in urban socio-ecology resilience planning and green innovation (Reguera, 2009). Renewable energy and technological innovation have been growing in the last years, but such technologies developed over time and are accompanied by decay and depreciation effects, need to be tracked to avoid their potential negative consequences (Su et al., 2022).

Organizational performance reflects the problems and dilemmas of the organizational practices linked to the state and urban anthropology in socio-ecological planning control and weighting the anthropological ethnography method used in organizational facilitation in professional practice. Urban anthropology in socio-ecological planning is a negotiation and mobilization process leading to the implementation of incremental action definition (de Melo, 2002). Urban neighborhood, as organizational anthropology in organizational culture exercise controls of symbolic culture which is manipulated as ideology aimed to control the urban anthropology or organizational living conditions and interrelationships of the social subjects.

Urban anthropology leading to socio-ecological planning processes from a centralized, sequenced, and systematic integral rational planning to a complete fragmentation and disjointed incrementalism (Lindblom, 1959). Strengthening the consolidation of the national systems of science, technology, and innovation has effects on competitiveness derived from incremental socio-ecological and green innovations in urban planning. Today it is important to include the covid in our urban anthropology study to understand how cities were being reshaped by the pandemic (Hyatt, 2021), it is imperative that the universities and governments start to study about it.

URBAN SOCIO-ECOLOGY RESILIENCE PLANNING AND URBAN GREEN INNOVATIONS AREAS

The evolution of scientific knowledge and socio-ecological and technological green innovation in urban planning is driven by the challenges proposed by the design of new goods, services and processes carrying out internal research and development activities in the firms and in cooperation with other agents in efforts to improve efficiency of regions with deficiencies in their scientific, technological and innovation systems. Firms seeking to innovate in urban socio-ecological planning face obstacles in finances.

Tax and market incentives leave decisions to firms about which urban development projects to finance (de Melo, 2001).

Urban planning of socio-technical and urban socio-ecological innovation in urban regions known as local industrial systems may constitute an example of an economy derived from industrial location self-reinforcing agglomerations of cumulatively socio-technical skills, investments and venture capital, infrastructure, suppliers, knowledge spillovers, services, etc. associated with informal information flows and proximity to research centers and universities (Saxenian, 1994). Urban planning impacts more variables that are related to the environment (Baschera and Schneider,2022).

To stimulate investments in green science, green technology and socio-ecological innovation systems for urban planning requires the necessary conditions to invest in research and development (de Melo, 2001). It is essential that government should stimulate enterprises to strengthen the research of environmental protection technologies (Su et al., 2023). Green technological innovation not tested has high risks and difficulties in evaluating the green high-tech goods and socio-ecological services market and green technology-based firms have difficulties in accessing capital sources. Investment activities in innovation activities have a high level of risk and uncertainty in investments in credit financing and equity financing in research & development of urban green planning innovation in technologies. It is important to consider that the sustainable urban development is essential since the expansion of city facilities need to be in line with the different aspects (social, economic, and environmental) (Gohari et al., 2023).

Universities provide technical training and development in urban planning and socio-ecological innovation. The university-industry interactions between the research institutes and centers for urban socio-ecological and environmental planning with firms interested in joint research programs sponsored by consortiums or individual government agencies and corporations to exploit results through patents and licenses. Through business incubators, universities and research institutions, technology parks, and science, green technology and socio-ecological innovation agents in university-based local green innovation systems, startups at earlier stages can have the opportunity to access government venture capital.

Government funding for research in green technological innovation for urban planning carried out by firms has a positive effect on business research and development spending (de Melo, 2001). Government authorities facilitate the formation of local green and urban socio-ecology resilience innovation systems from universities and research centers increasing public investing in foundations of science, green technology, and socio-ecological innovation. Government intervention in university-based local urban green innovation university oriented may regulate the incentives to promote green technology innovation training and strengthen emerging green technology-based firms. Green technology has been recommended as an effective solution for the growth of green spaces per capita and improving sustainability means of urban development's due to its diverse advantages (Gohari et al., 2023).

Local urban planning innovation systems for green socio-ecological areas integrate the definitions considered in Lundvall (1992) and the one used by MIT (Lester, 2005). They have some elements that interact with creation, dissemination, transference, and use of knowledge for economic value in specific regions (Lundvall, 1992), and show that urban planning impacts community vitality; ecological diversity and resilience; and most important, living standards of sustainable socio-economic development and environmental conservation. (Baschera and Schneider,2022). To make the balance between urban developments and environmental issues, and at the same time to combat climate change and global warming it is important to consider a sustainable development principle and an innovative solution (Gohari et al., 2023).

The context of urban green technological innovation transfers from the University-based to the firms is a factor of competitiveness of productive sectors of society. Massive investments are made in infrastructure and equipment by government agencies and institutions at research centers and universities for urban planning and green innovation (Porter, 2001). Creation of technical schools and incentive programs for implementation of urban planning policies and green innovative practices in industry. Several policies and programs have been instituted to manage climate change impacts but institutional capacity of agency systems to ensure the realization of the policies is limited (Asibey et al., 2022). One example is the green roofs developments. Green roofs have proven to be energy efficient, environment friendly, and economical in a long run. It has become necessary to implement green roofs as an effective and sustainable method that help in the achievement of urban sustainability (Alqahtany, 2023).

The design of a green roof system is a complex procedure that depends on several factors that include social, cultural, climatic, and environmental considerations (Williams et al., 2010). Understand the importance of the green roofs to alleviate environmental problems in cities and to mitigate the urban heat island effect is important to help to decrease air pollution and enhance urban species biodiversity. While green roof application has long been popular in Germany and Norway (Köhler and poll, 2010), There have recently been reports that addressed different advantages of green roofs in tropical areas (hot and humid climatic conditions) such as declining urban heat islands and reducing energy consumption for cooling.

Urban socio-ecology resilience planning and green innovation networks constitute a similar economy of connections which can be effective in the future of work, feeding the discussions of ideas, thoughts, and sentiments to uncover the native`s point of view in local perspective to obtain, analyze and interpret data. Some actors such as research laboratories, students and researchers' network, firms of the incubator, angel investors and venture capital, etc., participate in urban environmental planning processes, although those who are not involved on the governance of local urban planning and socio-ecological innovation systems are not necessarily less influential. Non-financial institutions in governments are hardly available to complement the green investment of private venture capitalists in green firms.

Declining government participation is a principle applied to mature local urban green innovation systems (OECD, 1997b; de Melo, 2001). Anchor organizations and companies focus on urban planning of local innovation systems university-oriented to cooperate in joint efforts of R&D projects directing the focus of cooperation for research to socio-ecological and green areas can get competitive advantage with a competent university. The analysis on morale and evolution of cooperation demonstrated that individual moral models tend to obey the rules and religious beliefs, predict intergroup cooperation and honest behavior (Purzycki et al. 2018). The participation takes greater relevance in these days, as it presents the challenge that countries must face to ensure that climate change and the heat island effect is increasing, as well as the obstacles that green innovation can confront, while the countries redesigning the laws in this new economic, political and social moment they are facing.

The set of green enterprises that participate in an articulating task to support mechanisms for the identification, selection and consolidation of urban green innovation projects through articulating instruments (Pimienta-Bueno, 1999). The local urban socio-ecological concentration of green enterprises allows economies of scale, agglomeration and gains from externalities (Audretsch, 1998). The urban green innovation technologies of technology-intensive green start-ups can be compared to those of mature firms in terms of achievements and performance. Start-ups based on green technologies have higher costs related to their initial phases, which are higher than those of small firms (OECD, 1997b). Nowadays, enterprises and industries must produce, import and upgrade traditional facilities with green technologies to optimize the energy structure (Su et al., 2022).

The urban imaginary phenomenon in the social participation processes can be treated from the anthropological organizational facilitation (Gravano, 2008a), aimed to the treatment of specific problems of these processes. The relationship between metaphorical mediation and interpretation deepens the analysis of urban anthropology contexts to challenge the societal construction of otherness.

The urban planning for green and socio-ecological innovation theory still was marginal during the 1970s until recently innovation acquired the multidisciplinary character encompassing knowledge in fields such as the sociology, economics, psychology, administration, geography, engineering, etc. (Schlapfer & Marinova, 2001). Throughout the 1980s and 1990s the economy was submitted to technological socio-ecological innovation restructuring and a process of economic internationalization and globalization of markets leading to replace the scientific and technology system to focus on the competitiveness of the industrial sector aimed to meet the international markets (González Alvarez, 1997).

Urban planning for socio-ecological and green innovation areas is the process that begins with the invention of a new element until its development for commercial use. The commercialization of urban socio-ecological planning green innovations generated in public universities and research institutions requires clear regulations for sharing and transferring intellectual property rights to private firms, new forms of association in public-private partnerships for technology development. Green innovation must be practical, possess commercial value and be socially acceptable (Sundbo, 1998). Knowledge sources for planning in urban green areas are embedded in institutional and human forms subject to financial capital and less to geographical movement (Schlapfer & Marinova 2001). Urban planning can adequately address the impacts of climate change while preparing urban areas to be more proactive and adaptive to climate change impacts through effective partnerships (Asibey et al., 2022). It is recommended that the local authorities initiate strong campaigns in favor of green technologies in the urban planning at least for the new buildings or the buildings under construction. They must educate people about the benefits these green roofs can provide. The authorities must facilitate the residents and house owners in designing, executing, and installing such systems (Alqahtany, 2023).

In urban socio-ecological planning, conceptual, theoretical and practical ruptures produce the transformation of participatory processes of urban socio-ecological planning and their relationships challenged by methodological research and facilitation. The approach to the professional practice of organizational facilitation and participatory processes presents theoretical-methodological challenges in urban metropolitan socio-ecological environmental planning (Gravano, 2007b) supported by the work of the organizational culture of organizational and economic agents and political actors, mainly the Manchester and Chicago schools have worked on urban anthropology (Cucó Giner, 2004; Gravano, 2005; Signorelli, 1999; Velho, 1978).

Work on entrepreneurship and entrepreneurial anthropology can be traced back to the study of ethnology in urban societies in the 1970s (Durrenberger, 2007). The expansion of university-built infrastructure through urban green innovation technology planning provides socioeconomic consequences results from the research on entrepreneurial alliances from the constitution of angel investors as the formal group of university employees and students who invest in companies in business incubators and innovation parks. The urban socio-ecological planning processes that occur with institutional and community participation address the circularity of management issues and exchange institutional power relations based on the anthropological object of cultural otherness (Gravano, 2006). Urban planning and architecture are demanded by specific actors and urban management agents linked to urban imaginaries weighted as anthropological object for the detection of meanings of living spaces for social actors. Administration and management practices have questioned effects on equality and wealth distribution.

Anthropological contributions to urban socio-ecological and environmental planning are based on community participation in institutional and organizational processes. Individual and corporate profits take precedence over vocations and values of local socio-ecological communities leading to the sacrifice of common good. Community participatory process enforced by law delimit the specific possibilities, difficulties and opportunities based on asymmetric rationalities at stake and the state action in the context or urban socio-ecological environmental planning. A specific situation of demand from the State for an anthropological contribution facilitates the community participatory process. A well-structured state can develop a local green and urban socio-ecology resilience innovation system to support activities of firms, institutions, universities and research centers, private economic sector, etc., It is important to understand that only when capital mismatch in the states is severe, and the low-end industrial structure is poor is the positive impact of green finance on the low-carbon economy highlighted based on scale and structural effects; the ability of green finance to contribute to the low-carbon economy through the technique effect has been more stable and significant (Zhu et al., 2022).

The process of urban planning and green innovation and green technology transfer begins with an idea that the research develops into a technology to be used and applied followed by a disclosure and evaluation of the potential economic value, discuss, and consult with specialists in concerning urban areas of discovery considering the criteria based on scientific and technical relevance, existent patents, and market potential. From this, it follows the registration of the patent and intellectual property. Reports indicate that urban planning efforts have yielded the maximum positive result in building the resilience and adaptive capacities of cities to climate change impacts (Asibey et al., 2022).

The impact of specific risk disasters in urban areas such as earthquakes, hurricanes, floods, public health emergencies, etc., lead to urban socio-ecology planning resilience to prevent the possible impact of these extreme events and provide a valuable diagnose of the weak links in urban resilience and improve the resistance and city resilience of future sustainable development of human society. The case was analyzed by Mu, Fang, Yang, & Guo (2022) based on the population mobility, centrality, and nodality between cities in China was significantly reduced affecting the urban resilience within the network during the COVID-19 epidemic.

The ancient Greeks developed critical thinking on urban socio-ecology resilience based on mutualistic relationships (Egerton, 2001). Socio-ecological change in societies is caused by exogenous factors, such as globalization, colonialism, capitalism, etc. Some colonial societies are compounded of heterogeneous elements with instability due to divergent interests (Radcliffe-Brown, 1940). On the other side, the Spanish applied anthropology concentrated more on the analysis of the legacy left on its colonies and directed to the colonized others by the Spanish empire. Corporations are considered and critiqued as an anthropological colony of the real opposition to knowledge transfer in socio-ecology of work (Mir & Mir, 2009). Individuals and groups with conflicting ecological interests encompasses incommensurable connections between ends and means (de Melo, 2002).

ANTHROPOCENE URBAN SOCIO-ECOLOGY PLANNING RESILIENCE

The definition and meaning of "Anthropocene" is largely disregarded and it is a topic of debate among scientists, the use of informal, flexible "Anthropocene" term can be used to describe the epochal transformation in which the effects of human technological activity and their cultural systems have profoundly altered the environment (Bagram, 2021). Socio ecosystems in urban areas are aimed to maximize the benefits of human needs in an Anthropocene era. The concept of Anthropocene is linked to socio-ecological aspects of climate change and to the analysis of the causes of the collapse of civilization and societies that can be learned from archaeological research when facing crises (Tainted, 1988; Tainted, 2008; McIntosh, et al., 2000; Trawick & Alf, 2015). Western science as a mode of existence is a source of prescriptions to climate change complemented with the interpretation and translational (LA tour, 2013). Communities merge global scientific explanations with regional and local knowledge blaming humans for the global warming and climate change that the world is facing (Sheng et al., 2021).

The global framework for climate services approach has provoked critical reflections on initiatives related to ethical implications, epistemic politics, colonial legacies, etc. (Haines, 2019b; Kraus, 2020; Nest, 2019). International policy frameworks underpinned discourses combined with the adverse effects of warming (Baer & Singer, 2014). Consistent with the anthropogenic climate change perspectives, humans may be accountable of weather. People give meaning and respond to change patterns of global warming and climate change (Barnes et al., 2013; Fiske et al., 2014; Peterson & Broad, 2009). The Anthropocene is much more than just a discussion of the impact of climate change and global warming, it is important to study it from the perspectives of sustainable development, regional security, or international collaboration (Chandler et al, 2022).

Urban resilience is more tangible than the invisible greenhouse gases with serious implications. Home insulation is associated with investment and frugality while organic food with risk and pastoral ideals (Press & Arnold, 2011). The ability of human beings to make decisions for nutritious and healthy food choices that increases survival chances (Birch *et al.*, 1998; Cashman, 1998). Pleasure, holistic health and purity are values of purchasing behavior in some European countries (Ditlevsen et al., 2019). Consumption is relational embedded in social relations. Consumers of traditional marketing goods and services are alienated (Lambert, 2019). Marx critics of consumer capitalism leads to naturalistic epistemology (Kramer, 2021; Stoner, 2021; Trujillo, 2021).

The social exchange theory is a dyadic approach in marketing that extends to incorporate philanthropic actors and multiple stakeholders. Marketing theory and practices replace the consumer fetish with the value creation by a collectively of stakeholders (Arnould & Cayla, 2015; Tsing, 2015; Atanasova 2021; Casey et al. 2020, Figueiredo & Scaraboto, 2016). Without replacement of the consumer construct, consumers can engage in sharing, gifting, bartering and other semiotic consumption processes (Wei, et al. 2021).

Consumption is socially learned and individually determined during the primary socialization The urban anthropology of consumption is the study, analysis and interpretation of consumption as social phenomenon that incorporates the acquisition, consumption and above all the appropriation of products and brands that serve as a person's identity reinforcer (Páramo and Ramirez, 2017).

Socio-technical approach has proved to be effective in some studies in coal mines and a project involving operating units at Royal Dutch Shell in England, in a textile factory in India, implying unions, employers and the government involved to improve the quality of working life in Norway and Sweden, and other socio-technical projects in Canada, Netherlands and Australia. Implementation of socio-technical urban projects are carried out in a continuous urban renewal process to achieve completeness as the consequence of revisions and modifications.

Technoscientific urban models of relationality, such as radioecology, enables to visualize the cross-boundary movements in urban areas. The sociotechnical approach is based on semi-autonomous groups with roots in the self-regulated projects and redundancy of functions (Trist, 1976; de Melo, 1997). The socio-technical support or urban systems based on promotions and rewards, promotions, must be congruent with activities and aligned with the objectives of the institution. Communication across the urban metropolitan boundaries of organizational departments links the control and supervision levels. Local urban autonomy and freedom of action is valued that what is prescribed by rules in the urban community. The socio-technical urban approach encourages the autonomy and initiative of employees. Variance must be controlled in sociotechnical urban criterion as the source for each person responsible for controlling learning from work and failures. The urban movements reveal the limits of the state to the post fallout urban socio-ecology resilience.

Locals consider that radioecology offers more than limited human relations and delimits an Eco semiotic of ineffable interrelated of living, decaying, and dying things. Application of Eco semiotics considers humans and other species co-constitute ecological phenomena (Maran & Kull 2014). Radioecology uses specific radioisotopes to investigate tracing and mapping connections radiological contaminants and their transboundary movements between the earth and the atmosphere (Odum, 1959). The resultant production of accumulative nuclear waste bringing urban socio-ecology resilience to collective recovery.

Global transformation has climate change at the heart of its concerns, and the sociological approach is concerned by unpredictability in the work research of Bauman and Beck and social acceleration by Hartmut Rosa (Hartmut Resonanz, 2016; W. S. et al., 2007). Evolution of human activities have resulted in temporal urban landscapes perceived as natural but shaped by evolution of generations (Meeus, 1995; Kuna & Dreslerová, 2007). Transformations of urban landscapes by human groups lead to the anthropogenic impacts on local socio-ecological equilibrium. The doughnut economics model proposes knowledges and practices for urban socio-ecology resilience, sustainability, and well-being to be applied everywhere (Strang, 2017).

Cooperative urban socio-ecological networks are characterized by established relationships of interdependence, specialization, communication, reciprocity, and trust (Nahapiet & Ghoshal, 1998). Consistent with the socio-ecological approach, the urban network integration mechanisms are also consisting of collaborating efforts to solve urban socio-ecological and resilience problems. The member´s objectives of the urban socio-ecology resilience network are complementary and not necessarily common integrated by mechanisms consisting of informal and voluntary relationships strengthened by a reticulation process.

DISCUSSIONS

The findings align with the body of existing research, the analysis of the relationships between the urban anthropology establishment of action is framed by the reflexive dialectic actor and author with specific disciplinary orientation in urban socio-ecology resilience planning and green innovation areas. The analysis of the relationships between the objectivity of inter-subjectivity of urban anthropology turn into methodological problems and dilemmas that require to be addressed in a specific way. Urban anthropological inter-subjectivity is an objective element of constructing reality without relativist ideology but with methodologies and practices based on valid assumptions for urban social-ecological resilience planning and green innovation.

Anthropological postmodernism considered as the anti-scientific narratives has disappeared but continue to prevent from access to scientific methods. Postmodernism argues that it is immoral to seek laws, generalizations, verifications and proofs, which dehumanize by objectifying people in urban planning and ecological innovation. (Barrett, 2009). Relativism is an escapist generalization perspective. Postmodernism in urban anthropology continues influencing academic development.

Anthropology will have to continue to be based on ethnography: the time-consuming, detailed, meticulous mapping of local life-worlds. It must be comparative, relating societies to one another in a bid to develop general theory. The new anthropology continues to be holistic in contextualizing ethnographic details within socio-cultural wholes (Hylland, 2022).

The urban socio-ecology resilience anthropology approach has shorter interactions in time between social groups aiming to urban socio-ecology resilience planning and green innovations. The development of urban anthropological motivation on creative abilities gives the conditions of the local innovation economy based on urban socio-ecological planning and green innovation areas. International institutions are engaged in studying sideways working with the groups of experts on urban socio-ecological planning and green innovation areas supported by urban anthropology (Hannerz, 2004; Holmes and Marcus, 2006, 2021; Ortner, 2010, 2013).

Change is not a continuous social life dimension considering stability as the norm. (Gluckman, 1968). A structural form for scientific purposes is not static but a dynamic continuity of an organic structure throughout a living body of an organism being constantly renewed and the social life that constantly renews the social structure (Radcliffe-Brown, 1940). Accounting practices on urban socio-ecology resilience and resilience stress accountability on verifiable data and information in annual reports. Changes in reporting require teamwork cooperation of the involved stakeholders. Several studies show that Urban planning remains an important tool in managing climate change impacts (Asibey et al., 2022).

Climate changes because of human activities and their cultural systems has gained attention in academia and set the agenda for many environmental anthropologists. Anthropology could play a central role, by offering methods to access the social, cultural and political processes that shape climate debates (Bejrami, 2021).

STUDIES OF CASE

The term and concept of Anthropocene is under debate, as the time of Anthropocene in human history as a formal official unit of geological time and the political and economic implications. Its definition and meaning is under scrutiny and probably will depend on the future development related to technology and human survival due to climate change (Bajrami, 2022), now there are currently some studies that are beginning to list the impacts generated, some of them are presented below.

The new anthropology is by necessity interdisciplinary, drawing on insights from disciplines as distinctive as history and biology. The impurity, hybridity and openness to other worlds of experience and nonhuman species on which anthropologists pride themselves, must be extended to other intellectual projects in academia as well (Hylland, 2022). In other hand, the Anthropocene has turned into a widely acknowledged concept in environmentalist discourse. As we become increasingly aware of mankind's destructive potential that threatens to make the earth unlivable (Land, 2019), that's why green innovations emerge, one of them are the green roofs. Most green roof studies have focused on reducing energy consumption. In developed countries, air conditioning in buildings consumes a large amount of electricity, these studies show that green roofs are an appropriate method of saving energy in hot climates, to improve the energy efficiency of indoor areas, the air quality of indoor buildings, and the air temperature of outdoor buildings and decrease the urban heat island effect, presents technological developments and how to use in urban green innovations spaces.

This topic takes greater relevance in these days, as it presents the challenge that countries must face to ensure that climate change and the heat island effect is increasing, as well as the obstacles that green roofs can confront, while the countries redesigning the laws in this new economic, political and social moment they are facing. For the sustainability concept to be useful, it must nevertheless be divested of its connection to an impossible idea of development as growth and should instead be defined as the ability of a system to reproduce itself indefinitely without undermining the conditions for its own existence. In this lies not only a societal challenge for humanity but also an intellectual project that will keep anthropologists busy in the foreseeable future (Hylland, 2022).

CONCLUSION

The analysis aimed to find some existing implications and relationships between urban anthropology and Anthropocene urban socio-ecology planning resilience, begins from the assumption that urban anthropology give support to create and develop any urban planning based on the Anthropocene urban socio-ecology resilience. This analysis concludes that urban anthropology fundamentals give support to create and develop urban planning based on the Anthropocene urban socio-ecology resilience.

Urban anthropological research positions between the boundaries of the global reach in connections and assemblages with urban anthropological socio-ecology and resilience and green innovation areas. The anthropoid dominance has led to results in resource shortages of the planet that may be a more acuity problem that climate change. The institutional support of urban anthropology to local urban planning and socio-ecological innovation system is a relevant factor to achieve successful results. Urban planning and green innovation institutions are characterized by their ability to create and develop social capital to facilitate relationships of cooperation and conflict.

There are several levels of urban anthropology applied to urban green planning and socio-ecological innovation. Technological urban green innovation park brings green firms seeking the benefits of agglomeration. Emergence of innovative green and socio-ecological ventures explore opportunities in urban planning innovation from main sources and drivers using the wealth of green knowledge. Generating scientific green knowledge and technological socio-ecological innovation, extension, and consulting activities must be structured through diverse mechanisms for the planning and formation of cooperative urban communities structured by interactions of deliberative processes to stimulate entrepreneurship and innovation.

Urban anthropology at the local dimension of the urban socio-ecological planning and green innovation system university-based must be supported by financial subsidies to guide the efforts of the governance structure. Other mechanism of urban socio-ecological planning and innovation is the indirect support from green infrastructures and constitutions of green assets used by flow firms and a set of projects. The green infrastructures, socio-ecological physical facilities, and green technologies of urban planning and innovation centers are green assets than can be financed.

An effort requiring shared objectives and collaboration supported by a governance structure capable to synchronize mission and objectives of agents and actors from the perspective analysis of urban anthropology to urban planning and socioecological innovation centers and parks. Deliberative actions on urban green innovation areas are triggered by normative and deliberative actions carried out by agents and actors. The involved actors focused on the urban green innovation technology park must have autonomy and flexibility and above all must avoid erosions caused by naturally occurring regulatory issues.

Organizational networks are vital to connecting each other in the space of a green movement. The environmental movements along the nongovernment organizations and environmental foundations lack of racial diversity despite the efforts to increase diversity and inclusion (Johnson, 2019) Residents also express their intentions to be involved in discussions and debates to develop a green urban framework. It is recommended that the local authorities initiate strong campaigns in favor of green technologies at least for the new buildings or the buildings under construction. They must educate people about the benefits these green roofs can provide. The authorities must facilitate the residents and house owners in designing, executing, and installing such systems (Algatamy, 2023).

Adding green components to technology can become sustainable green technologies able to avoid degradation and provide clean environment. Green roof technology is used to tackle social and environmental concerns to enhance climate resilience. Green roof is related to be linked to resilience, restorative, and accessible. Green nanotechnology represents opportunities and challenges in industry to encourage growth by supporting nanotechnology usage. Green roof technology has been recommended as an effective solution for the growth of green spaces per capita and improving sustainability means of urban development's due to its diverse advantages (Gohari et al., 2023).

Research on the mechanisms of urban socio-ecology planning resilience based on cross-field linkages must be strengthened.

LIMITATIONS AND FUTURE RESEARCH

It is necessary to deepen both theoretically and empirically on the association between urban anthropology and its impact on the Anthropocene urban socio-ecology planning resilience in different regions and countries, and their impacts in the new ecological innovations.

REFERENCES

Aldrich, H. (2014). *Stand up and be counted: Why social science should stop using the qualitative/quantitative dichotomy.* https://blogs.lse.ac.uk/impactofsocialsciences/2014/11/28/stand-up-and-becounted-social-science-qualitative-quantitative-dichotom

Alqahtany, A. (2022). Green Roofs as an Approach to Enhance Urban Sustainability: A Study of Public Perception in Riyadh, Saudi Arabia. *Buildings*, *2022*(12), 2202. doi:10.3390/buildings12122202

Alvarez, R. (2016). *Computational social science*. Cambridge University Press. doi:10.1017/CBO9781316257340

Arnould, E. J., & Cayla, J. (2015). Consumer fetish: Commercial ethnography and the sovereign consumer. *Organization Studies*, *36*(October), 1361–1386. doi:10.1177/0170840615580012

Asibey, M., Mintah, F., Adutwum, I., Wireko-Gyebi, R., Tagnan, J., Yevugah, L., Agyeman, K., & Abdul-Salam, J. (2022). Beyond rhetoric: Urban planning climate change resilience conundrum in Accra, Ghana. *Cities (London, England)*, *131*, 1–12. doi:10.1016/j.cities.2022.103950

Atanasova, A. (2021). Re-examining utopia in contemporary consumption: Conceptualization and implications for marketing. *AMS Review, 11*(1-2), 23–29. doi:10.100713162-021-00193-0

Audretsch, D. B. (1998). Agglomeration and the Location of Innovative Activity. *Oxford Review of Economic Policy, 14*(2), 18-28.

Baer, H., & Singer, M. (2014). *The anthropology of climate change: An integrated critical perspective*. Routledge. doi:10.4324/9781315818702

Bajrami, A. (2022). From nature to man: Environmental anthropology in the Anthropocene. *Journal of Biological Research (Thessaloniki), 95*(1), 10377. doi:10.4081/jbr.2022.10377

Baldassarri, D., & Abascal, M. (2017). Field experiments across the social sciences. *Annual Review of Sociology, 43*(1), 41–73. doi:10.1146/annurev-soc-073014-112445

Barnes, J., Dove, M., Lahsen, M., Mathews, A., McElwee, P., McIntosh, R., Moore, F., O'Reilly, J., Orlove, B., Puri, R., Weiss, H., & Yager, K. (2013). Contribution of anthropology to the study of climate change. *Nature Climate Change, 3*(6), 541–544. doi:10.1038/nclimate1775

Barrett, S. (2009). *Anthropology*. University of Toronto Press. doi:10.3138/9781442697362

Baschera, W., & Schneider, I. (2022). *Can sustainable urban planning determine people's happiness and well-being?* Brazilian Journal of Management / Revista de Administração da UFSM. doi:10.5902/1983465969433

Bernard, H. (2011). *Research methods in anthropology: Qualitative and quantitative approaches*. AltaMira Press.

Bilharz, J. A. (1998). *The Allegany Senecas and Kinzua Dam: Forced Relocation through Two Generations*. University of Nebraska Press.

Birch, L. L., Gunder, L., Grimm-Thomas, K., & Laing, D. G. (1998). Infants' consumption of a new food enhances acceptance of similar foods. *Appetite, 30*(3), 283–295. doi:10.1006/appe.1997.0146 PMID:9632459

Brady, H. (2019). The challenge of big data and data science. *Annual Review of Political Science, 22*(1), 297–323. doi:10.1146/annurev-polisci-090216-023229

Bunge, M. (1999). *Buscar la filosofía en las ciencias sociales*. Siglo XXI.

Casey, K., Lichrou, M., & O'Malley, L. (2020). Prefiguring sustainable living: An ecovillage story. *Journal of Marketing Management, 36*(17–18), 1658–1679. doi:10.1080/0267257X.2020.1801800

Cashdan, E. (1998). Adaptiveness of food learning and food aversions children. *Social Sciences Information. Information Sur les Sciences Sociales, 37*(4), 613–632. doi:10.1177/053901898037004003

Chandler, D., Rothe, D., & Muller, F. (2022). Relaciones Internacionales en el Antropoceno. *Relaciones Internacionales, 50*, 107–126. doi:10.15366/relacionesinternacionales2022.50.005

Connolly, C., Keil, R., & Ali, S. H. (2021). Extended urbanisation and the spatialities of infectious disease: Demographic change, infrastructure and governance. *Urban Studies (Edinburgh, Scotland)*, *58*(2), 245–263. doi:10.1177/0042098020910873

De Melo, A. (2001). *The Innovation Systems of Latin America and the Caribbean*. Inter-American Development Bank. Agosto.

De Melo, M. A. C. (1997). *Processo de Planejamento e as Inovações Tecnológicas e Sociais: uma perspectiva sócio-ecológica*. Anais do 5o. Seminário de Modernização Tecnológica.

De Melo, M. A. C. (2002). *Enriquecendo a Atuação de Incubadora de Emrpesas. Tecnologia e Inovação: experiências de gestão nas micro e pequenas empresas*. PGT/USP.

DeWalt, K., DeWalt, B., & Wayland, C. (1998). Participant observation. In H. R. Bernard (Ed.), *Handbook of methods in cultural anthropology* (pp. 259–299). Altamira.

Diamond, J., & Robinson, J. (2010). *Natural experiments of history*. The Belknap Press of Harvard University Press.

Díaz, D., Kristoff, J., Castro, M., Miceli, J., Castro, D., Quinteros, R., & Guerrero, S. (2007). *Exploraciones en antropología y complejidad*. Editorial Sb.

Ditlevsen, K., Sandøe, P., & Lassen, J. (2019). Healthy food is nutritious, but organic food is healthy because it is pure: The negotiation of healthy food choices by Danish consumers of organic food. *Food Quality and Preference*, *71*(January), 46–53. doi:10.1016/j.foodqual.2018.06.001

Dunning, T. (2012). *Natural experiments in the social sciences*. Cambridge University Press. doi:10.1017/CBO9781139084444

Durrenberger, P. (2007) The Anthropology of Organized Labor in the United States. *Ann Review of Anthropology, 36*, 73-88. https://www.annualreviews.org/doi/pdf/10.1146/annurev.anthro.36.081406.094433

Egerton, F. N. (2001). A history of the ecological sciences: Early Greek origins. *Bulletin of the Ecological Society of America*, *82*(1), 93–97.

Eriksen, T. H. (2016). *Overheating: An Anthropology of Accelerated Change*. Pluto. doi:10.2307/j.ctt1cc2mxj

Escalera, J., & Ruiz, E. (2011). Socioecological Resilience: Contributions and challenges from Anthropology. *Revista de Antropología Social*, *20*, 109–135. doi:10.5209/rev_RASO.2011.v20.36264

Escobar, A. (2020). *Pluriversal Politics: The Real and the Possible*. Duke University Press.

Figueiredo, B., & Scaraboto, D. (2016). The systemic creation of value through circulation in collaborative consumer networks. *The Journal of Consumer Research*, *43*(4), 509–533. Advance online publication. doi:10.1093/jcr/ucw038

Fiske, S. J., Crate, S. A., Crumley, C., Galvin, K., Lazrus, H., Lucero, L., Oliver-Smith, A., Orlove, B., Strauss, S., & Wilk, R. R. (2014). *Changing the atmosphere. Anthropology and climate change. Final report of the AAA global climate change task force*. American Anthropological Association.

Giner, C. (2004). J. Antropología urbana. *Ariel.*

Gluckman, M. (1968). The Utility of the Equilibrium Model in the Study of Social Change. *American Anthropologist, 70*(2), 219–237. doi:10.1525/aa.1968.70.2.02a00010

Gohari, A., Gohari, A., & Bin, A. (2023). Importance of green roof criteria for residential and governmental buildings: A multi-criteria decision analysis. *Environmental Science and Pollution Research International, 30*(2), 3707–3725. doi:10.100711356-022-22472-4 PMID:35953748

González Álvarez, M. D. (1997). Processos de Planejamento nos Pólos Tecnológicos: um enfoque adaptativo. Tese de Doutorado, Departamento de Engenharia Industrial, PUC-Rio, Rio de Janeiro.

Gravano, A. (1992). Antropología Práctica, muestra y posibilidades de la Antropología Organizacional. *Publicar en Antropología y Ciencias Sociales, Buenos Aires*, (1), 95–126.

Gravano, A. (2005). *El barrio en la teoría social.* Espacio Editorial.

Gravano, A. (2006). Imaginarios regionales y circularidad en la planificación: El caso del TOAR. *Intersecciones*, (7), 305–323.

Gravano, A. (2007b). Claves para la facilitación organizacional del proceso participativo en la planificación urbano-ambiental metropolitana. *Revista Regional de Trabajo Social, 21*(40), 9-22.

Gravano, A. (2008a) *¿Vecinos o ciudadanos? el fenómeno NIMBY (Not In My Back Yard) o SPAN (Sí, Pero No Aquí) del imaginario urbano en un proceso de participación social y su tratamiento desde la facilitación organizacional antropológica.* Trabajo presentado en II Congreso de la Asociación Latinoamericana de Antropología (ALA), Universidad de Costa Rica, Simposio "Imaginarios urbanos y participación social".

Grimmer, J., Roberts, M., & Stewart, B. (2021). Machine learning for social science: An agnostic approach. *Annual Review of Political Science, 24*(1), 395–419. doi:10.1146/annurev-polisci-053119-015921

Haines, S. (2019b). Managing expectations: Articulating expertise in climate services for agriculture in Belize. *Climatic Change, 157*(1), 43–59. doi:10.100710584-018-2357-1

Haller, T. (2019). The Different Meanings of Land in the Age of Neoliberalism: Theoretical Reflections on Commons and Resilience Grabbing from a Social Anthropological Perspective. *Land (Basel), 2019*(8), 104. doi:10.3390/land8070104

Hanners, U. (1986). *Exploración de la ciudad.* Fondo de Cultura Económica.

Hannerz, U. (2004). *Foreign New: Exploring the World of Foreign Correspondents.* University of Chicago Press.

Hartmut Resonanz, R. (2016). *Eine Soziologie der Weltbeziehung.* Suhrkamp.

Hirst, K. (2019). *Is anthropology a science?* ThoughtCo. https://www.thoughtco.com/isanthropology-a-science-3971060

Hofman, J., Sharma, A., & Watts, D. (2017). Prediction and explanation in social systems. *Science, 355*(6324), 486–488. doi:10.1126cience.aal3856 PMID:28154051

Hofman, J., Watts, D., Athey, S., Garip, F., Griffiths, T., Kleinberg, J., Margetts, H., Mullainathan, S., Salganik, M., Vazire, S., Vespignani, A., & Yarkoni, T. (2021). Integrating explanation and prediction in computational social science. *Nature*, *595*(7866), 181–188. doi:10.103841586-021-03659-0 PMID:34194044

Holling, C. S. (1973). *Resilience and Stability of Ecological Systems. Annual Review of Ecology and Systematics*, 4. https://www.jstor.org/stable/2096802

Holmes, D. R., & Marcus, G. E. (2006). Fast capitalism: para-ethnography and the rise of the symbolic analyst. In M. Fisher & G. Downey (Eds.), *Frontiers of Capital: Ethnographic Reflections on the New Economy*. Duke University Press.

Holmes, D. R., & Marcus, G. E. (2021). How do we collaborate? An updated manifesto. In D. Boyer & G. Marcus (Eds.), *Collaborative Anthropology Today: A Collection of Exceptions*. Cornell University Press. doi:10.7591/cornell/9781501753343.003.0002

Hyatt, S. (2021). *Teaching Urban Anthropology in a time of COVID*. National Association for the Practice of Anthropology. https://scholarworks.iupui.edu/bitstream/handle/1805/29883/Hyatt2021Teaching-pp.pdf?sequence=1

Hylland, T. (2022). The Sustainability of an Anthropology of the Anthropocene. *Sustainability*, *2022*(14), 3674. doi:10.3390u14063

Jorgensen, D. (2015). Participant observation. In R. Scott & S. Kosslyn (Eds.), *Emerging trends in the social and behavioral sciences* (pp. 1–15). Wiley. doi:10.1002/9781118900772.etrds0247

Kawulich, B. (2005). La observación participante como método de recolección de datos. *Forum Qualitative Social Research*, *68*(2), 43.

Köhler, M., & Poll, P. H. (2010). Long-term performance of selected old Berlin greenroofs in comparison to younger extensive green roofs in Berlin. *Ecological Engineering*, *36*(5), 722–729. doi:10.1016/j.ecoleng.2009.12.019

Kolb, Ch. (1998). Review of Bilharz, Joy A., *The Allegany Senecas and Kinzua Dam: Forced Relocation through Two Generations. H-AmIndian, H-Net Reviews*. https://www.h-net.org/reviews/showrev.php?id=2366

Kosova, R., Qendraj, D.H., Xhafaj, E. (2022). Meta-Analysis ELECTRE III and AHP in Evaluating and Ranking the Urban Resilience. *Journal of Environmental Management and Tourism, 3*(59), 756 - 768. doi:10.14505/jemt.v13.3(59).15

Kramer, J. M. (2021). From predator to parasite: On private property and our ecological disaster. *Journal of Economic Issues*, *55*(June), 416–422. doi:10.1080/00213624.2021.1908804

Krauß, W. (2020). Narratives of change and the co-development of climate services for action. *Climate Risk Management*, *28*, 100217. doi:10.1016/j.crm.2020.100217

Kuhn, T. S. (1962/1996). *The Structure of Scientific Revolutions* (3rd ed.). University of Chicago Press.

Kuna, M., & Dreslerová, D. (2007). Landscape archaeology and "community areas" in the archaeology of central Europe. In D. Hicks, L. McAtackney, & G.J. Fairclough (Eds.), Envisioning landscape: Situations and standpoints in archaeology and heritage. Left Coast Press.

Lambert, A. (2019). Psychotic, acritical and precarious? A Lacanian exploration of the neoliberal consumer subject. *Marketing Theory, 19*(3), 329–346. doi:10.1177/1470593118796704

Lapassade, G. (1986). *Grupos, organizaciones e instituciones: la transformación de la burocracia.* Gedisa.

Latour, B. (2013). *An inquiry into modes of existence: An anthropology of the moderns* (C. Porter, Trans.). Harvard University Press.

Latour, B. (2017). Anthropology at the Time of the Anthropocene: A Personal View of What Is to Be Studied. In M. Brightman & J. Lewis (Eds.), *The Anthropology of Sustainability*. Palgrave Studies in Anthropology of Sustainability. doi:10.1057/978-1-137-56636-2_2

Lester, R. K. (2005). *Universities, Innovation, and the Competitiveness of Local Economies A Summary Report from the Local Innovation Systems Project – Phase I.* Industrial Performance Center Massachusetts Institute of Technology 13 December 2005 MIT Industrial Performance Center Working Paper 05-010.

Lindblom, C. E. (1959). The Science of Muddling through. *Public Administration Review, 19*(2), 79–88. doi:10.2307/973677

Llanos, M. (2009). *Epistemología de las ciencias sociales*. UNMSM.

Lourau, R. (1979). *El análisis institucional*. Amorrortu.

Lundvall, B.-Å. (Ed.). (1992). National Systems of Innovation: towards a theory of innovation and interactive learning. Printer Publisher.

Malinowski, B. (1961). *Argonauts of the Western Pacific*. Routledge.

Maran, T., & Kull, K. (2014). Ecosemiotics: Main principles and current developments. *Geografiska Annaler. Series B, Human Geography, 96*(1), 41–50. doi:10.1111/geob.12035

Mayo, E. (1945). *The social problems of an Industrial Civilization*. Harvard University Press.

McIntosh, R. J., Tainter, J. A., & McIntosh, S. K. (Eds.). (2000). *The Way the Wind Blows: Climate Change, History, and Human Activity*. Columbia University Press.

Meeus, J. H. A. (1995). Pan-European landscapes. *Landscape and Urban Planning, 31*(1-3), 57–79. doi:10.1016/0169-2046(94)01036-8

Mir, R., & Mir, A. (2009). From the Colony to the Corporation: Studying Knowledge Transfer Across International Boundaries. *Group & Organization Management, 34*(1), 90–113. doi:10.1177/1059601108329714

Moore, H. L. (2017). What Can Sustainability Do for Anthropology? In M. Brightman & J. Lewis (Eds.), *The Anthropology of Sustainability*. Palgrave Studies in Anthropology of Sustainability. doi:10.1057/978-1-137-56636-2_4

Mu, X., Fang, C., Yang, Z., & Guo, X. (2022). Impact of the COVID-19 Epidemic on Population Mobility Networks in the Beijing–Tianjin–Hebei Urban Agglomeration from a Resilience Perspective. *Land (Basel), 2022*(11), 675. doi:10.3390/land11050675

Nahapiet, J. E., & Ghoshal, S. (1998). Social Capital, Intellectual Capital, and the Organizational Advantage. *Academy of Management Review, 33*(2), 242–266. doi:10.2307/259373

Nießen, D., Danner, D., Spengler, M., & Lechner, C. (2020). Big Five personality traits predict successful transitions from school to vocational education and training: A large-scale study. *Frontiers in Psychology, 11*, 1827. doi:10.3389/fpsyg.2020.01827 PMID:32903700

Nost, E. (2019). Climate services for whom? The political economics of contextualizing climate data in Louisiana's coastal master plan. *Climatic Change, 157*(1), 27–42. doi:10.100710584-019-02383-z

Odum, E. P. (1957). *Fundamentals of ecology*. W. B. Saunders Company.

OECD. (1997). Government Venture Capital for Technology-Based Firms. Ducument OCDE/GD (97) 201. Paris: OECD.

Olimid, A. P., Georgescu, C. M., & Gherghe, C. L. (2022). Influences of Covid-19 Crisis on Resilience Theories: An analysis of Community, Societal and Governance Resilience. *Revue des Sciences Politiques, 73*, 38–51.

Ortner, S. (2010). Access: Reflections on studying up in Hollywood. *Ethnography, 11*(2), 211–233. doi:10.1177/1466138110362006

Ortner, S. (2013). *Not Hollywood: Independent Film at the Twilight of the American Dream*. Duke University Press.

Páramo, D., & Ramírez, E. (2017). *Etnomarketing, la dimensión cultural del marketing*. Klasse Editorial.

Peterson, N., & Broad, K. (2009). Climate and weather discourse in anthropology: From determinism to uncertain futures. In *Anthropology and climate change: From encounters to actions* (pp. 70–86). Left Coast Press.

Pfeifer, J. (2006). Laws of nature. In S. Sarkar & J. Pfeifer (Eds.), The philosophy of science: An encyclopedia (pp. 439–444). Academic Press.

Pimienta-Bueno, J.A. (1999). *Parque de Inovação Tecnológica e Cultural da Gávea: A Visão da PUC-Rio*. Documento Interno.

Pizarro, C. (2014). La entrevista etnográfica como práctica discursiva: Análisis de caso sobre las pistas meta-discursivas y la emergencia de categorías nativas. *Revista de Antropologia, 57*(1), 461–496. doi:10.11606/2179-0892.ra.2014.87770

Porter, M. (2001). Clusters of Innovation: regional foundations of U.S competitiveness. Council on Competitiveness. Monitor Group.

Purzycki, B., Pisor, A., Apicella, C., Atkinson, Q., Cohen, E., Henrich, J., McElreath, R., McNamara, R., Norenzayan, A., Willard, A., & Xygalatas, D. (2018). The cognitive and cultural foundations of moral behavior. *Evolution and Human Behavior, 39*(5), 490–501. doi:10.1016/j.evolhumbehav.2018.04.004

Radcliffe-Brown, A. R. (1940). On Social Structure. *J. R. Anthropol. Inst. Great Br. Irel.*, *1940*(70), 1–12.

Reguera, A. (2009). *Empresarios de ayer, de hoy y de siempre. Un recorrido latinoamericano por sus formas espacio-temporales América Latina en la historia económica, num 32, jul-dic, México*. https://www.scielo.org.mx/scielo.php?script=sci_arttext&pid=S1405-22532009000200003&lng=es&nrm=iso

Reynoso, C. (1995). Hacia la perfección del consenso: Los lugares comunes de la antropología. *Intersecciones*, *1*, 51–72.

Roca, A. (2008). *La comprensión de la tecnología en la antropología*. Actas VII ESOCITE UFRJ, Jornadas Latino-Americanas de Estudos Sociais das Ciências e das Tecnologías.

Roethlisberger, F., & Dickson, W. (1939). *Management and the Worker*. Harvard Univ. Press.

Roni, S., Merga, M., & Morris, J. (2020). *Conducting quantitative research in education*. Springer. doi:10.1007/978-981-13-9132-3

Saxenian, A. (1994). *Regional Advantage: culture and competition in Silicon Valley and Route 128*. Harvard Business Press.

Schlapfer, A., & Marinova, D. (2001). Local Innovation Systems: nature, importance and role. *Conference Proceedings: International Summer Academy on Technological Studies: User Involvement in Technological Innovation*.

Schnegg, M., O'Brian, C. I., & Sievert, I. J. (2021). It's our fault: A global comparison of different ways of explaining climate change. *Human Ecology: an Interdisciplinary Journal*, *49*(3), 327–339. doi:10.100710745-021-00229-w

Signorelli, A. (1999). *Antropología Urbana*. UAM. doi:10.1080/0267257X.2014.925958

Snow, C. P. (1998). *The Two Cultures*. Cambridge University Press.

Stoner, A. M. (2021). Things are getting worse on our way to catastrophe: Neoliberal environmentalism, repressive desublimation, and the autonomous ecoconsumer. *Critical Sociology*, *47*(May), 491–506. doi:10.1177/0896920520958099

Strang, V. (2017). The Gaia complex: Ethical Challenges to an Anthropocentric 'Common Future. In B. Marc & L. Jerome (Eds.), *The Anthropology of Sustainability* (pp. 255–283). Palgrave. doi:10.1057/978-1-137-56636-2_12

Su, C., Liu, F., Stefea, P., & Umar, M. (2023). Does technology innovation help to achieve carbon neutrality? *Economic Analysis and Policy*, *78*, 1–14. doi:10.1016/j.eap.2023.01.010

Su, C., Pang, L., Tao, R., Shao, X. & Umar, M. (2022). Renewable energy and technological innovation: Which one is the winner in promoting net-zero emissions? *Technological Forecasting & Social Change,* 1 – 11. doi:10.1016/j.techfore.2022.121798

Sundbo, J. (1998). *The Theory of Innovation: entrepreneurs, technology, and strategy*. Edward Elgar.

Tainter, J. A. (1988). *The Collapse of Complex Societies*. Cambridge University Press.

Tainter, J. A. (2008). Collapse and Sustainability: Rome, the Maya, and the Modern World. *Archeological Papers of the American Anthropological Association, 24*(1), 201–214. doi:10.1111/apaa.12038

Tarantino-Curseri, S. (2018). Antropología, etología, sociología, y psicología, 4 disciplinas útiles en el ámbito empresarial. *Management Review, 3*(3). Advance online publication. doi:10.18583/umr.v3i3.126

Trawick, P., & Alf, H. (2015). Revisiting the Image of Limited Good: On Sustainability, Thermodynamics, and the Illusion of Creating Wealth. *Current Anthropology, 56*(1), 1–27. doi:10.1086/679593

Trist, E. L. (1976). A Concept of Organizational Ecology: an invited address to the three Melbourne universities. Academic Press.

Trujillo, M. L. N. (2021). Notes for a critical and ecological view of patriarchal capitalism in the web of life. *Capital and Class, 45*(March), 21–32. doi:10.1177/0309816820929115

Tsing, A. L. (2015). *The mushroom at the end of the world.* Princeton University Press.

Velho, G. (1978). Observando o familiar. In A aventura sociológica: objetividade, paixao, improviso e método na pesquisa social. Zahar.

W. S. (2007). JThe Anthropocene: Are Human Beings Now Overwhelming the Forces of Nature? *Ambio, 36*, 614–621. doi:10.1579/0044-7447(2007)36[614:TAAHNO]2.0.CO;2 PMID:18240674

Walker, B. (2019). *Finding Resilience: Change and Uncertainty in Nature and Society.* CSIRO Publishing.

Wei, X., Lo, C. K. Y., Jung, S., & Choi, T.-M. (2021). From co-consumption to co-production: A systematic review and research synthesis of collaborative consumption practices. *Journal of Business Research, 129*(May), 282–294. doi:10.1016/j.jbusres.2021.02.027

Welz, G. (2019). 'Environmental Orientations' and the Anthropology of the Anthropocene. *Anthropological Journal on European Cultures, 27*(1), 40–44. doi:10.3167/ajec.2018.270107

Williams, N., Rayner, J. & Raynor, K. (2010). Green roofs for a wide brown land: Opportunities and barriers for rooftop greening in Australia. *Urban Forestry & Urban Greening, 9*(3), 245–251. https://doi.org/.ufug.2010.01.005 doi:10.1016/j

Wolf, M. (2016). Rethinking urban epidemiology: Natures, networks and materialities. *International Journal of Urban and Regional Research, 40*(5), 958–982. doi:10.1111/1468-2427.12381 PMID:32336869

Yang, Y., Lili, W., & Hongchi, Z. (2022). Bibliometric research on the evolution of resilience theme from the perspective of Geographical Science. *2022 29th International Conference on Geoinformatics*, 1-9. 10.1109/Geoinformatics57846.2022.9963870

Zhu, Y., Zhang, J., & Duan, C. (2022). How does green finance affect the low-carbon economy? Capital allocation, green technology innovation and industry structure perspectives. *Economic Research*, 1-23. doi:10.1080/1331677X.2022.2110138

Chapter 9
Green Technology:
Exploring Its Impact on Indian Society

Ankit Mishra
Amity University, Raipur, India

ABSTRACT

In the domain of language, the term "green tech" encompasses a wide range, referring to innovative creations meticulously designed with environmental considerations woven into every stage, from inception to utilization. The swift depletion of Earth's natural resources, surpassing their regeneration, has spurred an unrelenting quest to refine such technology, making it more harmonious with our delicate ecology. The urgency to lessen reliance on finite resources, notably fossil fuels, aligns with the need to heal ecological wounds. These imperatives define green technology's core mission. Sectors like waste management, energy, and transportation ardently embrace this mission. Despite its benefits, green technology faces substantial hurdles before becoming mainstream. Over time, significant employment growth is observed in green tech, underscoring its growing relevance in our contemporary era.

INTRODUCTION

The nomenclature "green technology" embodies the conception and application of groundbreaking solutions that promote sustainable livelihoods and curtail detriments to the environment. This encompassing term encompasses a wide array of methodologies, artifacts, and systems that mitigate our dependence on finite resources, curbing emissions of greenhouse gases, while addressing pressing ecological concerns such as contamination, shifts in climate patterns, and the depletion of vital resources. Inclusive within this rubric are domains like eco-friendly transportation, environmentally conscious architectural practices, prudent waste management, and conservation strategies, each constituting mere fragments of the sprawling tapestry of green technology. Its armamentarium encompasses a plethora of innovations, including but not limited to solar energy, wind turbines, hydroelectric power, geothermal heating, verdant roofing, electric automobiles, and intelligent energy grids. For the pursuit of a more sustainable tomorrow and the amelioration of humanity's deleterious impact on its surroundings, the embrace of green technology becomes imperative. Its dividends are manifold: an economic upswing, societal well-

DOI: 10.4018/978-1-6684-9503-2.ch009

being, enhanced energy security, and a commensurate reduction in the carbon footprint. Nonetheless, the path to embedding green technology confronts its own set of impediments, from the weighty initial investments to resource scarcity, coupled with a public awareness shortfall and regulatory constrictions. It thus befits governments, corporations, and individuals to forge a symbiotic alliance, nurturing the promotion and funding of sustainable technological panaceas, thereby transcending these obstructions. The phrase "green technology" encapsulates the development and implementation of technology that harmonizes with the environment, embodying principles of sustainability. Its cardinal objective is to ameliorate the detrimental repercussions of human endeavors on the ecosystem while propelling the cause of sustainable advancement. This burgeoning realm spans a kaleidoscope of technologies, ranging from materials fostering ecological construction to renewable energy sources and energy-conserving modes of transit. The contemporary world is ensnared by an escalating environmental quandary, propelled by factors like climate upheaval. The utilization of fossil fuels and other finite assets engenders surges in greenhouse gas emissions, resulting in global temperature elevations – the underpinnings of climate alteration. Moreover, the Earth's biodiverse tapestry and intricate ecosystems bear the brunt of non-sustainable techniques and resources. The present juncture necessitates a resounding call for green technology, capable of delivering enduring antidotes to these ecological predicaments, thereby tending to these pressing exigencies (Banerjee, 2014).

DEVELOPMENT OF GREEN TECHNOLOGY IN VARIOUS COUNTRIES

The impetus behind the adoption of environmentally conscious practices, often referred to as "Green," has exhibited diverse motives across various sectors. Some sectors undertake this transition due to regulatory mandates (such as the energy sector), while others discern it as an opportunity to fortify their brand image among consumers (notably within the retail realm). This global trend has been embraced by governments worldwide, each formulating distinct strategies to develop robust green technology systems. These strategies encompass the promotion of concepts like green mining, eco-friendly purchasing, sustainable business operations, and the cultivation of green energy sources. A cornerstone of this movement is the concept of "green purchasing," which champions environmentally mindful management practices and the development of ecologically sensitive products within the market. The inception of the Green Purchasing Network in February 1996 in Japan, spearheaded by the Environment Agency, was instrumental in fostering this ethos. The network's mission is to propagate and actualize green purchasing principles, offering guidelines and essential information to facilitate its implementation (Bhardwaj, 2015).

In Europe, public purchasing carries substantial weight, accounting for nearly €1 trillion annually, which translates to over 14% of the GDP. Measures to promote green purchasing within the European context include an interpretative communication by the Commission in 2001 regarding community law applicable to public procurement, coupled with insights into integrating environmental considerations into the procurement process. Additionally, new public procurement directives issued in March 2004 elucidate the avenues for incorporating environmental concerns at different stages of the procurement process. A handbook, "Buying Green," published in August 2004, provides comprehensive guidance on environmental public procurement. Statistics on the status of green public procurement reveal significant adoption rates across EU countries, with percentages as follows: EU: 19%, Sweden: 50%, Denmark: 40%, Germany: 30%, Austria: 28%, UK: 23%.

In the United States, both the Federal government, with an annual expenditure of $500 billion, and State/Local governments, expending $400 billion, are heavily engaged in endeavors to promote sustainable practices. Initiatives such as the Energy Star program (1992), Comprehensive Procurement Guidelines (1995) advocating the purchase of recycled products, and the Environmental Preferable Purchasing (EPP) program, mandated by "Executive Order 13101" in 1998, exemplify efforts to prioritize environmentally friendly products and services (Butt, 2016).

In Japan, the government allocates nearly ¥14 trillion per year, while local governments disburse around ¥44 trillion, equivalent to 17.6% of the GDP, to advance green purchasing initiatives. Japanese efforts in this sphere comprise a Green Purchasing Network with over 2800 members and 15 purchasing guidelines, supported by awards, training, and seminars. The Eco Mark program, launched by the Japan Environment Association in 1989, has bestowed certification upon 5000 products spanning 45 categories, signifying their alignment with environmentally sustainable criteria (Hausfather, 2019).

The Indonesian Government has earnestly embraced the doctrine of Green Mining, thereby espousing a resolute commitment to sustainable resource extraction. Evidencing their dedication, they have orchestrated an intricate and far-reaching blueprint spanning from 2005 to 2025, underscored by a series of mission statements designed to:

- Astutely oversee and harness both renewable and non-renewable natural assets, aligning with principles of responsible resource management.
- Champion the safeguarding of invaluable water and energy reservoirs, as well as fostering the preservation of the rich tapestry of local biodiversity.
- Cultivate an enhanced proficiency in the realm of natural resource stewardship, buttressed by capacity-building initiatives.
- Exemplify rigorous management paradigms, undergirded by a robust commitment to environmental integrity.
- Vigilantly steer the reins in controlling environmental pollution and degradation, concurrently cultivating strategies for proficient natural disaster management.

These endeavors, undertaken with commendable diligence, have yielded tangible results, evidenced by a noteworthy uptick of 74% in the trade volume of eco-friendly products within the nation (Kenton, 2020).

THE GREEN INDIA MISSION

The Green India Mission (GIM), a pivotal facet of the National Action Plan on Climate Change (NAPCC), was inaugurated in February 2014. Its chief purpose is safeguarding the nation's biological riches and livelihoods from the threats of climate change. Stemming from an acknowledgment of the intricate interplay between forestry and ecological well-being, it aspires to shield, rejuvenate, and augment India's diminishing forest cover while fortifying the country's resilience against climate change through both adaptation and mitigation. The mission encompasses a comprehensive spectrum of green practices that surpass the mere act of tree planting. GIM's canvas extends to manifold ecosystem services, spanning biodiversity, water resources, biomass, and the conservation of vital habitats like mangroves and wetlands. It's a response not only to climate change but also to preserving and enhancing essential ecosystems. The mission hinges on augmenting carbon sinks within well-managed forests and ecosystems, aiding the adaptation of vulnerable species to the evolving climate, and bolstering forest-dependent communities. The mission's ambitions encompass enlarging forest and tree coverage, ameliorating the quality of existing forest cover, and elevating the wellbeing of both the environment and the communities dwelling in and around these ecosystems. This endeavor marries traditional ecological wisdom, contemporary scientific forestry, and advanced technology. In consonance with its holistic vision, GIM strives to synchronize with allied schemes and initiatives, thus creating a synergy that bolsters the development of forests and their fringes in a holistic and sustainable manner. This aligns with the mission's objective of surmounting challenges within the domains of environment, forest, and wildlife. Operating at both macro and micro levels, a multi-disciplinary assemblage from governmental bodies and non-governmental organizations facilitates the tactical planning and execution. The mission's emphasis on a landscape approach, treating regions as expansive interconnected units, resonates on both national and state levels, offering a promising pathway toward goal attainment. The mission adopts an integrated approach that extends across public and private lands, galvanizing grassroots organizations and local communities into the planning, decision-making, execution, and supervision processes. With keen awareness of the forces that contribute to deforestation, GIM is uniquely poised to combat "de-greening" trends, ensuring a cohesive and sustainable course toward ecological betterment (Richardson, 2018a).

INDIAN GOVERNMENT PLANS AND THEIR IMPLICATIONS

The quest for advancements in solar cell technology revolves around bolstering their efficiency, durability, and cost-effectiveness. This drive stems from the ambition to amplify the utilization of solar energy as a clean and renewable power source. In India, the demand for solar cells has surged, driven by several factors, including government-backed initiatives for renewable energy, heightened public awareness of solar benefits, and declining solar cell prices. By 2022, the Indian government aims to install 100 GW of solar energy capacity. Integral to this vision are programs like the Jawaharlal Nehru National Solar Mission (JNNSM), incentivizing solar energy adoption through tax breaks, subsidies, and favorable loan terms.

The surge in solar cell demand spans across residential, commercial, and industrial sectors. The residential domain has seen an influx of solar panel installations on rooftops, enabling homeowners to generate their own electricity and reduce reliance on the grid. Businesses, both commercial and industrial, have also embraced solar technology to lower energy costs and environmental impact. Furthermore, the agricultural sector is being encouraged to adopt solar power for activities like water pumping and irrigation.

India's rapid urbanization, industrialization, and growing population have heightened energy demands, propelling it to become the third-largest energy consumer globally. Despite fossil fuels being the predominant energy source, the country is actively transitioning towards renewable energy sources. To achieve this, India has set a goal of 175 GW of renewable energy capacity, including 100 GW of solar power, by 2022. The Energy Conservation Building Code (ECBC), a comprehensive framework for energy-efficient construction, is one initiative by the government to promote sustainability in buildings. Alongside this, the National Mission for Enhanced Energy Efficiency and the Energy Efficiency Services Limited (EESL) are pivotal undertakings aimed at fortifying energy efficiency in buildings (Richardson, 2018b).

THE FUTURE INDIAN MISSION

The mission concentrates on key domains:

- Solar Energy: Aiming for 100 GW of solar capacity by 2022.
- Wind Power: Doubling installed wind capacity to 60 GW by 2022.
- Biomass Energy: Aiming for 10 GW biomass capacity, promoting biomass usage for power.
- Small Hydro Power: Targeting 5 GW of small hydropower capacity, promoting small projects.
- Energy Efficiency: Striving for 20–25% energy intensity reduction by 2022 across industries.

The India Green Energy Mission propels the adoption of renewable energy, with 96 GW installed capacity in 2021, encompassing solar, wind, biomass, and small hydropower. The government, alongside initiatives like KUSUM, National Solar Mission, and National Wind Mission, is committed to advancing renewable energy.

ELECTRIC VEHICLES THE NEW ERA

The age-old automobile industry is poised for a monumental transformation driven by mounting concerns over fossil fuel prices and the environmental toll of emissions. This shift has steered the focus toward electric vehicles (EVs) as a more sustainable mode of transportation. EVs are powered by electric motors and rechargeable batteries, offering energy efficiency, lower greenhouse gas emissions, and reduced noise pollution. These vehicles come in distinct categories:

HEV (Hybrid Electric Vehicles): Combining fuel and electricity, these vehicles feature both an engine and an electric motor. Battery replenishment occurs through regenerative braking.

PHEV (Plug-in Hybrid Electric Vehicles): Similar to HEVs, these possess a smaller engine and larger batteries. Battery recharge happens either through braking or external electric charging points.

BEV (Battery Electric Vehicles): Completely reliant on electric motors, BEVs lack engines and utilize batteries for energy storage. Charging is accomplished via external power sources, making them also known as plug-in vehicles.

The transportation sector contributes a significant portion of greenhouse gas emissions, primarily from automobiles. Globally, China, the USA, and India are the leading contributors. In 2016, the Eighth Clean Energy Ministerial introduced the 'EV30@30' campaign, aiming for a 30% market share for EVs by 2030 across various vehicle categories.

India is committed to reducing pollution and carbon footprint. The government envisions a shift to EVs by 2030 to curtail oil expenses, emissions, and fuel import dependence. Battery swapping and leasing are potential models being explored to address adoption challenges, although adequate charging infrastructure remains a hurdle.

The shift to EVs is crucial for India due to urbanization-induced pollution. Figure 1 depicts EV projections for major nations, reflecting the growing global emphasis on sustainable transportation solutions (Popp, 2010).

Figure 1. Projection of electric vehicles
Source: Adopted from Bloomberg Opinion

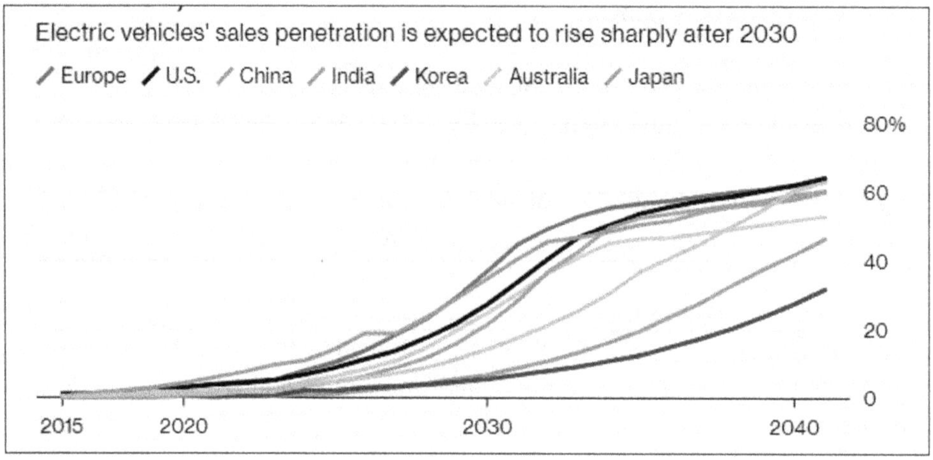

The Indian government has declared a mandate for all cars to be electric by 2030. The Society of Indian Automobile Manufacturers (SIAM) subsequently released a white paper projecting that EVs will comprise 40% of new car sales by 2030 and achieve full dominance by 2047, marking the nation's 100th year of independence.

GREEN TECHNOLOGY TRANSFER

The advent of emissions-reducing technologies like wind power, improved solar cells, and electric vehicles has sparked a new challenge: technology transfer. Developing countries often seek advanced technologies from multinational corporations, mostly based in developed nations, to balance their growing economies. However, the high costs associated with new technologies, reflective of substantial research and development investments, often surpass what developing countries can afford.

Technology transfer involves physically transporting high-tech equipment through international trade. The United Nations Conference on Trade and Development (UNCTAD) has introduced the International Code on the Transfer of Technology (ICTT) to guide this process. Technology transfer is influenced by push and pull factors. Market-based pull factors draw technology to markets with ample demand and resources. This transfer is a two-way street, requiring considerations from both the supply and demand sides. On the supply side, a conducive environment is needed in the recipient market, including production support, infrastructure, and favorable regulations. On the demand side, local demand must exist for green technologies to be effectively adopted. Developing countries must develop technological capacity and institutional structures to absorb, adapt, and enhance transferred technologies. Technology transfer is vital due to the rapid growth of developing countries, with major transfers happening in emerging economies like China, Brazil, and India. While various mechanisms for technology transfer exist, obstacles remain. Economic barriers such as tariffs and trade restrictions hinder imports, while social barriers stemming from foreignness or societal norms impede technology adoption. Overcoming these challenges is crucial for the successful transition to more sustainable green technologies.

CONCLUSION

The perilous threat of environmental degradation, including pollution, puts our planet at the brink of extinction. To foster the growth of green technologies, all barriers must be dismantled. With a heightened understanding of our environmental impact, the demand for green technology has surged, compelling companies to innovate and enhance their offerings. The surging popularity of electric vehicles stands as a testament to this demand. The Paris Agreement has spurred nations to ramp up their adoption of green technology, while businesses are investing to align with public sentiment and bolster their environmental stewardship image. This shift benefits the ecosystem, as green technology not only gains traction but also undergoes continuous refinement for greater ecological harmony.

REFERENCES

Banerjee. (2014). Advantages of green technology. *Recent Res. Sci. Technol., 6*, 97–100.

Bhardwaj. (2015). The Advantages and Disadvantages of Green Technology. *J. Basic Appl. Eng. Res., 2*, 1957–1960.

Butt, A. (2016). *What is Green Technology and Its Benefits?* https://usgreentechnology.com/green-technology/

Hausfather, Z. (2019). *Factcheck: How electric vehicles help to tackle climate change.* https://www.carbonbrief.org/factcheck-how-electric-vehicles-help-to-tackle-climate-change

Kenton. (2020). *Green tech.* investopedia.com

Popp, D. (2010). *The Role of Green Technology Transfer in Climate Policy*, https://www.resourcesmag.org/common-resources/the-role-of-green-technology-transfer-in-climate-policy/

Richardson, R. (2018a). *Depleting Earth's Resources.* https://msutoday.msu.edu/news/2018/depleting-earths-resources/

Richardson, R. (2018b). *Yes, humans are depleting Earth's resources, but 'footprint' estimates don't tell the full story.* https://theconversation.com/yes-humans-are-depleting-earths-resources-but-footprint-estimates-dont-tell-the-full-story-100705

Chapter 10
Green Knowledge in Green Roofs and Organizational Green Innovation

José G. Vargas-Hernandez
Tecnológico Superior de Jalisco Mario Molina Pasquel y Henríquez, Unidad Académica Zapopan, Mexico

Elsa Patricia Orozco Quijano
https://orcid.org/0000-0002-6616-3174
Laurentian University, Canada

Carlos A. Rodriguez-Maillard
https://orcid.org/0000-0003-2406-196X
Universidad Cristobal Colón, Mexico

ABSTRACT

This study aims to analyze the implications of green knowledge and technology in organizational green innovation, urban green innovation, and green roofs. Green roofs can be an effective tool for cities to improve the thermal environment, save energy, and combat climate change, and are an appropriate method of saving energy. The analysis is supported by the assumption that green technology is basic to organizational green innovation and urban green innovation areas practices, operations, and activities. Making the balance between urban developments and environmental issues, in consideration to sustainable development principle and its innovative green solutions, the methods employed are based on the analytical-reflective and descriptive supported with the review of theoretical and empirical literature. The analysis concludes that green knowledge sharing is relevant to create and develop the green technology with positive implications for organizational green innovation, urban green innovation areas, and green roofs.

DOI: 10.4018/978-1-6684-9503-2.ch010

INTRODUCTION

The COVID-19 pandemic, a major health crisis, has prompted numerous activities to become greener and more sustainable. Although the terms "sustainable" and "green" are often used interchangeably, they possess distinct meanings. "Sustainability" refers to the persistence and indefinite future of necessary and desirable characteristics of the human subsystem within the ecosystem (Hodge, 1997). Meanwhile, rapid urbanization in megacities has led to a decrease in green spaces due to the rise in construction activities. Hence, it is crucial to ensure that urban development is sustainable, aligning the expansion of city facilities with social, economic, and environmental aspects (Gohari et al., 2023).

Organizational environmental learning and green knowledge are linked to green technology, a key stimulant for environmental protection, organizational green innovation, urban green innovation areas, and green roofs. Sustainable development pertains to decision-making considering economic and social impacts (Wiering et al., 2020), while meeting the needs of the present without compromising future generations' ability to meet their own needs (Brundtland, 1987). Urban sustainability, including social and economic aspects, is one of the foremost features of cities (Dwift Consulting, 2017). Green trends such as smart and sustainable cities propose the use of green and eco-friendly technology, offering a more sustainable environment with zero gas emissions, along with other opportunities and challenges.

Individuals living in dense communities tend to use public urban green innovation spaces and parks more frequently for relaxation and leisure. They may also travel to the countryside, experimenting with and adopting greener practices within organizations focused on environmental sustainability. This enables the introduction of green innovations (Anderson et al., 2010; Stubbs & Cocklin, 2008). Organizations that adopt the concept of saving money through initiatives like recycling programs and thermostat monitoring can support non-governmental agencies involved in environmental preservation.

Frequent use of green spaces enhances individual satisfaction with public areas, improving social space and mental health (Hadavi, 2017). People attach meanings, identities, and psychological experiences to several types of green spaces, as described in the Place Identity Integrated Model and Environmental Representation (Bernardini & Irvine, 2007). Research on green organization identity has focused on the individual level (Chen & Chang, 2013). A green organizational identity strengthens tasks related to the organization's environmental activities and enhances its ability to cope with green-oriented conditions. However, landscape connectivity varies between urban green innovation roofs and urban open space management.

Organizations face challenges in complying with green sustainability strategies, a crucial measure against the negative impacts of industrialization on the environment. Organizational business sustainability encompasses green growth, green branding, and green sustainable reporting. Branding sustainability enhances an organization's appeal to environmentally conscious customers.

Green sustainable development, coupled with economic growth, social justice, progress, and environmental security, are pertinent research issues in areas such as green entrepreneurship in organic farming (Gupta & Vegelin, 2016; Ihnatenko & Novak, 2018; Kucher, 2019; Mohd & Norhidayah, 2016; Savickiene & Miceikiene, 2018; Shevchenko & Petrenko, 2020; Skydan et al., 2020). Achieving a balance between urban development and environmental concerns, while combating climate change and global warming, requires the consideration of sustainable development principles and innovative solutions (Gohari et al., 2023).

Organizations undertake green initiatives to develop sustainable competitive advantages (Galdeano-Gómez et al., 2008; López-Gamero et al., 2008; Wysocki, 2021). These sustainable components offer organizations a competitive edge (Bansal, 2005; Barnet, 2007; Chang & Fong, 2010; Namkung & Jang, 2013). Organizational business systems form the foundation for any sustainable business, formulating and implementing green initiatives to achieve sustainability and competitiveness. The ecological partnership between organizations and society facilitates sustainable competitiveness.

Organizational ecological and sustainable procedures follow a strategic orientation (Fernandes & Solimun, 2017). Although sustainability was previously considered an additional cost for organizations, it is now perceived as a strategy and tool to derive value and foster innovation (Van Holt et al., 2020). A shared organizational vision promotes the development of green behaviors, contributing meaningfully to the organization's overall goals (Aragon-Correa et al., 2008; Renwick et al., 2008).

BACKGROUND

Numerous studies, including one by Castro & Orellana (2019), explore the concept of green roofs. They examine how the introduction of green roofs can enhance the sustainability and creativity of cities. According to these researchers, green roofs can help establish urban green innovation zones, aimed at promoting sustainable urban development through the integration of green infrastructure and innovation. The paper includes a literature review discussing the benefits of green roofs, such as their incorporation into urban green innovation spaces and their ability to mitigate the impacts of urban heat islands, improve stormwater management, and boost biodiversity.

Green roofs have demonstrated their energy efficiency, environmental friendliness, and long-term economic viability. The implementation of green roofs has become imperative as an effective and sustainable approach that aids in achieving urban sustainability (Alqahtany, 2023). The design of a green roof system is a complex process involving numerous factors, including social, cultural, climatic, and environmental considerations (Williams et al., 2010).

There are diverse types of green roofs. These include the extensive green roof system, the semi-intensive living roof planted with horizontal cover plants, and the intensive green roof which includes growth medium, plantation, and irrigation and must be established over a concrete floor base (Karachaliou et al., 2016). The innovation in creating these roofs has become significant in recent years. Klemm et al. (2019) investigate the role of green roofs as a part of urban green infrastructure and their potential to contribute to sustainable and innovative cities. The authors discuss the challenges and obstacles in implementing green roofs, such as the lack of knowledge and expertise, and the need for supportive policies and regulations. They conclude that green roofs are a crucial part of urban green infrastructure for innovative cities, and their full potential should be realized through supportive policies and regulations.

Additionally, Hemsley-Brown & Laing (2019) focus their research on potential hindrances to the adoption of green roofs as a method for sustainable urban development. They identify several potential barriers such as a lack of understanding and knowledge about green roofs, high initial installation costs, the absence of encouraging policies and regulations, and a lack of incentives for property owners to install green roofs, all derived from a literature review. The study also examines social and cultural variables, including aesthetic preferences and perceptions of the value of green space, which could influence the adoption of green roofs. The authors propose that to overcome these obstacles, a multifaceted strategy involving collaboration between various stakeholders, including policymakers, researchers, practitioners, and the public, is needed.

Palomino & Durán (2020) provide a comprehensive analysis of the study on the impact of green roofs on the urban heat island effect. The authors discuss the urban heat island effect—a phenomenon where urban areas experience higher temperatures than their rural surroundings due to the built environment and human activity—and its causes. The study underscores the potential of green roofs to reduce the urban heat island effect and improve thermal comfort, offering insights for policymakers, practitioners, and scholars interested in the design and implementation of green roofs in urban settings.

MAIN FOCUS

Wu & Jim (2021) provide an extensive review of the literature on the roles green roofs play in sustainable urban development. They highlight the benefits of green roofs, including their capacity to reduce the urban heat island effect, enhance biodiversity, improve stormwater management, and provide social benefits, all in line with the broader goals of sustainable urbanization. The authors posit that green roofs are an integral part of urban green infrastructure—a network of green spaces and technologies offering numerous social and environmental benefits—and that they hold immense potential to support sustainable urban growth. In the same vein, Jamei et al. (2023) suggests that installing green roofs reduces heat exchange between indoor and outdoor environments, thereby lowering building energy consumption. This demonstrates the importance of understanding and considering the impacts of green roofs, which stem from a combination of knowledge, technology, and green innovation.

Green roofs present a promising alternative to traditional rooftop materials, realigning urbanized spaces with the natural environment. A green roof typically consists of a roofing membrane, drainage layer, geotextile filter, growth substrate, and vegetation suitable for the regional climate. These components together aid in stormwater retention, urban heat island mitigation, and reduction in building energy consumption (Gunn et al., 2022). Understanding the value of green roofs in addressing environmental issues in cities, including mitigation of the urban heat island effect, is crucial for efforts to decrease air pollution and enhance urban species biodiversity.

Green Knowledge and Technology

Today's business ecosystem presents uncertainties and volatility in the development of green organizational and business models, challenging corporate risk management and decision making. Technological knowledge about the economic, social, and environmental benefits of constructing green roofs promotes their efficient construction and maintenance. Organizational green knowledge shares issues adopting a multilevel framework, contributing to green technology. For environmental protection and to stimulate green innovation, organizational management must formulate green knowledge and green technology. Knowledge management of environmental issues serves as a resource for learning, problem-solving, and creating competitiveness (Liao et al., 2004). Green knowledge management is a potent tool for organizational improvement in a transformative environment (Raudeli ¯ unien˙ et al., 2018). Environmental knowledge learning and green technology form the dynamic complexity of organizational green innovation (Li et al., 2021; Redman, 2014).

Knowledge sharing forms the foundation of knowledge management (Zhang & Sundaresan, 2010), and in the context of sustainable development goals like eliminating pollution, environmental protection, and Sustainable Development Goals (SDGs), sharing knowledge with other organizational members is beneficial (Bock et al., 2005). Green Knowledge Sharing (GKS) refers to the behavior of organizational members who disseminate information, data, and knowledge about green issues to others, thereby creating new knowledge and learning opportunities (Chen & Chang, 2013; Norton et al., 2015). Green property psychological ownership and green knowledge sharing are influenced by the green organizational identity. Knowledge sharing of individuals under time pressure is reduced (Chang et al., 2020). Organizational knowledge sharing focuses on fostering an atmosphere of environmental protection to create green knowledge sharing related to green issues.

Green technology is a socio-environmental healing technology that can minimize environmental damages and degradation while conserving natural resources. Sustainable green technologies used in processes and applications leave no footprint (Aithal & Shubhrajyotsna, 2016). Urban green innovation technology can be transferred by public government and private businesses aiming for inclusive green growth and ensuring social integration by eradicating poverty among vulnerable groups and keeping humanity's footprint within ecological boundaries. Building an inclusive and green organization of available and retained talent is a formidable task (Brands & Fernandez-Mateo, 2017). Green roof technology is employed to address social and environmental concerns and enhance climate resilience. Green plant-based green chemistry principles are used for green synthesis, nanomaterials, and nanoparticles (Maas & Hox, 2005). Green and eco-friendly nanotechnology based on green chemistry techniques seeks to decrease the risks associated with industrial applications.

Green eco-friendly nanotechnology solutions used in sustainable development goals reduce the technification threat. Environmental degradation and climate change pose threats to global sustainable development and the Sustainable Development Goals. The idea of achieving sustainable development goals through the use and development of proven green eco-friendly nanotechnology processes furthers the proliferation and growth of many other systems and devices. Green nanotechnology is based on miniature-sized communication and computational devices, nano sensor devices, high-density memory chips, etc. Ensuring ecological communication among stakeholders discloses the relevance of environmental concerns (Abimbola et al., 2010). Sustainable green nanotechnology principles and processes in the primary sector of the economy relate to the extraction of raw materials from nature.

Long-term effects of human food are required by green manufactured nanomaterials designed according to principles of green nanotechnology, complemented with current regulations aimed at fostering sustainable development of green technology. Increased instances of illnesses have been linked to the consumption of manufactured and processed inorganic food, leading to concerns about food safety. Green design and development of health and environment using nanoscale materials is a concern aimed at improving sustainable solutions. Green synthesis methods can mitigate adverse effects of nanomaterials on the environment and user health. The complementarity of mixed methods helps detect contradictions, improve the results of one method using the results of the other, and enhance the scope and breadth of different components of research (Greene et al. 1989).

Green nanotechnology in the service industry is intangible in nature and affects directly and indirectly. Green nanotechnology fosters innovation in educational technology through high-quality, low-cost, and ubiquitous online education. ICCT technologies working with green nanotechnology provide intelligent services in artificial intelligence, the internet of things, cloud and quantum computing, 3D printing, etc. Green nanotechnology supports ICCT technologies in the development of super-intelligent machines and human beings. Green nanotechnology enhances the durability, speed, and reach of digital entertainment instruments (Aithal & Shubhrajyotsna, 2016).

Organizational Green Innovation

The principles, values, and norms of economic concern, societal parity, and ecological accountability underpin organizational green business innovations (Alawattage & Fernando, 2017). Green organizational innovations can be generated either internally or externally, driven by green context-oriented factors. Based on green organizational identity, organizations have attributed more significance to environmental issues. This identity enhances organizational green innovation, competitiveness, and ensures organizational competitive advantages (Geraie 2015). Theories of organizational identity and psychological ownership are utilized to construct a conceptual model.

Organizational green process innovation is eco-friendly, developing goods manufacturing processes in accordance with their life cycles (Kam-Sing 2012). Organizational green innovation and actions are conceived to mitigate adverse environmental effects. The green innovation behaviors carried out under a strategy establish a relationship between organizational green innovation strategy and actions. Motivational mechanisms stimulate innovation to facilitate organizational green innovation. Prosocial motivation, identified as a mechanism linking contextual factors and innovative outcomes, motivates organizational stakeholders to pursue collective efforts towards green innovative outcomes with their teammates, leading with values of organizational green innovation (Norton et al., 2015; Liu et al., 2016).

Potential risks and challenges of implementing innovative resources and inputs supported by green efficacy are addressed by organizational green innovation (Liu, et al. 2015). Green organizational innovation is associated with individual tasks of organizational environmental activities to strengthen the organization. Green organizational identity enhances organizational creativity and innovation and ensures sustainable competitive advantages (Geraie, 2015). This leads to social acceptance, enabling the organization to develop sustainable organizational green competitiveness capability.

The green competitive advantage describes the conditions necessary for a firm to hold a position in sustainable and ecological management and innovation, which cannot be imitated by others (Lin & Chen, 2017). The resource-based view of organizational research focuses on the characterization of organizations (Pereira & Vence, 2012) and their resources and capabilities (Fong & Chang, 2012; Marchi, 2012). This view includes factors such as organizational redundancy, innovation capabilities, network embedding, strategy orientation, and green innovation strategy as drivers of green innovation practices. To enhance environmental sustainability, organizations rely on intangible resources (Singh et al., 2020). A green pioneer organization in a resource-intensive sector must have an organizational model (Elkington,1994) with a social and environmental impact.

Human resources green management deals with valuable organizational resources and assets considered based on environmental sustainability. Evidence indicates that the workforce strongly engages with environmental sustainability development and become more committed and satisfied organizationally. Human resource green management is engaged in organizational environmental sustainability management (Shoeb & Nisar 2015; Wirtenberg et al., 2007). Green dynamic capabilities lie within organizational processes and are consistently reliable in organizations (Chen, 2008).

Urban Green Innovation Spaces and Green Roofs

The rapid growth of the urban population, coupled with the low rate of economic development due to the pandemic, is leading to significant civilization challenges. These include the scarcity of urban green innovation spaces, pollution, and health issues. Smart growth, fostered by sustainable and inclusive growth, is based on the adoption of greener, efficient technologies and job creation to support social cohesion (Gunn & Mintrom, 2016). To remain inclusive, organizations need to embrace elements like equity and social justice for goal creation, strategy formulation, and public commitment development (Johnson, 2017).

Organizational green innovation emphasizes the cultivation of an organizational green culture, environmental pollution prevention, and waste reduction to enhance sustainability. Urban green innovation spaces are expanding in green metropolises, driven by the preservation and expansion of sustainable environmental development (Kalandides, 2021).

Urban green innovation spaces should be delineated by construction and protection limits. People's interaction with these spaces can be public or private (Vijayaraghavan, 2016). Public and private urban green innovation spaces can coexist and include green roofs, multifunctional open spaces framed within effective landscape management. These spaces, despite having different perceptions (Coolen & Meesters 2012), should ideally coexist (Mesimaki, et al. 2017).

People's perceptions of private and green urban spaces vary, with a focus on social identity and environmental sustainability. Green behaviors, defined as employee actions that contribute to environmental sustainability, extend beyond job tasks, and are considered part of organizational citizenship (Ones & Dilchert, 2012). The loss of green urban areas due to residential and commercial expansion, exacerbated by urban sprawl, leads to a decrease in habitats, biodiversity, recreational areas, natural resources, and elements (Burszta-Adamiak & Fiałkiewicza, 2019). Urban green spaces with high social and environmental vulnerability, especially urbanized areas with high land temperature, should be prioritized for the implementation of urban green innovation areas and green roofs.

Traditional streets can transform into green urban connection areas, serving as an open space system and parkland. An urban green innovation belt can prevent urban sprawl by merging with nearby settlements. The greenbelt concept conserves open space accessibility along roadways, connecting to a regional trail. However, the greenbelt has not successfully prevented urban expansion and sprawl, but it has facilitated access to open spaces (Maruani & Amit-Cohen 2007). Investments to develop accessible green spaces, designed to meet economic, social, and leisure needs, contribute to community-building and alleviate construction pressures. Urban green innovation areas are open to enhance urban green innovation space networks and extend green infrastructure functions, including green roofs (Fung, 2018).

Green infrastructure comprises a network of green open spaces that provides economic, social, and environmental benefits to city, town, and village residents. It creates an appealing environment, increases urban resilience benefits by managing stormwater, mitigates the urban heat island effect, and purifies the air. Green roofs, regarded as passive open green spaces, offer environmental benefits, and provide open spaces for greater urban livability (Mesimaki, et al. 2017). New city infrastructures and facilities should incorporate eco-roofs. Multi-functional urban green innovation infrastructure should allocate funding for green development initiatives.

Scholars have used quantitative and qualitative analysis to understand people's needs and perceptions of green roofs. The environmental effectiveness of green roofs can enhance public health and meet the social needs of vulnerable populations (Vijayaraghavan 2016). Green roofs, solar roofs, and living roofs, considered multi-benefit assets, can bring significant economic benefits beyond stormwater management, energy efficiency, and air quality improvement. They can also reduce the urban heat island effect. Loss of green spaces intensifies the heat island effect and stormwater runoff. Hence, the development of urban green innovation spaces and roofs should comply with regulations on stormwater control. An incentive program for the installation of eco-roofs can provide better stormwater management.

Historically, green roofs have evolved from the Hanging Gardens of Babylon, providing aesthetic value, and fostering human interaction, and even contributing to insulation. Le Corbusier integrated green roofs into modern architecture (Berardi et al., 2014). Countries like Germany, France, Switzerland, the USA, Canada, and Japan have researched and implemented green roofs. For instance, in Portland in 2005, authorities required city-owned buildings to incorporate green roofs. Similarly, Toronto passed a bylaw requiring all newly developed buildings to include green roofs. In Tokyo, all new buildings must have green roofs (Vijayaraghavan 2016).

Urban green innovation spaces should facilitate green infrastructure solutions, such as building green roofs, which positively influence the urban environment. Green infrastructure's operational stage involves maintenance solutions, focusing more on effective stormwater management and ecological landscape conservation than social and economic benefits (Mell, 2010). Urban green innovation roofs, integrated into open space management, offer multifunctional opportunities to develop green infrastructure and benefit urban communities and neighborhoods. Both private and public green roofs, as open spaces, should serve as multifunctional green infrastructure and meet residents' urban open space needs (Meerow & Newell, 2017).

A well-developed urban green innovation infrastructure requires a multifaceted approach, with increasing awareness of green roofs to alleviate global and local environmental impacts. Vegetated green roofs naturally sequester carbon dioxide through photosynthesis (Rowe, 2011). Studies confirm the capacity of green roofs to retain rainwater, reduce air pollutants, and improve the microclimate, positively influencing the urban heat island effect (Burszta-Adamiak et al., 2019; Shafi que et al., 2018). There are several types of intensive and extensive green roofs, each incorporating different soil depth, weight, irrigation, plant species, and maintenance. Urban green innovation projects include urban gardening, permaculture, guerrilla gardening, smart city, and building architecture. Productive urban green innovation combines urban agriculture, allotment gardens, and a culture of self-sufficiency, all ecologically motivated (Chen et al., 2020).

Greenhouse gas emission mitigation techniques to improve agriculture's environmental impact need to consider animal welfare, the largest anthropogenic contributor (Kucher et al., 2018; Llonch et al., 2017; Reisinger & Clark, 2018). Unused spaces can transform into productive greenery and develop into natural and green innovation areas that preserve and increase ecological value. They can also serve as joint creation and learning spaces (Green, 2020; Nachhaltiges, 2020).

Energy optimization in building rehabilitation should consider creating new alternative green spaces, transformed into green innovation areas for recreational purposes. Green roofs offer leisure spaces, enhance the aesthetic value of buildings (Sutton, 2014), reduce the urban carbon footprint (Ugai, 2016), and manage the direct water footprint (Fialkiewicz et al., 2018). Identifying urban places that need urban green innovation areas and suitable buildings for green roofs is critical for community and neighborhood implementation. Administrative and technical support is essential in designing, constructing, developing, and maintaining urban green innovation areas and green roofs. This task enhances the allocation of urban green innovation areas based on the structural design of buildings and surrounding environmental characteristics in the community. Green roof design and construction should consider the selection of plant species and the development of small.

Urban green innovation areas and green roofs are sustainable resources contributing to the attractiveness of the city. The integration of the open green areas in the city is creating a natural structure where green roofs in the private properties contribute to the natural structure of the cities (Corporation, 2016). Urban green innovation spaces and green roofs contribute to the increase of the efficiency and effectiveness of energy consumption in urban areas (Lee et al., 2018). The green infrastructure maintenance depends on the available space and configuration and the different purposes of green spaces and should consider the social benefits to the local community.

In the green economic innovation model, urban green innovation areas and green roofs are crucial parts of the open green spaces (Lee et al., 2018; Yang, et al., 2018). The economic performance of companies in urban green innovation areas and green roofs are based on the sustainability, environmental friendliness, and energy efficiency (Fam et al., 2018). Urban green innovation spaces and green roofs are part of the green economic innovation model for green economic growth that should develop social benefits and the welfare of the community.

The use of urban green innovation areas and green roofs as natural areas and as multi-functional open green spaces in cities, towns, and villages is highly recommended. Green roofs and urban green innovation spaces are significant for sustainable environmental management and sustainable urban development. The strategies and policies for green growth must consider the balance between economic development and environmental sustainability to make the cities more livable and resilient.

Urban Green Innovation Spaces and Green Roofs

The rapid growth of the urban population and the low rate of economic development due to the pandemic, are leading to serious civilization problems, among others the lack of urban greener innovations spaces, pollution, and lack of health. Smart growth is supported by sustainable growth based on adoption of greener, efficient technologies, inclusive growth, fostering employment and social cohesion (Gunn & Mintrom, 2016). Organizations must take an approach to stay in inclusive green organizations embracing elements such as equity and social justice needed to create goals, formulate, and implement strategies and develop public commitment (Johnson, 2017). Organizational green innovation refers to the innovation that emphasizes the implementation of organizational green culture, environmental prevention of pollution and waste reduction to enhance organizational sustainability. Urban green innovation areas are increasing in green metropolis driven by the preservation and expansion of sustainable environmental development (Kalandides, 2021).

Urban green innovation spaces must be defined by the limits of construction and protection. The interactions of people and urban green innovation spaces are divided in public and private (Vijayaraghavan, 2016). Public and private urban green innovation spaces co-exist and can be incorporated green roofs as open green spaces aimed to use multi-functionality framed in an effective landscape management. these spaces have differences in perception (Coolen & Meesters 2012) although they should co-exist (Mesimaki, et al. 2017).

People have different perceptions on private and green urban spaces with social identity and environmental sustainability. Green behaviors are defined as the actions and behaviors on which employees engage linked with and contribute to environmental sustainability (Ones & Dilchert, 2012). This concept includes voluntary green behaviors beyond the job tasks referred as organizational citizenship, also considered as task-related environmental behavior (Bissing-Olson et al., 2013; Boiral & Paillé, 2012). The disappearance of green urban areas due to residential and commercial purposes, enhanced by the urban sprawl, leads to loss of habitats, biodiversity, and recreation areas, and decrease of natural resources and elements (Burszta-Adamiak &, Fiałkiewicza, 2019). Urban green spaces with high social and environmental vulnerability are the most urbanized areas with high land temperature might be prioritized for the implementation of urban green innovation areas and green roofs.

Traditional streets can be transformed into green urban connecting areas considered as an open space system and park lands. An urban green innovation belt prevents urban expansion and urban sprawl by merging with nearby settlements. Greenbelt conserves accessibility of open spaces along roadways and connect to a regional trail. Nevertheless, greenbelt was not successful to prevent urban expansion and sprawl but facilitates the access to open spaces (Maruani & Amit-Cohen 2007). Intensification of investments to develop accessible green spaces conceived to cover economic, social and leisure needs, and activities contribute to create the sense of community and respond to the pressures of construction. Urban green innovation areas are open to enhance urban green innovation space networks and extend the green infrastructure with the functions of green roofs (Fung, 2018). Green infrastructure is the urban physical environment in a network of green open spaces that brings economic, social, and environmental benefits to residents of communities in cities, towns, and villages.

Green infrastructure is a way to create an attractive environment, increase the urban resilience benefits by better managing stormwater, mitigating the urban heat island effect, purifying air. Green roofs are considered passive open green spaces that have environmental benefits and provide open spaces for more urban livability (Mesimaki, et al. 2017). Eco roofs are required for new city infrastructures and facilities. Multi-functional urban green innovation infrastructure should allocate funding for green initiatives of development.

Scholars have applied quantitative and qualitative analysis to address the needs and perceptions of people on green roofs. The environmental effectiveness of green roofs should improve public health and connect the social needs of vulnerable people (Vijayaraghavan 2016). Green roofs and solar and living roofs considered a multi-benefit asset bringing significant economic benefits beyond stormwater management, energy efficiency for cooling, improving air quality, reducing urban heat island. Loss of green spaces intensify heat island effect and storm water runoffs. The development of urban green innovation spaces and roofs are required to comply with regulations on stormwater control. A motivation program for the installation of eco roofs provides incentives to better management of the stormwater.

Green roofs have historically evolved from the hanging gardens in Babylon embracing the aesthetic values and facilitating the human interactions and increase the insulation. Le Corbusier applied green roofs to modern architecture. (Berardi et al., 2014). In Germany followed by France and Switzerland occurred the research and implementation of green roofs, in USA, in Portland in 2005 the authority requires the city-owned buildings to build green roof, in Canada in Toronto the authority passed a bylaw requiring all the new developed buildings meanwhile in Japan in the city of Tokyo, the new constructed buildings must have green roofs (Vijayaraghavan 2016).

Urban green innovation areas should provide spaces for green infrastructure solutions such as the building green roofs, which have beneficial influence on the urban environment. The stage of operation of green infrastructure is the maintenance solutions. The green infrastructure is single function derived from the ability to effectively manage stormwater and focusing on ecological landscape conservation than social and economic benefits (Mell, 2010). Urban green innovation roofs integrated into open space management have multifunctional opportunities to develop green infrastructure and benefits in urban communities and neighborhoods. Private and public green roofs as open spaces a multi-functional green infrastructure and should accommodate the urban open space needs of residents (Meerow & Newell, 2017).

Well-developed urban green innovation infrastructure needs to be supported by a multi-faceted approach. The increasing awareness of green roofs to alleviate global and local environmental impacts. Vegetated green roofs reduce carbon dioxide in the atmosphere naturally sequestered through the photosynthesis processes (Rowe, 2011). It has been confirmed the capacity of green roofs to retain rainwater, reduce air pollutants and improve microclimate that have a positive influence in urban heat island effect (Burszta-Adamiak et al., 2019; Shafi que et al., 2018). There are diverse types of intensive and extensive green roofs which include vegetation with different soil depth and weight, irrigation, plant species and maintenance. Urban green innovation projects include urban gardening, permaculture, guerrilla gardening, smart city and building architecture. Productive urban green innovation combines urban agriculture, allotment gardens and self-sufficiency culture ecologically motivated (Chen et al., 2020).

Mitigation of greenhouse gas emission technique to improve the environmental impact of agriculture needs to consider animal welfare the largest anthropogenic contributor (Kucher et al., 2018; Llonch et al., 2017; Reisinger & Clark, 2018). Wasteland spaces can be converted into a place of productive greenery and developed into natural and green innovation areas that preserve and increase the ecological value and be used also as place for joint creation and learning, (Green, 2020; Nachhaltiges, 2020).

Rehabilitation of buildings using energy optimization should consider the creation of new alternative green spaces transformed into green innovation areas for free recreational purposes. Green roofs provide leisure spaces and enhance aesthetical values of buildings (Sutton, 2014) reduce carbon footprint in urban areas (Ugai, 2016) and direct water footprint (Fialkiewicz et al., 2018). Matching urban places needing urban green innovation areas and green roofs in suitable buildings to determine opportunities to implement in communities and neighborhood. Administrative and technical support are necessary in designing, constructing, developing, and maintaining urban green innovation areas and green roofs. This task recommendation enhances the allocation of urban green innovation areas based on the structural design of buildings and the surrounding environmental characteristics in the community. The design and construction of green roofs need to consider the plant species and the development of small fauna.

The construction and application of green roofs can be achieved in large urban areas are located near another urban constructed areas and should not be single investments. Stovin et al. (2012) calculate that the surface of roofs accounts for almost 50% of the sealed urban areas with the potential for urban green innovation roofs. Green roofs decrease the costs for discharging snowmelt and stormwater. The infrastructure of green storm water initiative reduces greenhouse gas emissions and improves the air quality. The development initiatives of green roofs are linked with storm water management to meet the flood management control required. Local building codes must allow green roofs exempted from the building floor area and related to open spaces. The incentives for zoning to encourage the implementation of the urban green innovation roof should include the development of parking landscapes as one measure adopted in a stormwater management plan that may provide solutions (Murray, 2017).

The development of green roofs value investments as an additional incentive for potential customers. Meeting the ecological needs of customers (Jain & Kaur, 2004). There is a trend of organizations to shift operations towards the direction of environment-friendly processes despite the challenges of customer cynicism regarding organizational green actions (Kumar & Christodoulopoulou, 2014). Co-financing and local legal regulations for the construction of green roofs are most often used incentives. Green development finance has been assigned to Multilateral Development Banks, alongside public-private partnerships (PPPs) and new forms of finance, blended finance, bond instruments, green bonds, social impact bonds and development impact bonds. Urban green innovation roofs are the result of the implementation of motivational tools and other incentives for the different stakeholders.

The existence of types of barriers encourages local authorities to improve the use of incentives for the creation and functioning of urban green innovation areas and urban green innovation roofs and the most commonly use of incentives are the direct ones and legal regulations for the construction of regulations (Mentens et al., 2006). Incentives must be translated from sufficient interest into motivation for the excessively strict requirements for the construction of green roofs such as the minimum surface area, to become eligible for co-financing and tax allowance. Market demand-related incentives such as promotional instruments, are tools to determine demand for buildings with greenery in comparison to traditional. The implementation of financial and non-financial incentives is a crucial factor for the green roofs as a relevant element of urban environment (Brudermann & Sangkakool, 2017). Financial incentives include subsidies or donations granted to reimburse costs of investments that are aimed to support potential investors.

The development of rural and urban green innovation tourism close to nature and based on settlements is becoming more relevant for the impact on green economy and ecology, with the self-employment of residents to provide environmentally accommodation and food to the visitors (Kolomiets & Tomashuk, 2021; Mazur & Tomashuk, 2019; Tomashuk & Baldynyuk, 2021). The green economy development and the pursuit of environmental protection is bringing profits to the organizations (Chen et al., 2006).

Case Studies

Green roof applications have long been popular in Germany and Norway (Köhler and Poll, 2010). Recently, reports have emerged highlighting the different advantages of green roofs in tropical areas, characterized by hot and humid climatic conditions. These advantages include the mitigation of urban heat islands and reduced energy consumption for cooling. There are now studies beginning to enumerate the impacts generated by green roofs, with details provided below.

Green roof studies have primarily focused on reducing energy consumption. In developed countries, air conditioning in buildings consumes a large amount of electricity. These studies show that green roofs are an effective method for saving energy in hot climates, improving the energy efficiency of indoor spaces, enhancing indoor air quality, reducing outdoor air temperature, and mitigating the urban heat island effect. They also present technological developments and their applications in urban green innovation spaces.

This topic has become increasingly relevant, as it presents the challenges that countries must face regarding climate change and the increasing heat island effect. It also explores the obstacles green roofs might confront, as countries revamp their laws in response to the new economic, political, and social circumstances they face.

CONCLUSION

This study analyzes how organizational environmental and green knowledge learning is linked to green technology for environmental protection, stimulating organizational green innovation, urban green innovation spaces, and green roofs. Green roof technology can be considered an effective solution for sustainable urban development in megacities due to its numerous economic, environmental, and social benefits. However, the importance of these benefits varies for diverse types of buildings, warranting further investigation.

Organizations transform to mitigate and neutralize environmental impact and to adapt to environmental sustainability. Green transformation is a multifaceted process involving interconnected and overlapping processes, posing several managerial challenges. Environmental community involvement promotes and improves the natural environment, aiming for the sustainable growth of society and inspiring voluntary participation in social activities. The green government should ensure access to affordable, sustain-

Table 1. Main articles on green roofs innovations and urban spaces

Article	Author(s)	Main Topic
Evaluation of cool roof and vegetations in mitigating urban heat island in a tropical city, Singapore	(Li & Norford, 2016)	"The result reveals that the city-wide deployment of cool roofs can significantly reduce near-surface air temperature and surface skin temperature during the daytime, especially at noon, although the effect is less pronounced during nighttime in Singapore"
Green Roof: Its Awareness Among Professionals and Potential in Malaysian Market	(Rahman et al., 2013)	"It discusses the growing awareness of green roof technology among professionals in Malaysia's built environment industry. A total of thirty local green roof projects have been identified, indicating a growing tendency to deploy green roof technology in Malaysia across various professions, including construction, landscape architecture, urban planning, project management, and academia.
Using Green Roofs for Social Housing to Improve Energy Consumption in New Cities. (An Applied Study of Social Housing in Egypt's New Cairo City)	(Goda et al., 2023)	"The study shows the improve in energy consumption of cooling and heating energy in social housing in new urban cities in Egypt, reduce high temperatures inside and outside building, and decrease the urban heat island effect, the study suggests a method that is based on modifying the traditional roof of residential construction in Egypt by converting it to a green roof. An integrated analysis model including a strategy shows the environmental and climatic benefits of using green roofs to provide the external outdoor temperature of social housing buildings. The result revealed that green roofs, which are used by 50% of green roofs in traditional buildings, were shown to be efficient in reducing annual energy consumption by 12% and decreasing the average temperature inside buildings by 2.44°C, whereas the average temperature outside was reduced by 3°C."
Green Roofs as an Approach to Enhance Urban Sustainability: A Study of Public Perception in Riyadh, Saudi Arabia	(Alqahtany, 2023)	Most of the population of Riyadh understands the benefits related to green roofs, and they have shown a positive response to a few factors. Almost 94% of people agreed with the fact that green roofs enhance the aesthetics of the building, and the same number of people agreed that they play a role in controlling the air quality. Similarly, 93% of people that green roofs provide additional recreational space in the building. On the other hand, 91% of the respondents identified the climate of the area as the biggest challenge in implementing green roofs in the buildings
The use of green roofs in urban green innovation spaces: A case study in Barcelona.	(Nogue & Salas, 2018)	The study offers a comprehensive perspective on the motivations, advantages, and challenges of implementing green roofs in urban green innovation spaces. It investigates the reasons for installing green roofs in these areas, the benefits, and challenges associated with it, and the role stakeholders play in the planning and maintenance of these green roofs.
How to design green roofs to enhance urban green innovation space? Evidence from Beijing	(Gong et al., 2020)	The study provides a more in-depth analysis of the design features and effectiveness of green roofs. It further examines the design elements of green roofs and their efficiency in enhancing urban green innovation spaces. The study also offers suggestions on designing and deploying green roofs to support sustainable urban development and innovation, providing valuable insights for policymakers and practitioners.
The application of green roofs in urban green innovation spaces: A case study in Jinan City, China	(Li et al., 2021)	The case study focuses on the application of green roofs in urban green innovation spaces in Jinan City, China, providing a comprehensive analysis of their design and implementation in these spaces. It assesses the effectiveness of green roofs in yielding various environmental and social benefits, such as diminishing the urban heat island effect, improving air quality, and providing community activity spaces.
Green roof innovation in Singapore: Policies, performance, and potential	(Tan et al., 2019)	A case study on green roof innovation in Singapore presents a broader perspective on the policies and regulations that have facilitated the widespread implementation of green roofs throughout the city. The study examines the laws and policies that have streamlined the installation of green roofs in Singapore, evaluating their efficacy in providing a host of environmental and social benefits, while also identifying any existing obstacles to their utilization. Furthermore, the study offers recommendations on how to design and implement green roofs not only in Singapore but also in other urban contexts, providing valuable insights for policymakers and practitioners.

Source: Own elaboration based on references included.

able housing in urban green innovation spaces, and open buildings to create an urban social cohesion process, consolidated in urban social movements. The design of an effective green roof system is a complex procedure that depends on several sustainability factors, including social, cultural, climatic, and environmental considerations.

Organizational networks are vital for connecting each other in the green movement space. Despite efforts to increase diversity and inclusion, environmental movements, along with non-government organizations and environmental foundations, lack racial diversity. Residents express their intention to participate in discussions and debates to develop a green urban framework. It is recommended that local authorities initiate strong campaigns in favor of green technologies, particularly for new or under-construction buildings. They must educate people about the benefits green roofs can provide and facilitate residents and homeowners in designing, executing, and installing such systems.

Incorporating green components into technology can create sustainable green technologies capable of preventing environmental degradation and promoting a cleaner environment. Green roof technology is used to address social and environmental concerns and enhance climate resilience. Green nanotechnology offers both opportunities and challenges in the industry to foster growth by supporting nanotechnology usage. Green roof technology has been recommended as an effective solution for increasing per capita green spaces and improving sustainability means of urban development due to its diverse advantages.

Gaining competitive advantage through implementing a sustainable organizational strategy is crucial for survival. The organizational green shared vision should include sustainable business goals. Long-term organizational green innovation strategy must encourage followers to implement green operations. Organizational socio-ecological strategies are formulated and implemented to achieve green competitiveness. Despite significant benefits, other factors, including high construction costs and maintenance issues, need to be considered for future global applications.

Most green roof studies have focused on reducing energy consumption. In developed countries, air conditioning in buildings consumes a large amount of electricity, these studies show that green roofs are an appropriate method of saving energy in hot climates, to improve the energy efficiency of indoor areas, the air quality of indoor buildings, and the air temperature of outdoor buildings and decrease the urban heat island effect, presents technological developments and how to use in urban green innovations spaces.

This topic takes greater relevance in these days, as it presents the challenge that countries must face to ensure that climate change and the heat island effect is increasing, as well as the obstacles that green roofs can confront, while the countries redesigning the laws in this new economic, political, and social moment they are facing.

REFERENCES

Abimbola, T., Lim, M., Hillestad, T., Xie, C., & Haugland, S. A. (2010). Innovative corporate social responsibility: The founder's role in creating a trustworthy corporate brand through "green innovation.". *Journal of Product and Brand Management, 19*(6), 440–451. doi:10.1108/10610421011085758

Aithal, P., & Shubhrajyotsna, A. (2016). Nanotechnology Innovations and Commercialization – Opportunities, Challenges & Reasons for Delay. *Proceedings of National Conference on Changing Perspectives of Management, IT, and Social Sciences in Contemporary Environment, 14*, 1-12. 10.5815/ijem.2016.06.02

Aithal, P., & Shubhrajyotsna, A. (2018). Nanotechnology based Innovations and Human Life Comfortability –Are we Marching towards Immortality? *International Journal of Applied Engineering and Management Letters, 2*(1), 1–8. doi:10.5281/zenodo.1485048

Alawattage, C., & Fernando, S. (2017). Postcoloniality in corporate social and environmental accountability. *Accounting, Organizations and Society*, *60*, 1–20. doi:10.1016/j.aos.2017.07.002

Alqahtany, A. (2022). Green Roofs as an Approach to Enhance Urban Sustainability: A Study of Public Perception in Riyadh, Saudi Arabia. *Buildings*, *12*(12), 2202. doi:10.3390/buildings12122202

Anderson, R., Amodeo, M., & Harzfeld, J. (2010). *Changing business cultures from within.* 2010 State of the World Report. The World Watch Institute. doi:10.1016/j.jenvman.2006.12.032

Aragon-Correa, J. A., Hurtado-Torres, N. E., Sharma, S. & García-Morales, V. J. (2008). Environmental strategy and performance in small firms: A resource-based perspective. *Journal of Environmental Management*, *86*(1), 88–103. doi:10.1016/j.jenvman.2006.12.032

Bansal, P. (2005). Evolving sustainably: A longitudinal study of corporate sustainable development. *Strategic Management Journal*, *26*(3), 197–218. doi:10.1002mj.441

Barnet, M. L. (2007). Stakeholder influence capacity and the variability of financial returns to corporate social responsibility. *Academy of Management Review*, *33*(3), 794–816. doi:10.5465/amr.2007.25275520

Barney, J. (1991). Firm Resources and Sustained Competitive Advantage. *Journal of Management*, *17*(1), 99–120. doi:10.1177/014920639101700108

Berardi, U., GhaffarianHoseini, A. H., & GhaffarianHoseini, A. (2014). State-of-the-art analysis of the environmental benefits of green roofs. *Applied Energy*, *115*, 411–428. doi:10.1016/j.apenergy.2013.10.047

Berlin, G. (2020). *Capital of Green Trends: How Berlin leads the Way in Urban Sustainability*. Available online: https://about.visitberlin.de/en/greenberlin

Berlin, N. (2020). *Stadt der grünen Trends-Wie Berlin den Weg der urbanen Nachhaltigkeit geht*. Available online: https://about.visitberlin.de/nachhaltiges-berlin

Bernardini, C. & Irvine, K. N. (2007). The 'nature' of urban sustainability: Private or public green spaces. *Sustainable Development and Planning III*, *2*, 661-674. doi:10.2495/SDP070642

Bissing-Olson, M. J., Iyer, A., Fielding, K. S., & Zacher, H. (2013). Relationships between daily affect and pro-environmental behavior at work: The moderating role of pro-environmental attitude. *Journal of Organizational Behavior*, *175*(2), 156–175. doi:10.1002/job.1788

Bock, G. W., Zmud, R. W., Kim, Y. G., & Lee, J. N. (2005). Behavioral intention formation in knowledge sharing: Examining the roles of extrinsic motivators, social-psychological forces, and organizational climate. *Management Information Systems Quarterly*, *29*(1), 87–112. doi:10.2307/25148669

Boiral, O., & Paillé, P. (2012). Organizational citizenship behaviour for the environment: Measurement and validation. *Journal of Business Ethics*, *109*(4), 431–445. doi:10.100710551-011-1138-9

Brands, R. A., & Fernandez-Mateo, I. (2017). Leaning out: How negative recruitment experiences shape women's decisions to compete for executive roles. *Administrative Science Quarterly*, *62*(3), 405–442. doi:10.1177/0001839216682728

Brudermann, T., & Sangkakool, T. (2017). Green roofs in temperate climate cities in Europe - An analysis of key decision factors. *Urban Forestry & Urban Greening*, *21*, 224–234. doi:10.1016/j.ufug.2016.12.008

Brundtland, G. (1987). *Our Common Future: The World Commission on Environment and Development*. Oxford University Press.

Burszta-Adamiak, E., & Fiałkiewicza, W. (2019). Review of green roof incentives as motivators for the expansion of green infrastructure in European cities. *Scientific Review – Engineering and Environmental Sciences, 28*(4), 641–652. DOI doi:10.22630/PNIKS.2019.28.4.5

Burszta-Adamiak, E., Stańczyk, J., & Łomotowski, J. (2019). Hydrological performance of green roofs in the context of the meteorological factors during the 5-year monitoring period. *Water and Environment Journal : the Journal / the Chartered Institution of Water and Environmental Management, 33*(1), 144–154. doi:10.1111/wej.12385

Castro, A. J. L., & Orellana, F. (2019). Green roofs as a strategy for urban green innovation spaces. *Sustainable Cities and Society, 44*, 697–705. doi:10.1016/j.scs.2018.10.034

Chang, N. J., & Fong, C. M. (2010). Green product quality, green corporate image, green customer satisfaction, and green customer loyalty. *African Journal of Business Management, 4*(13), 2336–2344.

Chang, T. W., Chen, Y. S., Yeh, Y. L., & Li, H. X. (2020). Sustainable consumption models for customers: Investigating the significant antecedents of green purchase behavior from the perspective of information asymmetry. *Journal of Environmental Planning and Management, 64*(9), 1668–1688. doi:10.1080/09640568.2020.1837087

Chen, Y. S. (2008). The driver of green innovation and green image–green core competence. *Journal of Business Ethics, 81*(3), 531–543. doi:10.100710551-007-9522-1

Chen, Y. S., & Chang, C. H. (2013). The determinants of green product development performance: Green dynamic capabilities, green transformational leadership, and green creativity. *Journal of Business Ethics, 116*(1), 107–119. doi:10.100710551-012-1452-x

Chen, Y. S., Lai, S. B., & Wen, C. T. (2006). The influence of green innovation performance on corporate advantage in Taiwan. *Journal of Business Ethics, 67*(4), 331–339. doi:10.100710551-006-9025-5

Chen, Y. S., Lin, S. H., Lin, C. Y., Hung, S. T., Chang, C. W., & Huang, C. W. (2020). Improving green product development performance from green vision and organizational culture perspectives. *Corporate Social Responsibility and Environmental Management, 27*(1), 222–231. doi:10.1002/csr.1794

Consulting, D. (2017). *12 mal Berliner Leben, 12 mal Berlin Erleben. Konzept für einen stadtverträglichen Berlin-Tourismus 2018+*. VisitBerlin. https://about.visitberlin.de/sites/default/files/2018 02/Tourismuskonzept_Berlin_Studie_2017.pdf

Coolen, H., & Meesters, J. (2012). Private and public green spaces: Meaningful but different settings. *Journal of Housing and the Built Environment, 27*(1), 49–67. doi:10.100710901-011-9246-5

Elkington, J. (1994). Towards the sustainable corporation: Win-win-win business strategies for sustainable development. *Cal. Management Review, 36*(3), 90-100.

Fernandes, A. A. R., & Solimun. (2017). The mediating effect of strategic orientation and innovations on the effect of environmental uncertainties on the performance of business in the Indonesian aviation industry. *International Journal of Law and Management, 59*(6), 1269–1278. doi:10.1108/IJLMA-10-2016-0087

Fialkiewicz, W., Burszta-Adamiak, E., Kolonko-Wiercik, A., Manzardo, A., Loss, A., Mikovits, C., & Scipioni, A. (2018). Simplified direct water footprint model to support urban water management. *Water (Basel), 10*(5), 630. doi:10.3390/w10050630

Fong, C. M., & Chang, N. J. (2012). The impact of green learning orientation on proactive environmental innovation capability and firm performance. *African Journal of Business Management, 6*(32), 727–735.

Fung, K. L. (2018). *Expanding the green network on rooftops: A study of integrating green roofs as a part of urban green innovation space planning* (Order No. 10932095). Available from ProQuest One Academic. (2124999351). Retrieved from http://wdg.biblio.udg.mx:2048/login?url=https://www.proquest.com/dissertations-theses/expanding-green-network-on-rooftop s-study/docview/2124999351/se-2?accountid=28915

Galdeano-Gómez, E., Céspedes-Lorente, J., & Martínez-del-Río, J. (2008). Environmental performance and spillover effects on productivity: Evidence from horticultural firms. *Journal of Environmental Management, 88*(4), 1552–1561. doi:10.1016/j.jenvman.2007.07.028 PMID:17825476

Geraie, M. S., & Rad, F. M. (2015). Mediator role of the organizational identity green in relationship between total quality management and perceived innovation with sustainable competitive advantage. *International Journal of Biology, Pharmacy and Allied Sciences, 4*, 266–276.

Goda, Z., Foda, M. & Elsayyad, N. (2023). Using Green Roofs for Social Housing to Improve Energy Consumption in New Cities. (An Applied Study of Social Housing in Egypt's New Cairo City). *Future Cities and Environment, 9*(1), 2, 1–11. doi:10.5334/fce.165

Gohari, A., Gohari, A., & Bin, A. (2023). Importance of green roof criteria for residential and governmental buildings: A multi-criteria decision analysis. *Environmental Science and Pollution Research International, 30*(2), 3707–3725. doi:10.100711356-022-22472-4 PMID:35953748

Gong, Y., Li, X., & Li, Z. (2020). How to design green roofs to enhance urban green innovation space? Evidence from Beijing. *Habitat International, 107*, 102–113. doi:10.1016/j.habitatint.2020.102113

Greene, J., Caracelli, V., & Graham, W. (1989). Toward a conceptual framework for mixed-method evaluation designs. *Educational Evaluation and Policy Analysis, 11*(3), 255–274. doi:10.3102/01623737011003255

Gunn, A., & Mintrom, M. (2016). Higher Education Policy Change in Europe: Academic Research Funding and the Impact Agenda. *European Education, 48*(4), 241–257. doi:10.1080/10564934.2016.1237703

Gunn, P., Burak, H., Van Geel, P., & Baldwin, C. (2022). Development of resistor-capacitor and finite difference models to evaluate green roof thermal performance. *Building and Environment, 226*, 1–15. doi:10.1016/j.buildenv.2022.109700

Gupta, J., & Vegelin, C. (2016). Sustainable development goals and inclusive development. *International Environmental Agreement: Politics, Law and Economics, 16*(3), 433–448. doi:10.100710784-016-9323-z

Hadavi, S. (2017). Direct and indirect effects of the physical aspects of the environment on mental well-being. *Environment and Behavior, 49*(10), 1071–1104. doi:10.1177/0013916516679876

Hemsley-Brown, T., & Laing, A. (2019). Green roofs and sustainable urban development: An exploratory review of potential barriers to adoption. *Sustainable Cities and Society*, *50*, 101648. doi:10.1016/j.scs.2019.101648

Hodge, T. (1997). Toward a conceptual framework for assessing progress toward sustainability. *Social Indicators Research*, *40*(1/2), 5–98. doi:10.1023/A:1006847209030

Huang, J., Kong, F., Yin, H., Middel, A., Liu, H., & Meadows, M. E. (2023). Green roof effects on urban building surface processes and energy budgets. *Energy Conversion and Management*, *287*, 1–13. doi:10.1016/j.enconman.2023.117100

Ihnatenko, M., & Novak, N. (2018). Development of regional programs for the development of agrarian enterprises with organic production based on the European and international experience. *Baltic Journal of Economic Studies*, *4*(4), 126–133. doi:10.30525/2256-0742/2018-4-4-126-133

Jain, S. K., & Kaur, G. (2004). Green marketing: An Indian perspective. *Decision (Washington, D.C.)*, *31*(2), 161–209.

Jamei, E., Chau, H., Seyedmahmoudian, M., Mekhilef, S., & Alzahraa, F. (2023). Green roof and energy – role of climate and design elements in hot and temperate climates. *Heliyon*, *9*(5), 1–16. doi:10.1016/j.heliyon.2023.e15917 PMID:37215798

Johnson, S. K. (2017, August 17th). What 11 CEOs have learned about championing diversity. *Harvard Business Review*.

Johnson, S. K. (2019). *Leaking Talent How People of Color are Pushed Out of Environmental Organizations How People of Color are Pushed Out of Environmental Organizations*. https://diversegreen.org/research/leaking-talent/

Kalandides, A., & Grésillon, B. (2021). The Ambiguities of "Sustainable" Berlin. *Sustainability (Basel)*, *2021*(13), 1666. doi:10.3390u13041666

Kam-Sing Wong, S. (2012). The influence of green product competitiveness on the success of green product innovation: Empirical evidence from the Chinese electrical and electronics industry. *European Journal of Innovation Management*, *15*(4), 468–490. doi:10.1108/14601061211272385

Karachaliou, P., Santamouris, M., & Pangalou, H. (2016). Experimental and numerical analysis of the energy performance of a large scale intensive green roof system installed on an office building in Athens. *Energy and Buildings*, *114*, 256–264. https://doi.org/.enbuild.2015.04.055 doi:10.1016/j

Klemm, W., Zinngrebe, Y., & Heidenreich, S. (2019). Green roofs as part of urban green infrastructure for innovative cities. *Environmental Innovation and Societal Transitions*, *31*, 50–56. doi:10.1016/j.eist.2018.11.003

Köhler, M., & Poll, P. H. (2010). Long-term performance of selected old Berlin greenroofs in comparison to younger extensive green roofs in Berlin. *Ecological Engineering*, *36*(5), 722–729. doi:10.1016/j.ecoleng.2009.12.019

Kolomiets, T. V., & Tomashuk, I. V. (2021). Entrepreneurship and development of rural areas in Ukraine. *Colloquium-journal*, *9*(96), 29–42.

Kucher, A. (2019). *Sustainable soil management in the formation of competitiveness of agricultural enterprises.* Academic Publishing House «Talent». doi:10.13140/RG.2.2.19554.07366

Kucher, L., Heldak, M., & Orlenko, A. (2018). Project management in organic agricultural production. *Agricultural and Resource Economics, 4*(3), 104–128. doi:10.22004/ag.econ.281753

Kumar, V., & Christodoulopoulou, A. (2014). Sustainability and branding: An integrated perspective. *Industrial Marketing Management, 43*(1), 6–15. doi:10.1016/j.indmarman.2013.06.008

Li, J., Yin, Z., Li, Y., Sun, S., Li, M., & Li, Y. (2021). The application of green roofs in urban green innovation spaces: A case study in Jinan City, China. *Environmental Science and Pollution Research International, 28*(3), 3525–3537. doi:10.100711356-020-11544-2 PMID:32892283

Li, X., & Norford, L. (2016). Evaluation of cool roof and vegetations in mitigating urban heat island in a tropical city, Singapore. *Urban Climate, 16*, 59–74. doi:10.1016/j.uclim.2015.12.002

Liao, S. H., Chang, J. C., Cheng, S. C., & Kuo, C. M. (2004). Employee relationship and knowledge sharing: A case study of a Taiwanese finance and securities firm. *Knowledge Management Research and Practice, 2*(1), 24–34. doi:10.1057/palgrave.kmrp.8500016

Lin, Y. H., & Chen, Y. S. (2017). Determinants of green competitive advantage: The roles of green knowledge sharing, green dynamic capabilities, and green service innovation. *Quality & Quantity, 51*(4), 1663–1685. doi:10.100711135-016-0358-6

Liu, D., Jiang, K., Shalley, C. E., Keem, S., & Zhou, J. (2016). Motivational mechanisms of employee creativity: A meta-analytic examination and theoretical extension of the creativity literature. *Organizational Behavior and Human Decision Processes, 137*, 236–263. doi:10.1016/j.obhdp.2016.08.001

Liu, J., Chen, J., & Tao, Y. (2015). Innovation Performance in New Product Development Teams in China's Technology Ventures: The Role of Behavioral Integration Dimensions and Collective Efficacy. *Journal of Product Innovation Management, 32*(1), 29–44. doi:10.1111/jpim.12177

Llonch, P., Haskell, M. J., Dewhurst, R. J., & Turner, S. P. (2017). Current available strategies to mitigate greenhouse gas emissions in livestock systems: An animal welfare perspective. *Animal, 11*(2), 274–284. doi:10.1017/S1751731116001440 PMID:27406001

López-Gamero, M. D., Claver-Cortés, E., & Molina-Azorín, J. F. (2008). Complementary resources and capabilities for an ethical and environmental management: A qual/quan study. *Journal of Business Ethics, 82*(3), 701–732. doi:10.100710551-007-9587-x

Maas, C. J., & Hox, J. J. (2005). Sufficient sample sizes for multilevel modeling. *Methodology: European Journal of Research Methods for the Behavioral and Social Sciences, 1*(3), 86–92. doi:10.1027/1614-2241.1.3.86

Marchi, V. D. (2012). Environmental innovation and R&D cooperation: Empirical evidence from Spanish manufacturing firms. *Research Policy, 41*(3), 614–623. doi:10.1016/j.respol.2011.10.002

Maruani, T., & Amit-Cohen, I. (2007). Open Space planning models: A review of approaches and methods. *Landscape and Urban Planning, 81*(1–2), 1–13. doi:10.1016/j.landurbplan.2007.01.003

Mazur, K. V., & Tomashuk, I. V. (2019). Governance, and regulation as an indispensable condition for developing the potential of rural areas. *Baltic Journal of Economic Studies*, *5*(5), 67–78. doi:10.30525/2256-0742/2019-5-5-67-78

Meerow, S., & Newell, J. P. (2017). Detroit, Spatial planning for multifunctional green infrastructure: Growing resilience in. *Landscape and Urban Planning*, *72*, 62–75. doi:10.1016/j.landurbplan.2016.10.005

Mell, I. C. (2010). *Green infrastructure: concepts, perceptions, and its use in spatial planning* [Doctoral Thesis]. School of Architecture, Planning and Landscape, Newcastle University.

Mentens, J., Raes, D., & Hermy, M. (2006). Green roofs as a tool for solving the rainwater runoff problem in the urbanized 21st century? *Landscape and Urban Planning*, *77*(3), 217–226. doi:10.1016/j.landurbplan.2005.02.010

Mesimaki, M., Hauru, K., Kotze, D. J., & Lehvavirta, S. (2017). Neo-spaces for urban livibility? Urbanities' versatile mental image of green roofs in the Helsinki metropolitan area, Finland. *Land Use Policy*, *61*, 587–600. doi:10.1016/j.landusepol.2016.11.021

Mohd, H. A. & Norhidayah, S. (2016), Sustainable food production: insights of Malaysian halal small and medium sized enterprises. *International Journal of Production Economics, 181*(part B), 303–314. . doi:10.1016/j.ijpe.2016.06.003

Murray, N. (2017). Urban disaster risk governance. A systemic review. PPI-Centre. UCL Institute of Education.

Namkung, Y., & Jang, S. (2013). Effects of restaurant green practices on brand equity formation: Do green practices really matter? *International Journal of Hospitality Management*, *33*(2), 85–95. doi:10.1016/j.ijhm.2012.06.006

Nogué, S., & Sala, M. (2018). The use of green roofs in urban green innovation spaces: A case study in Barcelona. *Urban Forestry & Urban Greening*, *33*, 19–29. doi:10.1016/j.ufug.2018.02.008

Norton, T. A., Parker, S. L., Zacher, H., & Ashkanasy, N. M. (2015). Employee green behavior: A theoretical framework, multilevel review, and future research agenda. *Organization & Environment*, *28*(1), 103–125. doi:10.1177/1086026615575773

Ones, D. S., & Dilchert, S. (2012). Employee green behaviors. In D. S. S. E. Jackson (Ed.), *Managing human resource for environmental sustainability*. Jossey-Bass.

Palomino, J. C., & Durán, P. (2020). The influence of green roofs on the urban heat island effect: A review of the current state of knowledge. *Urban Climate*, *34*, 100696. doi:10.1016/j.uclim.2020.100696

Pereira, Á., & Vence, X. (2012). Key business factors for eco-innovation: An overview of recent firm-level empirical studies. *Cuadernos de Gestión*, *12*, 73–103. doi:10.5295/cdg.110308ap

Rahman S, Ahmad H & Rosley M. (2013) Green roof: its awareness among professionals and potential in Malaysian market. *Procedia Social and Behavioral Science, 85*, 443–453. https://doi.org/.sbspro.2013.08.373 doi:10.1016/j

Raudeliuniene, J., Davidaviˇciene, V., & Jakubaviˇcius, A. (2018). Knowledge management process model. *Entrepreneurship and Sustainable Issues*, *5*, 542–554.

Redman, C. (2014). Should sustainability and resilience be combined or remain distinct pursuits? *Ecology and Society*, *19*(2), 190–202. doi:10.5751/ES-06390-190237

Reisinger, A., & Clark, H. (2018). How much do direct livestock emissions contribute to global warming? *Global Change Biology*, *24*(4), 1749–1761. doi:10.1111/gcb.13975 PMID:29105912

Renwick, D. W. S., Redman, T., & Maguire, S. (2008). Green human resource management: A review and research agenda. *International Journal of Management Reviews*, *10*(1), 1–18. doi:10.1111/j.1468-2370.2011.00328.x

Rowe, D. B. (2011). Green Roofs as a means of pollution abatement. *Environmental Pollution*, *159*(8-9), 2100–2110. doi:10.1016/j.envpol.2010.10.029 PMID:21074914

Savickiene, J. & Miceikiene, A. (2018). Sustainable economic development assessment model for family farms. *AgriculturalEconomics – Czech, 64*, 527–535. doi:10.17221/310/2017-AGRICECON

Shafique, M., Kim, R., & Rafiq, M. (2018). Green roof benefits, opportunities, and challenges– A review. *Renewable and Sustainable Energy Reviews, 90*, 757-773.

Shevchenko, A., & Petrenko, O. (2020). Current state of micro and small agribusiness in Ukraine. *Agricultural and Resource Economics*, *6*(1), 146–160. doi:10.51599/are.2020.06.01.10

Shoeb, A., & Nisar, T. (2015) Green Human Resource Management: Policies and practices, *Cogent Business & Management, 2*(1), 1030817-103. https://www.tandfonline.com/doi/full/10.1080/23311975.2015.1030817 doi:10.1080/23311975.2015.1030817

Singh, S. K., Del Giudice, M., Chierici, R., & Graziano, D. (2020). Green innovation and environmental performance: The role of green transformational leadership and green human resource management. *Technological Forecasting and Social Change*, *150*, 119762. doi:10.1016/j.techfore.2019.119762

Skydan, O., Nykolyuk, O., Pyvovar, P., & Martynchuk, I. (2020). Methodological approach to the evaluation of agricultural business system flexibility. *Management Theory and Studies for Rural Business and Infrastructure Development*, *41*(4), 444–462. doi:10.15544/mts.2019.36

Stovin, V., Vesuviano, G., & Kasmin, H. (2012). The hydrological performance of a green roof test bed under UK climatic conditions. *Journal of Hydrology (Amsterdam)*, *414*, 148–161. doi:10.1016/j.jhydrol.2011.10.022

Stubbs, W., & Cocklin, C., (2008). Conceptualizing a sustainability business model. *Organizational Environment, 21*(2), 103-127.

Sutton, R. (2014). Aesthetics for green roofs and green walls. *The Journal of Living Architecture*, *1*(2), 1–20. doi:10.46534/jliv.2014.01.02.001

Tan, P. Y., Wong, N. H., Tan, T. K., & Wong, K. W. (2019). Green roof innovation in Singapore: Policies, performance, and potential. *Journal of Environmental Management*, *233*, 128–135. doi:10.1016/j.jenvman.2018.12.060

Tomashuk, I. V., & Baldynyuk, V. M. (2021). Identification of problems and prospects of rural infrastructure development of Ukraine. *Colloquium Journal,* 13(100), 58-70. doi:10.24412/2520-6990-2021-13100-58-70

Ugai, T. (2016). Evaluation of sustainable roof from various aspects and benefits of agriculture roofing in urban core. *Procedia: Social and Behavioral Sciences*, *216*, 850–860. doi:10.1016/j.sbspro.2015.12.082

Van Holt, T., Statler, M., Atz, U., Whelan, T., van Loggerenberg, M., & Cebulla, J. (2020). The cultural consensus of sustainability-driven innovation: Strategies for success. *Business Strategy and the Environment*, *29*(2), 3399–3409. doi:10.1002/bse.2584

Vijayaraghavan, K. (2016). Green roofs: A critical review on the role of components, benefits, limitations, and trends. *Renewable & Sustainable Energy Reviews*, *57*, 740–752. doi:10.1016/j.rser.2015.12.119

Wang, Y., Hu, H., Dai, W., & Burns, K. (2021). Evaluation of industrial green development and industrial green competitiveness: Evidence from Chinese urban agglomerations. *Ecological Indicators*, *124*, 107371. doi:10.1016/j.ecolind.2021.107371

Wiering, M., Liefferink, D., Boezeman, D., Kaufmann, M., Crabbé, A., & Kurstjens, N. (2020). The Wicked Problem the Water Framework Directive Cannot Solve. The Governance Approach in Dealing with Pollution of Nutrients in Surface Water in the Netherlands, Flanders, Lower Saxony, Denmark, and Ireland. *Water (Basel)*, *12*(5), 1240. doi:10.3390/w12051240

Williams, N., Rayner, J. & Raynor, K. (2010). Green roofs for a wide brown land: Opportunities and barriers for rooftop greening in Australia. *Urban Forestry & Urban Greening*, *9*(3), 245–251. https://doi.org/.ufug.2010.01.005 doi:10.1016/j

Wirtenberg, J., Harmon, K. D., Russell, W. G., & Fairfield, K. D. (2007). HR's role in building a sustainable enterprise. *Human Resource Planning*, *30*, 10–20.

Wu, W., & Jim, C. Y. (2021). Urban green infrastructure and sustainable urbanization: A comprehensive review of the roles of green roofs. *Landscape and Urban Planning*, *212*, 104097. doi:10.1016/j.landurbplan.2021.104097

Wysocki, J. (2021). Innovative green initiatives in the manufacturing SME sector in Poland. *Sustainability (Basel)*, *13*(4), 2386. doi:10.3390u13042386

Zhang, Z., & Sundaresan, S. (2010). Knowledge markets in firms: Knowledge sharing with trust and signalling. *Knowledge Management Research and Practice*, *2010*(8), 322–339. doi:10.1057/kmrp.2010.22

APPENDIX

QUESTIONS

1. Why and how technological developments can minimize the occurrence of environmental problems through the adoption of green innovation and green roofs in urban areas?
2. What are the main technological factors that this study aims to identify in the use of urban green innovation spaces?
3. What are the main factors that this study aims to identify in the use of technology in green roofs?
4. What are some of the technological factors that drive the adoption of green roofs in horizontal property by means of a technology acceptance of the place integrated identity model.
5. Under what assumptions are sustained the technological dynamics of the urban green innovation?
6. How can be described the perceived use of a technology in urban green innovation areas when addressing concerns such as attitude toward use, intention to use, perceived ease of use, and perceived usefulness?
7. Which methodology you may suggest being used to determine the perceived use of technology in green roofs?
8. For an efficient use of technology in urban green innovation areas and green roofs, which criteria suggest helping to minimize the chaos, the occurrence of environmental breakdowns, such as air pollution, increased temperature, urban heat islands, etc.
9. How could technology help to address such socio-ecological problems through the implementation of simple strategies?

Chapter 11
Green Marketing Among Z-Generation (Z-Gen.)

Abhinav Agrawal
Agrasen Mahavidyalaya, India

Minal Shah
Bhilai Institute of Technology, India

Dharmendra Kumar Gangeshwer
Bhilai Institute of Technology, India

ABSTRACT

Environmental problems have become prime global challenges, and in developing countries like India, there is a need for high attention to curb these problems. These problems bring unsustainable economic and environmental development, and global sustainable growth has been a major concern of G-20 countries. For this purpose, in every economy, the business organizations, producers, and consumers can play their roles to protect the environment through their business and buying activities. Manufactures and businesses can play vital roles through production of green product and green marketing activities, and buyers can play their role through demanding such product as these products are proven to be good for the environment as well as for the health for consumers and also attract maximum profit for the business organizations. Marketers can target young consumers aged between 11-26 called "Z-Generation" consumers because they are kind of consumers who are more aware of the environmental issues and express their will to protect mother nature.

INTRODUCTION

The problem of increasing global environment has become a matter of concern today; if we couldn't get rid of with this problem today, then it will bring negative consequences in the future. All nations striving for curbing this problem. For a developing country like India, it has become even more important that it does not allow this problem to grow; as per Environmental Performance Index 2022 in environment

DOI: 10.4018/978-1-6684-9503-2.ch011

performance India has ranked lowest among 180 countries due to existence of many problems in country including climate change, water pollution, global warming, unorganized waste disposal system etc. Hence it is necessary that these problems should be resolved on priority basic. In some extent to overcome these problems business organizations and consumers can play their vital role through producing green product and purchase of green products respectively; Business and manufacturers can practice green marketing and shifting their product to green products and by creating demand for such products through advertisement and various green promotional activities. So that by using green advertising, packaging etc., sustainable development can be done in the country.

Green marketing has become essential for demand todays demand creation for marketers as many business organizations have adopted this method of marketing which has become key competition factor. To survive in business it is now basic requirement for all businesses to go green with production of green product and its marketing, some of the brand like PATANJALI and MAMAEARTH have made strong brand image among the buyers mind with its herbal and ecofriendly characteristics, and also proven tough competitors for non-green products and made them helpless to follow the trends for instance since its launch PATANJALI Herbal toothpaste gain tremendous popularity among its consumers resultant the big brand COLGATE has to launched its herbal toothpaste too and plenty of such instance are there. MAMAEARTH believe in delivering products which are gentle as with usage of natural, plant-based ingredients and this practice of this brand has made it most popular and used brand whereas there was top existing brands already had strong market and now they are also shifting towards green marketing. It exhibits how important is to go green for marketers for long run survival and targeting "Z-Generation" as segmented population can be more beneficial for any economy as Z-Generation is in the driver's seat of sustainable environmental and economic development vehicle. So that marketers must focus green marketing towards the Z-Generation consumers along with the all age group consumers but the prime focus must be this smart generation "Z-Generation".

Marketers can go green in marketing strategies that positively influence consumers' perceptions of such activities (McDonald, S., & Oates, C. J., 2006). Under green marketing, those items are marketed and sold which can contribute to the protection of the environment and to improve the environment, so by changing its current marketing activities with green marketing marketers can boost their profitability with contribution towards environmental protection. Under green marketing, green Promotion of green products and changes in advertising, packaging, and production of green products etc. are included.

Along with time interests of the consumers of each age may vary. This variation can also be seen in their today's education level, there has been a lot of change and developments witnessed in the level of education and the quality has increased in the field of education, whose direct effect can be seen in today's generation thinking. "Z-GENERATION" is receiving its education with technological advancements and gaining a lot of knowledge from its former generation and this generation is such a generation which has the ability to determine the future demand also, so it is necessary that by targeting these generations Marketers should develop their products so that they can increase the demand for their products in the present as well as in the future and the increasing demand for green products can be seen as an opportunity in front of the business. Businesses can attract the people of the new generation through the development of green products and can create demand for green products through its various activities. Along with being educated, this generation understands their responsibility for the protection of the environment very well; there has been a significant change in the thinking of the present generation. These changes accelerated post corona pandemic the effect of the pandemic has made people more aware towards their immunity and now they want to choose such product which will be good for their health in future also.

SOME OF GREEN PRODUCTS AND SERVICES

Products

Clothes made of recycled plastic, Reusable Cloth or Cotton Shopping Bags, Stainless-steel Water Bottle, Reusable Cloth or Cotton Shopping Bags, LED bulbs, E-vehicles, Eco-friendly House Cleaners, Furniture's made of recycled cork, Solar Water Heaters, Energy Saving Electronic appliances, Reusable Bamboo Utensil, Eco-friendly Herbal FMCG products, Rechargeable batteries, Organic products, Ayurveda Medicine, Eco-Friendly Paint made by cow dung etc.

Services

Solar panel (Renewable energy), Agro-tourism, Car sharing, Usage of Public transport, Rain water harvesting, etc.

Among above listed green product and services so many product and services have got tremendous growth and demand among the consumers some of them are LED bulbs, E-vehicles, Rechargeable batteries, Furniture's made of recycled cork, Solar panel etc. it exhibits the future growth and demand for all the existing green product and services and also upcoming new green product and services through innovation and research and development activities of the businesses.

WHAT IS GREEN MARKETING?

Green Marketing also called "Eco-Marketing" concerned with the marketing of environmental friendly products on the basic of their importance on environmental sustainability.

In other word Green Marketing is a type of eco-friendly marketing, where product and services are promoted on the basic of their environmental benefits.

Green Marketing Includes

- Manufacturing Eco-Friendly products.
- Packing the products in sustainable manner.
- Following sustainable business Practices.
- Promoting the green benefits of existing products being produced by the company.
- Using recycled materials in production.
- Usage of green energy.
- Product waste Reduction.
- Usage of Eco-friendly methods.

According to the American Marketing Association, green marketing is the marketing of products that are presumed to be environmentally safe. Thus green marketing incorporates a broad range of activities, including-

1. Product modification,
2. Changes to the production process,
3. Packaging changes, as well as

Green marketing must be more than a green way of marketing of the so-called green products. Green has to refer both to the method and to the product. This is why the very idea of green marketing needs considerable development and analysis, with rules and integrity in economic, scientific, academic and ethical terms. (Kingis)

Green or Environmental Marketing consists of all activities designed to generate and facilitate any exchanges intended to satisfy human needs or wants, such that the satisfaction of these needs and wants occurs, with minimal detrimental impact on the natural environment. (Polonsky, 1994)

Green Marketing is broader concept to study, it is associated with consumer goods, services as well as industrial goods, green marketing is correlated with the words like recyclable, eco-friendly and refillable.

HISTORY, IMPORTANCE, BENEFITS, AND PROBLEMS OF GREEN MARKETING

History of Green Marketing

Green marketing concept assumed to be introduced in the 1970s and the idea of green marketing come into force in 1980s. It was started in Europe in early 1980s when it is found that some of the manufactured goods are harmful to the environment. From that time green marketing has passed through three phases.

First phase of green marketing is called "Ecological" green marketing, under this phase, various activities of marketing focused on providing solutions of various environmental problems. Producers started using green marketing tool to fulfill consumers' needs and wants. They assumed that consumers will buy green eco-friendly products and by this the organizational credibility will be increased and they will be able to capture large share in the market, but this idea didn't work because of Green washing, it was only shown by businesses that they are highly engaged on only showing that their products are eco-friendly but actually it was not a ground reality.

The second phase started when backlash was seen by the marketers, the second phase was recognized as "Environmental" green marketing. The focus point during this phase moved towards clean technology, which was related to designing new products which were not harming the natural environment.

In mid-1990s, people become aware about their role for environmental protection and preservation. People started considering about the environmental problems. This introduced third phase. It termed

"Sustainable" green marketing later on. Buyers were more interested to buy product or avail services which were less harmful for the environment; organizations had to change their behaviors of selling.

GREEN PRODUCTS AND GREEN CONSUMERS

In developed western countries people have become more conscious about the protection of environment. Asian regions also following this trend. So that consumers all over the world are becoming more conscious about the protection of environment. Green products are kind of products which never degrade the environment and use the natural resources and can be recycled. Recently the demand for green products are increased. The prevalence of Ayurveda and herbal stores and Ayurvedic knowledge in India has forced many foreign companies to make herbal products. For instance Wide range of herbal products PATANJALI run by Well-known personality Yog Guru Baba Ram Dev have given tough competition to many foreign companies FMCG products and forced them to make herbal products too. This clearly shows people's belief and trend on green products has increased. For long run survival of any business it has become important to go green and have some responsibility toward the Mother Nature with their produced and this move by the business can take them towards high rate of success.

TYPES OF GREEN CONSUMERS

There are several types of consumers to motivate the business to perform marketing activities for the green products. And it is important for the business to understand their consumers for the success.

There Are Four Types of Green Consumers

Behavioral Greens Consumers

This type of customer buys those items which have a positive image in their eyes. They do not want to buy all those products which pollute the environment. In their daily life they follow green practices and inspire others to buy products that are right for protecting the environment.

Think Greens Consumers

Consumers who buy green product when they find it affordable, and if it is out of budgeted item they easily skip buying that product.

Potential Greens Consumers

They are the kind of consumers when they are influenced or motivated by their family, friends, collogues to buy green products so they pretend like green product buyer or some time they buy green product for the sake of the satisfaction of their peer group. Such kind of consumers can become future regular consumer of eco-friendly products.

True Brown Consumers

These kind of consumers are less active consumers and does not show their responsibility towards the environment and just ignore the product which is environment friendly may be due not enough knowledge about the future risk if environment are not protected.

GREEN CONSUMER ATTITUDE AND BUYING BEHAVIOR

As green marketing has developed, so have the consumers changed their buying behavior. Now more and more green products and services are being demanded by the consumers. Consumers are converting their concern for the protection of their environment into green buying behavior. Today, many consumers, especially young consumers, understand their responsibility towards the environment very well and are behaving accordingly. But this habit has not been fully developed in many consumers. Many consumers are motivated to buy green products only partially or occasionally. There is a need to identify these types of consumers and motivate them more.

The income of the consumer also affects their buying behavior. Consumers with high income are able to spend extra money to buy any product, while having low income, the consumer considers it appropriate to take those things which come in their budget, and in such a situation those green products which are a bit expensive are ignored by the consumers. Apart from income, the effect of education is also seen in the behavior of the consumer. A highly educated person understands his responsibility towards the environment better and executes his consumer behavior accordingly.

IMPORTANCE OF GREEN MARKETING

Today, any organization can increase its profit potential by green marketing; every organization can bring a new revolution today by using various limited resources of production and promoting green products. As we are living in an era of recyclable, non-toxic and eco-friendly goods and services availability. Green marketing has become a new tool for businesses to fulfill consumer needs and wants and generate maximum profits. There is a need to recognize this new opportunity and ethically choose and produce the product while protecting the environment and human health. Thus they can lead in the competition as well. Today, the trend of natural products among consumers is increasing very fast, an epidemic like Corona has taught everyone that if we want to keep our immunity strong, then we need the right food and drink healthy. Therefore, an organization needs to realize the importance of green product which can increase its profit by producing healthy eco-friendly products.

Statistical Data That Supports Importance of Green Marketing

According to Business Research Insights

"The global Green Marketing Market is expected to grow from 49250.7 million in 2021 to USD 60811.48 million by 2027, at a CAGR of 3.58% from 2021 to 2027."Green Marketing Market Provides Such by

Type-Broker, Reseller, Manufacture, such by Application-Health, Food & Agriculture, Natural Resources & Environment, Industrial Processing,

According to Shiksha.com

1. India has provided 100% electricity country's villages, decreased 38 million tonnes of annual CO2 emissions. By 2030, it is planned by government plans to create 450GW of renewable energy and also restoration of 26 million hectares of degraded land by the year 2030.
2. Many FMCG (fast-moving consumer goods) brands are also focusing environment-friendly practices which exhibits the importance of green marketing these companies includes
    - Colgate-Palmolive is Offering recyclable toothpaste tubes.
    - Amazon India has Remove single-use plastic usage in packaging and replaced it with paper cushions.
    - Tata Metaliks Limited has Increased greenery in areas where mining and metal industries are located, through natural fertilizers, water-neutral operations, etc
    - LG Electronics runs the "Green Program Plus (GP Plus)" which is a sustainability centered supply chain management program. Any supplier of LG must go through its rigorous sustainability test.
    - Another FMCG brand, The Body Shop, has been focusing on reducing its carbon footprint for decades. The cosmetics brand from England claims that it has never tested products on animals. If you didn't know, 88% of the 50 largest cosmetic companies in the world fund animal testing

Greening the Organization

An organization can "green" themselves by three ways

1 Value-adding Processes – Under value-adding processes the organizations can update itself with adoption of new technology, or introducing new technologies for reducing the negative impact of the environment.

2. Management System- Adopting management system which can create the environment for reducing the negative impact of environmental on value-adding processes.
3. Products- Using eco-friendly raw materials to produce any product can bring high level of satisfaction among society, so that careful selection of product is the key factor for going green for any organization.

GREEN MARKETING MIX

The problem of availability of limited means of production and their various uses is a common problem of every economy. Degradation of natural resources calls for a proper marketing mix. Along with this, the consumers are provided with such products at reasonable prices which are not harmful to their health as well and protection of nature has also become an essential aspect. In the present and future, only those products will attract people towards itself, which are safe for them as well as for Mother Nature.

Elements of Green Marketing Mix

Price

Price is the consideration received by seller in the terms of offering goods or services on the behalf of sales. For the success of any product or services price plays a vital role. If the price is charged too high to the competitive product then gaining consumers can be difficult and unachievable task for any organization; except the product is differentiated one than the competitors' products. For green product sale the price factor can play a key role as the buyers find most of the green products costly and if the price penetration strategy is practiced by the marketers when they introduce any new green product then they can enjoy competitive advantage in the market.

Product

Product is called the physical goods which are offered to fulfill the need or wants of consumers; and the green products are those products which is harmless for the environment as well as for the human health and body. Most of the green product carries recyclable characteristics. They not only protect the environment but have full capacity to satisfy the needs and wants of the consumers. Green marketers can use eco-friendly packaging for protection of green product to make it fully eco-friendly product.

Package

It is an activity which protects the product from any damage and also helps to carry the product from one place to other. The packaging also should be harmless for the environment as well as product itself. Packaging should not include use of plastic bag or any box made by plastics instead eco-friendly packing material should be opted by the marketers.

Promotion

Promotion is a tool to influence, motivate and inform the prospective buyers for positive buying decision making. Promotion includes the audio-visual advertisement, promotion through internet, websites, social media platform etc. for promotion selection of advertising agency which also promote and believe on going green initiatives is better because the image of promotional indirect tool can also put impact on the organizational product. If promotion is performed through superstars who personally understand and actively participate on environmental protection activities can be appointed as a brand ambassador for the product promotion. It can be more influential among the targeted market.

People

It is mandatory for the organization to teach its manpower with the all aspect of environmental issues and their responsibilities to protect it. The man power must have knowledge about the eco-friendly product being produced and offered by the organization. The practice for going green can be performed by the employees in organizations by their daily operational activities like using less paper using electronic means for various activities.

Distribution

The distribution channel of green products should create positive image of organization among the buyers mind. Harmful means for distribution of goods should be avoided for placing the goods for sale. Currently electrical vehicles are getting popularity so the organizations can use such vehicles for local distributions and if possible it can be used for distanced distribution activities too. This practice can create positive image of the organization as well as the products being offered by the organizations among prospective consumers mind.

After Sales Service

After-sales Services are referred with providing services after performing sales. If consumers encountered with any issue on product then it is the responsibility of the organization to get the product repaired or if possible replace it with the new one. And providing required guideline for the usage of product, if required then demo of using the product must also be provided for making good brand image.

BENEFITS OF GREEN MARKETING

Many organizations have opted go green policy because of changing trends and human psychology. They focus on gaining competitive advantages by producing innovative eco-friendly products and catching attraction of new customers. Especially young consumers as they are considered more environmentally educated and conscious for protecting the environment. Some important benefits of green marketing are as follows:

1. Sustainable and log-term growth

For the sustainable and long-term development of any organization, it is necessary that they recognize their role towards the environment and society, and through their work they can acquire the image of an organization that guarantees the protection of the environment and the safety of its customers in all respects. Through green marketing the organization can gain trust. Post Corona Pandemic, people have become very conscious about the environment and their health, and have also started demanding the available green products. By increasing this demand further through green marketing, each organization can increase its profit manifold by contributing in the interest of the environment.

2. Helps in capture new market

Through green marketing, the organization can find new markets and gain market share. As it has become clear from many researches that in future only those firms will earn more profit which will keep doing something new. Through green marketing today and in the future too, by bringing new products in the market, the organization can achieve its new position in the market.

3. Helps in gaining customer loyalty

Through green marketing, the organization can achieve loyalty among its customers, continuous provision of eco-friendly products by the organization and adopting such a policy, which can make the all-round development of the country and society possible, must be successful in creating loyalty in the minds of its consumers. In India many products have been able to create a good image, one of the most popular brand among them is PATANJALI, which provides wide range of herbal products getting high popularity among the buyer since it is launched. The company has created loyalty among its consumers which influence so many organizations to go green policy.

4. Helps in increasing brand image

Brand image is considered with the image or reputation of the product being offered. It is about consumers experience about the product and what they feel about the brand. And the consumer's experiences are influenced by various factors, such as quality of the product and the impression that company made through advertisement and customer services provided by the company. Brand image can change time to time because the opinion about the brand changes time to time. It is green product which can help to organization to build a positive brand image among the consumers because the environmental and human conscious characteristics. And positive brand image is something which can bring high profit potentiality for the organization.

5. Competitive advantages

Competitive advantage is a kind of favorable factor for an organization which leads them to be more profitable than their competitors. And to enjoy competitive advantage, an organization has to be unique and innovative on its marketing strategies and product features. If any organization offers its product for a price less than their competitors and adding some new features than its rival's products then they can have more advantages in the term of profit and customer loyalty. Offering green marketing is something which is unique and emerging concept and if any organization actively adopts it then they can definitely enjoy competitive advantage and become more profitable.

6. Self-satisfaction

Green marketing can bring self-satisfaction for the organization. Offering high quality product in affordable price range with environmental friendly characteristics brings self-image as organization knows they are offering product which can create high value in consumers mind, and all organization wants to gain loyal consumers. Satisfied consumers motivate others also through positive mouth publicity. And this increases the market share and profit of organization so that organization can feel more self-satisfied.

PROBLEM OF GREEN MARKETING

1. Less consumer awareness on green products

The biggest problem in green marketing is that even today there is a lack of awareness among people about green products, people still do not express their willingness to accept green products, because many people believe that Regular products are better than green products, they are better for whatever they are using, so a problem in green marketing is the awareness of the products that can be seen among the people. Many customers get information about green products from many places, even though they have not been able to include green products in their daily purchases; the reason is also their habit, which makes them unable to choose and use a new product. For example, today we all know about the harm caused by plastic to the body, even after this, many customers do not use steel bottles and still use plastic bottles, in addition to this in daily use. Plastic is used in many works, it is a big problem that even knowing that we are not able to use green products.

2. Resources problem

One of the problems before green marketing is also the problem of resources, because today many organizations want to include green products in their production work, but there are many resources which are not available, due to their non-availability, Green products cannot be produced. As we know that the means of production are limited and its usage are unlimited, so the problem always remains before a producer that how to use these limited means of production optimally. It would also not be wrong to say that only green product alone cannot satisfy the needs of the consumers because today there are many such products which are not possible to be produced as green products because the means of production are also not available for this. One of the biggest challenges before green marketing always remains.

3. Heavy investment requirements

The production of green product requires a lot of investment, which appears as a problem for a producer. Producers want to upgrade their production and give place to the green product, but for this they need heavy investment, which is sometimes that not possible for the producer to spend in highly expensive machines, technology etc. It may not be possible for all the producers; which appear in the form of a problem of green marketing.

4. Price centric consumers

The price of any product has a direct impact on the demand; price consciousness negatively affects purchase intentions (Sun, Y., and Wang, S., 2019). The effect of the price of green products can also be seen on the consumers. The price of green products is usually slightly higher than non-green products, which prevents consumers from purchasing organic products. It also remains a serious problem before the producer as to how to sale the product to the customers at a given price and create demand for it. Due to this, many consumers avoid purchase of green products even if they want to.

5. Competitive pressure

Green product as it is emerging as a new concept, therefore it has to face tough competition to make its place in the market as the available product has made its mark in the eyes of the consumers and the consumers have been persuaded to purchase the product. Competition is a formidable problem, so facing competitors remains a big challenge for green products and green marketing work, so it is necessary that through green marketing, more and more awareness should be increased among the people that how green products can be profitable for them as well as with product differentiation the green product can be differentiated from the competitors product so that existing competition can be faced.

6. Availability of substitute product

Increasing demand for a product is hindered by the availability of an alternative product because the customers feels that the product may not be able to satisfy their need or its price is higher than the price of the substitute product or there is a difference in quantity or quality. So the customer usually turns towards to substitute products, which is a common problem. To overcome with this problem, it is necessary that the importance of green products in human life should be explained to the human beings and for their good health and sustainable environmental development.

WHAT IS "Z-GENERATION" (Z-GEN)?

According to Pew Research 11–26-year-old age group of people fall under the category of "Generation- Z". They are more technology-run generation grown up with access to the internet and electronic devices such as- Smart-phone, Notepad, Lap-top, Computer etc. This generation is faster than its previous generation and having more innovative and creative characteristics. Gen-Z teenagers are more concerned with their academic performance and moral values and self-image, and enjoying their adulthood with having high buying power in present as well as in future first generation.

CHARACTERISTICS OF "Z-GENERATION"

Technology-Driven Generation

The member of "Generation Z" is more driven by Technology and electronic devices say- Smart-phone, I-phone, laptop, computer, electronic book reader, notepad, laptop etc.

Future Economic Contributor

"Z-GENERATION "carries the characteristics of present as well as future buyers their present as well as future contribution for any economic development cannot be neglected. If they are targeted by advertising then it can be very much positively impactful for the growth of any product or services.

Educated and Well-Behaved

With growing advanced education system whether private or government the level of education has grown significantly keeping overall development of students moral and academic aspect, one subject Environmental studies has always been integral part of each class which have made this generation more concerned and personally conscious towards the environmental protection.

Self-Motivated

"Z-GENERATION" falls under young and teen aged consumers who believe in their own values and self-concept, they are highly influenced by marketing activities of organization and by their peer group; they do what they think appropriate to fulfill their need and wants. Days are gone when parents used to decide what product will be bought for their children. Today's generation is smart enough to take this decision as they plan their purchase according to their self-need.

Upbringing

Upbringing of "Z-GENERATION" population mostly done by their parents with extra care as they early 21st century parents with high moral and education level, impact of upbringing can be seen among "Z-GENERATION" as they are more concerned about the cleanliness and hygiene for choosing any product.

Digitally Literate

"Z-GENERATION" is called the smart generation as they are more digitally literate. Many school's teaching learning methodology is now based on ICT which makes compulsory the usage of Lap-top, Notebook, smart-phone and these devices have made learning too easy among "Z-GENERATION" they can access any knowledge or information with one click on smart devices, days are gone when for information or knowledge one has to depend on high cost and inaccessible books with access of internet and world wide web now students can get any information in no time.

Social Media Platform Users

Use of social media has become key part of daily lives of ""Z-GENERATION"". By mobile internet technology they stay connected virtually with their friends, Relatives through various social media platform as- Facebook, Instagram, Whatsapp, Twitter etc.

E-Shopper

"Z-GENERATION" are extensively engaged in online shopping, they are called smart generation as this innovative shopping platform are introduced and got pick up in their era so that they are more confortable to use online shopping through various online shopping platform like- Flipkart, Amazon, Ajio, Myntra etc. through these platforms they also read various reviews given by users already bought and used the product, also they can check the rating of the product which they are willing to buy.

Influencer

With their creative and innovative thinking they have power to influence others. "Z-GENERATION" has to capacity to influence their friends, family as they are more conscious about the environmental protection as they have been studying the subject in their school and college which motivates them to influence others through their knowledge and concern for the environment.

WHAT MARKETERS NEED DO UNDERSTAND ABOUT "Z-GENERATION" CONSUMERS?

Environmental Knowledge of "Z-Generation"

The marketer should more focus on advertising which hit the knowledge of "Gen-Z" there is strong need to understand the knowledge of "Z-GENERATION" about their knowledge on environmental issues, Their concern over the product safety and the level of information they possesses about environmental friendly products.

Self-Image or Green BRAND IMAGE

Self-image concerned with the personal image of product or services in the mind of buyer. There is need to understand the self-image of product among "Z-GENERATION" whether the product being offered or going to offered suits the personal image of "Generation -Z", there concern over the protection from using the product, self-concept, showing their role towards becoming environmental friendly organization through introducing environment friendly product so that they can stay positive in "Z-GENERATION" mind image.

Environmental Concern

Environment concern means the role and alertness towards the responsibilities for protecting environment, thinking about maintaining the balance with the nature for surviving. What "Generation -Z" concern for environment is needs to understand and concern can be created through advertisement. The best example for the creation of concern by on of leading green product is namely "MAMA EARTH" the company created the concern in the mind of parents for their babies by showing the importance of green baby product with no harming characteristics on baby skin. And the product got high growth and success among the parents.

Labeling and Branding

It is assumed to be understood the nature of "Generation -Z" while selecting or opting the product, means are they willing to buy product with recycling label. If yes then marketers marketing should be more focused on highlighting the eco-label and eco-brand characteristics and if "Generation -Z" does not check or concern about the labeling and branding then for gaining competitive advantage this characteristics can be highlighted through the advertisement so that importance of green product can be stored in "Generation -Z" mind.

Green Product Purchase Intention

Learning "Z-GENERATION" intention driving variables or factors like their thinking about the green products, less anti-pollution characteristics, spending power and role of their peer-group, community, friends, teachers, family etc. are advisable for the marketers for generating future demand.

GENERATION-Z INFLUENTIAL FACTORS FOR GREEN PRODUCT PURCHASE

Green Product Awareness

Green products are produced through natural process and in non-toxic manner. Awareness of such product is important for consumers as these products are environmental friendly and good for human health point of view. Awareness of such products can be created through advertisement, Packaging and labeling. Existing consumers know the importance of green product for the conservation of environment and their role for the product for their better health. But there is need to aware large number of prospective and long lasting buyers for continuous demand for this kind of life saving products. Marketers can target young "Generation -Z" consumers keeping in mind their dual characteristics of present and future buyer and long lasting demand creator of such goods. There is high need to convert these "Generation -Z" consumers into green consumers if any nation wants sustainable overall development; awareness of such product can significantly and positively influence all category of consumers specially the seed("Generation -Z") consumers, who will become tree in future with its wide branches.

Price of Green Product

Criteria of price are the key hindrance on green product purchase, but consumers can get ready to pay premium price for any product if they find it useful for them and their family. Consumers can react positively towards the eco-friendly product if it fulfill their need and want equally like non eco-friendly product and can pay even extra as they know the importance of that product for their health and for environment. For "Generation -Z" price matters a lot but if they realize the value of product for them then they can also feel comfortable to pay premium price for the product.

Availability of Green Product

Awareness is the first task before the availability of products. Various studies show that the awareness rate of green products is higher than its availability. Consumer decision to buy any product depends upon its availability, green product availability can motivate the consumers for buying decision making and availability also creates a positive impact among the users. If there is a weakness in the availability of green products then it will not attract green consumers and it will lead to no demand for such products. No availability leads to no purchase, so green product availability is a key factor in "Generation -Z" consumers' attention for the purchase of green products. If demanded or required product is available in the market then it can influence "Z-GENERATION" consumers on their intention to buy green products.

Green Product Value

Green product value is the imagination of consumers on the overall benefits from usage of green products. Green products values are highly correlated with the intention to purchase green products. Consumers are attracted to buy product if they feel and assume that the product will bring overall high value for them. Value of any prospective product are evaluated by consumers through deep research before taking decision for buying that product so that the marketers must reflect in its promotional activities about the values of product that consumers can get and why they should go for green product purchase. So that "Generation -Z" can also have positive influence on their purchase intention by the value they get from green products.

Green Product Quality

The product being advertised as a green and eco-friendly product must felt green by its users or consumers also by its quality. Quality is the key intention maker factor for green consumers on the purchase decision. If eco-friendly consumers feel that the usage of product certainly helpful on environment conservation along with providing utility and satisfaction on their usage then it can positively influence them to be a regular buyer of such products. So that it can be said that green perceived quality can positively influence the "Generation -Z" consumers on their green purchase intention.

WHY IS THERE A NEED FOR GREEN MARKETING AMONG "Z-GENERATION"?

India aimed to boost up G20' contribution for achieving 50 percent reduction in the term of degraded land following the Biodiversity and Land Degradation them along with enhancing global efforts to foster development and growth with pursuing strong climate ambitions. To curb environmental degradation and climate change issue there is strong need to educate the people specially young generation about the importance of our environment in their life, marketers through their marketing and advertising activities can attract the young consumers belongs to "Z-GENERATION" who considered as present and future buyers. Green marketing activities can enhance profitability as well as can be helpful for overall sustainable development. "Z-GENERATION" consumers are also called smart generation consumers who are smart enough to use technologies and their daily activities. Through social media advertising large number of such group of consumers can be attracted to create demand for green product and through various short videos platform like YOUTUBE short, INSTAGRAM Reels since this generation is highly active in these platforms thorough these platforms they can be motivated to play their role to protect the environment thorough making changes in their purchase habit.

Green Marketing Strategies to Influence Z-Generation Consumers

Along with offering green product the business owners can adopt following practice under their marketing strategy to influence and attract the Z-Generation Consumers':

- Educating Z-Generation about green product importance social advertisement.
- Adopting eco-friendly packing materials.
- Using green colour packing as a symbol of eco-friendly product.
- Practicing systematic waste disposal system.
- Eco-labeling.
- Social media campaign.
- Distributing free sample of green products to the Z-generation and motivating them to use it and recommend other about its importance.
- Adding green philosophy in modern marketing concepts by satisfying Z-generation with harmless products.

CONCLUSION

Green marketing or Eco-marketing connected with manufacturing of eco-friendly products, sustainable packing, sustainable business practice, promotion of green product, usage of recycled material in production activities, using green energy and following all eco-friendly process and methods to introduce product in the market to attract large number of buyers along with contributing towards sustainable environmental and economic development and this green marketing activities can be more efficient if performed to the segmented population of "Z-GENERATION", consumers as they are the seed generation and present and future buyer and influencer to the other. Through their Generation -Z focused green marketing activities business organizations can bring long lasting growth and profitability along with protecting environment and providing health benefitted products and services to the consumers. Since the global concern is now towards the sustainable development so that the marketers also need to change their business strategy for long-term sustainability and growth. Now the time has come to cope with the changing global business environment for future success and existence, acceptance of green marketing and green practice is the demand of time and every business organization has need to realize it. And what they target today will decide future and "Z-GENERATION" consumers are present and future consumers so that business organization should always primarily consider these age group consumers in their business strategy.

REFERENCES

Alsmadi, S. (2007). Green Marketing and the Concern over the Environment: Measuring Environmental Consciousness of Jordanian Consumers. *Journal of Promotion Management*, *13*(3-4), 339–361. doi:10.1080/10496490802306905

Ansu-Mensah, P. (2021). Green product awareness effect on green purchase intentions of university students': An emerging market's perspective. *Futur Bus J.*, *7*(1), 48. doi:10.118643093-021-00094-5

Guerreiro & Pacheco. (2021). *How Green Trust, Consumer Brand Engagement and Green Word-of-Mouth Mediate Purchasing Intentions*. doi:10.3390/su13147877

Joshi, Y., & Rahman, Z. (2015). Factors Affecting Green Purchase Behaviour and Future Research Directions. *International Strategic Management Review*, *3*(1-2), 128–143. doi:10.1016/j.ism.2015.04.001

Kautish, P., Paul, J., & Sharma, R. (2019). The moderating influence of environmental consciousness and recycling intentions on green purchase behavior. *Journal of Cleaner Production*, *228*, 1425–1436. doi:10.1016/j.jclepro.2019.04.389

Kim, Y. (2011). Understanding green purchase: The influence of collectivism, personal values and environmental attitudes, and the moderating effect of perceived consumer effectiveness. *Seoul Journal of Business*, *17*(1), 65–92. doi:10.35152nusjb.2011.17.1.003

Levine, D. S., & Strube, M. J. (2012). Environmental Attitudes, Knowledge, Intentions and Behaviors among College Students'. *The Journal of Social Psychology*, *152*(3), 308–326. doi:10.1080/00224545.2011.604363 PMID:22558826

Liao, Y.-K., Wu, W.-Y., & Pham, T.-T. (2020). Examining the moderating effects of green marketing and green psychological benefits on customers' green attitude, value and purchase intention. *Sustainability (Basel), 12*(18), 7461. doi:10.3390u12187461

McDonald, S., & Oates, C. J. (2006). Sustainability: Consumer Perceptions and Marketing Strategies. *Business Strategy and the Environment, 15*(3), 157–170. doi:10.1002/bse.524

Mohd Suki, N., & Mohd Suki, N. (2019). Examination of peer influence as a moderator and predictor in explaining green purchase behaviour in a developing country. *Journal of Cleaner Production, 228*, 833–844. Advance online publication. doi:10.1016/j.jclepro.2019.04.218

Mukonza, C. (2021). *Green Marketing in Emerging Markets Strategic and Operational Perspectives.* Springer Nature Switzerland AG. doi:10.1007/978-3-030-74065-8

Pagiaslis, A., & Krontalis, A. K. (2014). Green Consumption Behavior Antecedents: Environmental Concern, Knowledge, and Beliefs. *Psychology and Marketing, 31*(5), 335–348. doi:10.1002/mar.20698

Schmuck, D., Matthes, J., Naderer, B., & Beaufort, M. (2017). The Effects of Environmental Brand Attributes and Nature Imagery in Green Advertising. *Environmental Communication, 12*(3), 414–429. doi:10.1080/17524032.2017.1308401

Sun, H., Teh, P.-L., & Linton, J. D. (2018). Impact of environmental knowledge and product quality on student attitude toward products with recycled/remanufactured content: Implications for environmental education and green manufacturing. *Business Strategy and the Environment, 27*(7), 935–945. Advance online publication. doi:10.1002/bse.2043

Sun, Y., & Wang, S. (2019). Understanding consumers' intentions to purchase green products in the social media marketing context. *Asia Pacific Journal of Marketing and Logistics, 32*(4), 860–878. doi:10.1108/APJML-03-2019-0178

Wang, H., Ma, B., & Bai, R. (2019). How Does Green Product Knowledge Effectively Promote Green Purchase Intention? *Sustainability (Basel), 11*(4), 1193. doi:10.3390u11041193

Wang, S.-T. (2014). Consumer characteristics and social influence factors on green purchasing intentions. *Marketing Intelligence & Planning, 32*(7), 738–753. doi:10.1108/MIP-12-2012-0146

Yadav, R., & Pathak, G. S. (2016). Young consumers' intention towards buying green products in a developing nation: Extending the theory of planned behavior. *Journal of Cleaner Production, 135*, 732–739. doi:10.1016/j.jclepro.2016.06.120

Chapter 12
Gross National Happiness:
Bhutan's Model for Sustainable Development – Its Relevancy in Current Times

Bhuwan Shrivastava
https://orcid.org/0009-0004-2114-3606
Royal Thimphu College, Bhutan & Amity University, Raipur, India

Sumita Dave
Amity Business School, Amity University, India

ABSTRACT

Bhutan, known to the outside world as the land of happiness, has attracted attention due to its unique way of transforming the lives of its citizen through its approach towards Gross National Happiness and Sustainable Development Goals. This chapter explores Bhutan's way to treat the concept of Gross National Happiness. This chapter provides an overview of GNH philosophy by examining its different pillars. This chapter also takes into consideration the criticism and shortcoming which GNH has over the most common indicator of growth, the GDP. Furthermore, the chapter also explores the development of concept of GNH and its presence and relevancy in current times. It tries to understand the attainment of Sustainable Development Goals via GNH thereby understanding further Bhutan's model of SDG attainment and how it acts as a model for other nations aspiring for Sustainable Development Goals. Overall, this chapter is an attempt to understand, explore, and present Bhutan's way to reach the target of SDG.

INTRODUCTION

Bhutan, geographically located between two of the world's biggest economy, India and China, finds itself, sometimes, in the middle of tussle between these two biggies. Bhutan, which claims to have resolved the most daunting issue within the region, the Doklam crisis (Financial Express, 2023), which almost brought the two big neighbors, on the verge of war, is probably known to the outsider world as the land

DOI: 10.4018/978-1-6684-9503-2.ch012

of "Gross National Happiness" (GNH). With the core value of "Collective Happiness" and emphasis on "Community Wellbeing", Bhutan has successfully extended the process of modernization without harming the environment or its relationship (Walcott, 2011). This policy was developed against the drawback of most common indicator of growth "Gross Domestic Product", which only consider the variables of production and consumption as a measure of growth of an economy without realizing its impact on the ecology and environment and thus its effect on human wellbeing (Masaki, K., & Tshering, J. (2021). Though the countries, especially European, which have adopted the concept of GNH from Bhutan, have shown some of the positive results with respect to their wellbeing and happiness. This is clearly evident from the latest list of countries and their ranking in the World Happiness Report, 2022 (World Happiness Report 2022). The list has eight countries amongst the top ten countries from Europe and Bhutan is nowhere in top 20 even. The reason may be, perhaps, there are issues which makes the measurement of GNH in Bhutan difficult. With deprivations in education and increase in domestic violence, many, even Bhutanese, feel that only lip service is being done to implement GNH properly across all areas and domains of life and economy (Luechauer, 2013).

There are certain doubts with regard to the application of GNH philosophy, with the current trend of economy all across the world. To what extent the GNH can be considered as genuine replacement or alternative to the conventional and most prevailing model of economic development? What is the possibility of GNH getting success in the modern globalized world? (Helena See, 2022).

Finland, for example has always remained in the top five of the world's happiest countries. The reason, that they have taken the happiness as free education, generous parental leave and a healthy work-life balance which ensured that people have the time and the means to pursue their pleasures, no matter how mundane (The Economist, Mar 26th 2018). That is their way of taking the happiness in the real time. They have renovated their own way of happiness, which is though not different from the basic philosophy of GNH. In recent times, they have also included the immigrants in their survey of happiness to have the holistic understanding. And the finding was that the happiness of immigrants stems from a strong support network and integration policies (The Economist, Mar 26th 2018). Which clearly means that, happiness of people is very subjective and very specific to the circumstances and situations. It is not that world's most developed nations also comes as world's happiest country. If we look the ranking of world's most developed nations like US and UK, we will find that they are not finding their place in even top ten. Though United States have doubled their per capita income in the last 40 years, but they may have not sufficiently invested in the wellbeing on people.

This chapter is an attempt to understand the philosophy of Gross National Happiness vis a vis its impact on the development of country in economic terms. This chapter will try to unfold the relevance of GNH in current times in Bhutan. It will study the impact of application of GNH philosophy in some of the important sector of Bhutanese economy i.e. Tourism and also it will try to understand the impact in management of natural resources.

This chapter will try to serve the Bhutan's example to the world. In the other words, it will analyze critically the implementation of the policy and steps which could have taken to make this policy better.

This chapter will end with recommendations for future research and policy development in Bhutan's economy and sustainability.

GNH, THE PHILOSOPHY: ITS DEVELOPMENT

Bhutan's policy of Gross National Happiness is no more a new term. It is equally adapted by some countries and praised by the leaders worldwide. According to Bhutan's Gross National Happiness Commission, "...the philosophy of GNH is an effort to have a harmonious balance between material wellbeing and the spiritual, emotional cultural need of our society" (GNH Centre Bhutan). It was developed on the premise that while counting the development index of a nation, it is not only the money should be counted but also the people's happiness and satisfaction should be taken into consideration.

The concept of GNH introduced to Bhutan by, His Majesty Jigme Singye Wangchuck, the fourth king of Bhutan. He emphasized the importance of ensuring the overall happiness of the people without being sidetracked with the statistics of Gross Domestic Product (GDP). The ultimate goal of GNH is happiness and to ensure a good quality life for the people through people centric development approach (Royal Bhutanese Embassy, India website).

There are literatures which traces the development and introduction of GNH into Bhutan since the formation of Bhutan in the year 1651.

The first evidence of happiness policy being imparted in governance of Bhutan dates back to the year 1652 just one year after the Bhutan has become a nation. Unlike political guidelines of other part of world, this guideline provided a broad framework for the good governance and emphasized that whole nation of Bhutan should move towards becoming an enlightened nation. (Givel, Michael & Figueroa, Laura, 2015). Thus the "happiness" became the part of Bhutan's governance from the very beginning of its existence. It is based on Mahayana Buddhist values.

This framework of governance of Bhutan is highlighted in legal codes. The prominent of them is the legal code of 1729 which states that 'if the government cannot create happiness for its people, there is no purpose for the government to exist' (Ura 2010). Thus, after Bhutan becoming the constitutional monarchy in the year 2008, it laid in very clear terms in the Article 9 of constitution that state has 'to promote those conditions that will enable the pursuit of Gross National Happiness.' And hence Gross National Happiness became the integral part of Bhutan's governance and developmental policy.

Over the years GNH has taken many shapes. Everyone has understood the concept according to their environment.

As such there is no universal definition of GNH available, but according to GNH center, Bhutan, the GNH means, "...a holistic and sustainable approach to development, which balances material and non-material values with the conviction that humans want to search for happiness. The objective of GNH is to achieve a balanced development in all the facets of life that are essential for our happiness." (GNH Center website). Further GNH center of Bhutan has come out with current understanding of GNH as a "multidimensional development approach seeking to achieve a harmonious balance between material well-being and the spiritual, emotional and cultural needs of society." The keywords being harmonious balance – that may be achieved by balancing the needs of the body with those of the mind (GNH Center website).

GNH is based on 4 pillars. These are:

- Good Governance
- Sustainable Socio-economic Development
- Preservation and Promotion of Culture
- Environmental Conservation

A brief understanding of all the above pillars is as below.

Good governance provides the foundation for government policies to be framed as per the GNH principles. It makes sure that GNH policy with the help of various tools and processes, should be employed in such a way that it is reflected from the social policy so framed by government.

Sustainable Socio-economic Development lays importance to overall development of society which includes families. It emphasizes that sufficient leisure time should be given so that it can act as catalyst for families and society to develop and hence adding the values to Gross National Happiness.

Preservation and Promotion of Culture is another important pillar for GNH in Bhutan. It lays emphasis on preservation of culture and norms. It says that culture is a way to preserve the identity.

Environmental conservation is another key pillar to the GNH philosophy. Its importance lies on the preservation and conservation of environment. It has identified environment as a important part of human's life since it gives not only valuable assets but also adds aesthetic stimulus.

To have the clear and better application of all the above 4 pillars, GNH policy is further divided into 9 domains, which presents the GNH as more elaborate tool to understand the basic elements that are involved in measuring the "happiness" in human life.

These 9 domains of GNH policy are:

- Living standards
- Education Health
- Environment
- Community
- Vitality
- Time-use
- Psychological well-being
- Good Governance
- Cultural resilience and promotion

These nine domains are not to be dealt in aloof. They are so thought that all of these domains are interrelated to each other. They effect the human life not only as single factor but as connected web of elements.

For instance, one of the domains "living standard" is directly related to "Education" status of a country and further these are related to the "Environment" of a country.

Hence, the interrelationship between the material and non-material elements in the human life is responsible for happiness and this is the only factor which differentiate the GNH from GDP.

Keeping in line with these nine domains, Bhutan has developed 38 different sub-indices, 72 indicators and 151 variables that can be used to measure the happiness of Bhutanese people. (gnhcentrebhutan.org)

Though GNH has attracted global interest as an alternative to the GDP for the overall growth of human being at large, the success and sustainability of GNH in Bhutan is itself under doubt (Helena See, 2022). The doubts comes in the form of two major broad questions, first being that to what extent the GNH philosophy is genuine and best alternative to the prevalent scale of measuring economic development and second being its probability of survival in the long run (Helena See, 2022). In the next section, we will try to understand the philosophy of GNH vis - a – vis Bhutan's economic policy.

GNH AND BHUTAN'S ECONOMIC POLICY

Bhutan's economic policy or Bhutan's economic environment is attributed to Mahayana form of Buddhism. Bhutan's economy is not only the one being guided by the Buddhism. There are other countries as well which are equally influenced by Buddhism. They are Thailand, Bhutan and Yogyakarta, the southern part of Indonesia (Kershaw, Roger, 2001). All of these three economies are predominantly Buddhist and have evolved their own way to design their economy. They are "Sufficiency Economy" of Thailand, "Unification of King and people" in Indonesia and "Gross National Happiness" of Bhutan. There is one peculiarity about these "Buddhist" economies is that, they directly come from the king. It is the king who plays a prominent role in developing the basic economic model specific to their kingdoms / countries. According to Buddhism, king has to take the responsibility in the form of spiritual guidance and high degree of social responsibility (Drechsler, Wolfgang, 2019).

For the sake of clarity, when Buddhism says that it has to followed a "middle path", that doesn't mean that it is trying to be skeptic with the worldly affairs or beings. But it is trying to have a balance between the two extremes of life, by giving the due importance and respect to both the side. This is clearly evident from all of the Buddhist Economies which we have pointed out earlier.

Out of all the Buddhist Economy one thing which is common is that, all of them talk of sustainability as their core idea. This is evident from the belief that Buddhism itself talks about the "Right Livelihood", which directly or indirectly points out towards the village economy or emphasizing on SMEs or traditional farming and hence sustainability. It is evident from all three Buddhist economies, the importance they put in sustainability. "Sufficiency Economy" of Thailand, which according to the king His Majesty King Bhumibol Adulyadej, "…The important thing is for us to have a sufficient economy. A sufficient economy means to have enough to support ourselves…we have to take a careful step backward…each village or district must be relative self-sufficient." (Chaipattana Foundation website). Whereas in case of "Unification of King and People" of Indonesia which emphasizes on "food sovereignty" through high focus on agriculture and hence the self-sustainability of people and the economy. And also, if take the example of GNH philosophy of Bhutan we find the importance of sustainability and maintaining the environment in its area of focus. Thus, all three of the Buddhist economy has certain thing in common, sustainability and peace of mind / nature. These principals have guided in some or other way all of these economies. These philosophies have one ultimate aim which is "Happiness". "Happiness" here according to Singer et al, 2015 is different from the happiness which comes from the consumption. This consumption ultimately acts as the most common guide stick of growth as the concept of "Gross Domestic Product (GDP)". Rather "Happiness" has no direct connection with the material belongingness or consumption. It's all about the inner happiness of people and sustainability of economy. While GDP emphasizes on the continuous rising income as the sign of prosperity and growth on the other hand Buddhist economy focuses on inner happiness and satisfaction is the key to the economic wellbeing. It further states that "wanting less" is in interest of both individual and economy as a whole (Drechsler, Wolfgang, 2019).

As far as Bhutanese economy is concerned, the evolution and adaptation of GNH was needed more because of its position in the world political map. As said earlier, being a small country located between the two super powers, GNH acted as a tool to give them an identity (Helena See, 2022). Since then, Bhutan has taken a new approach towards the economic development by inculcating the concept of five-year plans, GNH has been a guiding principle in all these plans. In every of the plan, the GNH had its impact.

Bhutan started the five-year plan policy from the year 1961. Since the beginning, and even to the extent today also, Bhutan has majorly been a closed and isolated economy driven by the philosophy of

GNH. As we have dealt earlier in this chapter that, out of various area of focus of GNH one of such area is preserving the culture and another is preserving the environment. While deciding the five-year plan, the four pillars of GNH philosophy are strongly adhered to.

Bhutan's first five-year plan had a direct reflection of GNH philosophy. While at the end of first five-year plan 108 schools were opened up all across Bhutan, the major emphasis was still not the setting up the industry and getting the employment numbers up, the major focus was going with special impetus on preservation of nature and culture. This is clear from the total budget of Rs. 1747 lakhs out of which Rs. 100 lakhs approximately were kept for health, forest, agriculture and animal husbandry and for industry it was Rs. 11 lakhs only (Gross National Happiness Commission, Royal Government of Bhutan). Though this figure also provided a positive side of the policy makers of Bhutan, that they didn't leave the importance of education at peril. The alone budget for education was Rs. 100 lakhs (Gross National Happiness Commission, Royal Government of Bhutan). Tha major area where country worked upon the animal husbandry, public health and forest. Here the basic idea was to preserve the balance of country's cultural natural wealth with that of growth. The authorities believed that this policy will generate the slow growth in short run but in long run this strategy will definitely achieve the sustainable growth objective as sighted under the GNH philosophy (International Monetary Fund, 1995).

Till further couple of decades, the country kept the focus only on resource-based industries (International Monetary Fund, 1995). This objective was accompanied with the achievement in preserving the forest cover which was one of the pillars of GNH policy. One industry was identified which can bring the addition to the country's economy without harming the natural heritage, was hydropower. Because of country's natural advantage of geographical making, the generation of hydropower came as natural option to the country's economy. Having limited impact to the country's natural wealth, these hydropower project can even run on river schemes, they didn't need any of the dams to be built, hence benefitting both the economy and fulfilling the GNH's idea of sustainable development.

Even with the low investment in the industrial sector, the power generation industry brought a major change in the development of Bhutan's economy. With the revenues from these projects and generous

Table 1. Bhutan's economy

Particulars	Bhutan	Comparable Income Group in Asia*
GDP	US$255 million	...
GDP per capita	US$380	US$310
Population	0.675 million	...
Population growth rate	3.1 percent	2.1 percent
Life expectancy at birth	66 years	60 years
Access to safe water (percent of population)	58	69.5
Infant mortality (per thousand live birth)	70	84
Population per physician	3,822	2,850 persons
Daily calorie supply	...	2,416 calories per person
Adult illiteracy (percent)	46	54
Primary school enrollment (percent of school-age population)	70	106

Sources: Data provided by the Bhutanese authorities; and World Bank, Social Indicators of Development, 1995. Estimates are for 1994 or most recent available data. *Asia includes Afghanistan, Bangladesh, India, Maldives, Nepal, Pakistan, and Sri Lanka.

Figure 1. Bhutan: Composition of GDP
Source: IMF Staff Country Reports 1995, 135; 10.5089/9781451806212.002.A001

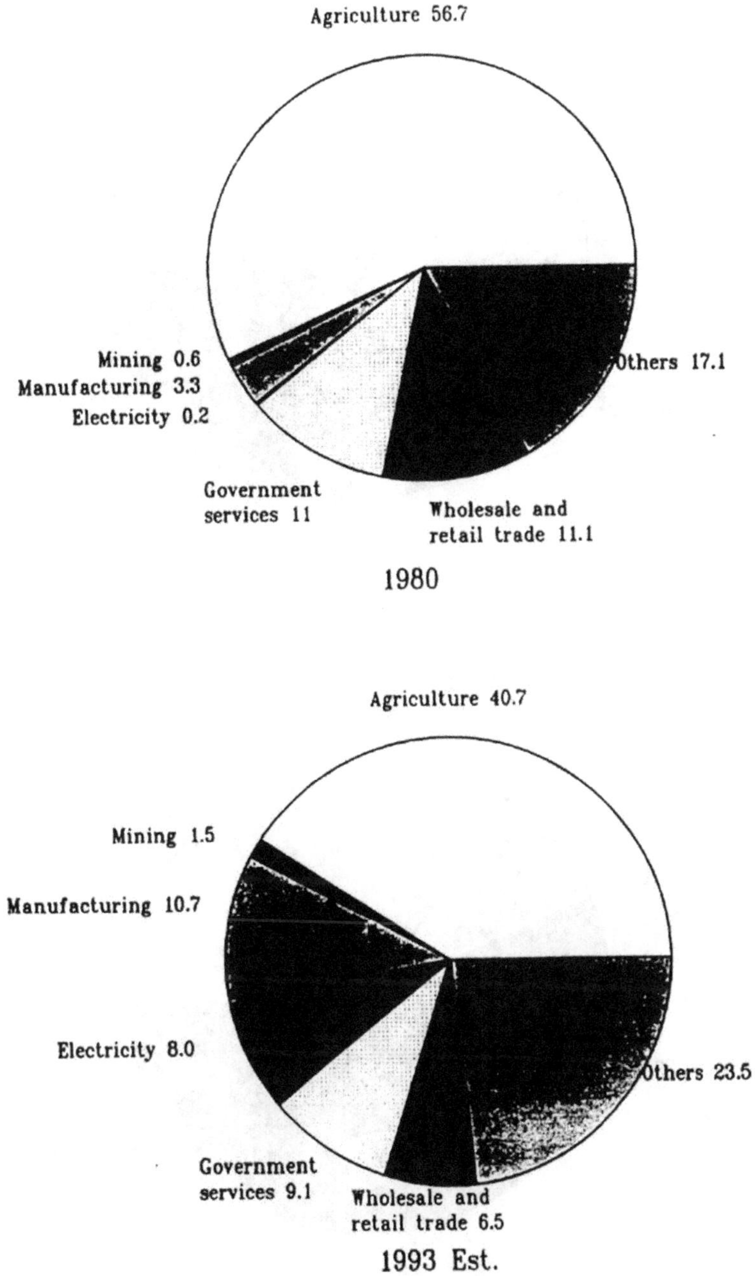

assistance, country saw a jump in the number of villages and towns getting connected with roads and other infrastructural facility. Medical infrastructure was strengthened further with a greater number of hospitals and clinics opening up across the country. With virtually no roads in 1960s, till 1990s country had almost 3100 km of road network. The number of schools were also increased from just 11 in 1950s to almost 300 in 1990s (International Monetary Fund, 1995). With all these things in place, Bhutan saw

Figure 2. Bhutan: Composition of export
Source: IMF Staff Country Reports 1995, 135; 10.5089/9781451806212.002.A001

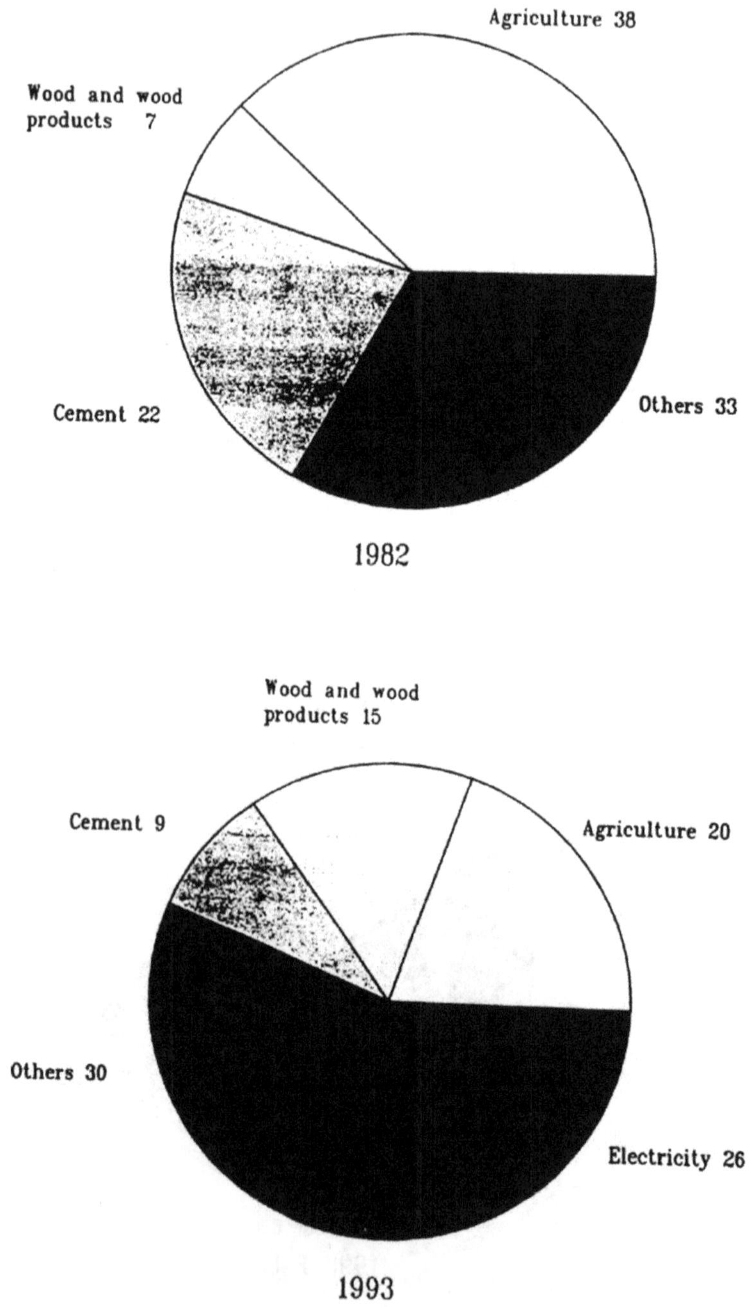

a better standard of life in terms of per capita income and other important indicators of growth in terms of GDP, than the neighbors combined.

From the Table 1 it was clear that Bhutan's economy was moving in the direction as desired and visioned in GNH policy. It had better per capita income with respect to other countries of Asia. And also, in terms of health sector Bhutan had better numbers than rest of Asian countries combined.

One of the major concerns for the policy makers which came at the end of this decade of 1990s was the lack of labour and managerial talent, which rose the number of unemployment in the coming decade. This was further deepened by the strict adherence to the cultural and religious belief that made the employment and investment scenario in Bhutan bit difficult thus having the major impact on GNH's sustainable developmental policy. With the decade 1980s and 1990s saw a major shift of labour force from agriculture to urban based source of employment like industries and retail trade, the percentage of contribution of agriculture declined during this couple of decades and which was in contrast to the basic idea of GNH philosophy.

An aspect that contradicted the concept of GNH was that, with the decline in the contribution of agriculture in GDP and increasing the contribution of industry, the export also got effected. This decade saw the decline in agriculture-based export with the corresponding increment in the export of electricity to the neighboring countries. Another portion which was making significant impact on the export scenario of Bhutan was increment in the export of wood and wood related products which was again in exactly opposite to what was the basic idea of GNH was, the preservation of natural vegetation and forest.

It is worth mentioning the introduction of television to this far Himalayan kingdom of Bhutan; which had more of sociological impact than economical. But while considering the importance of GNH to the economy of Bhutan, this also has to be taken into consideration. One of the pillars of GNH lays the importance of preservation of culture. This means that is one is having respect for his/her culture then that person will definitely strive for the inner happiness through mutual respect to others. But with the pressure from the mass, the government on 2 June 1999 opened the Bhutan's air for television, of course with certain apprehensions (https://www.reuters.com). Now when Bhutan is on the cusp of development and when everything is set for country's economy to stride forward, the introduction of television was seen as hinderance in attainment of overall objective of GNH. One of the doubts which policy makers had that with television, country's culture and belief which was preserved as it is for so longer a period, may get diluted. This was something according to some was getting reflected in the increment in crime and instances of theft in the urban centers. But the reasons were more attributed to the rising unemployment rate in the city because of migration of work force from agriculture to the cities in search of urban employment opportunities. With some restrictions in channels and broadcasters, keeping in line with the basic idea of GNH, now television is playing the crucial role in shaping the country's generations.

Further decades of 2000s and beyond were the decisive ones. These decades brought some of the major shift in the country's economic and political policy which had major impact on the way the concept of GNH was being handled and hence has the effect on sustainable economic development of the country. The year 2008 saw the first general elections and the country's transition from being a monarchy to democratic country took place. Now country became the constitutional monarchy with prime minister and its team of ministers as head of the policy making bodies.

With strong support and aspiration from public, the first prime minister Mr. Jigme Thinley assumed the responsibility of keeping the country's growth pace at par with rest of Asian nations strictly in lines with drawn by GNH policy. During his tenure from 2008 to 2012, the economic indicators were concerning. Economy had a jump in GDP growth rate and per capita income, the point of concern was the unemployment rate which rose to almost 4 percent at the end of 2009 (RMA, 2009). The balance which government wanted to make here is of maintaining the basic essence of GNH, thereby taking the country towards the sustainable development with the fast-changing economy of the world. Though the year 2008 saw the global crises (Kose et al., 2020), the Bhutanese economy did not have that impact which the rest of the world saw. One of the important sectors, which was also the sector of focus, i.e., tourism got a dent from this global phenomenon and also some of the manufacturing sector. The consequence to this effect is clearly visible in the downfall of the employment rate (RMA, 2009).

Since the period of 2008 – 2012 was the period of the first tenure of democratically elected government, the aspiration of people was high with respect to their employment generation and confining the inflation rate to the permissible limits. Royal Monetary Authority of Bhutan, country's topmost monetary authority, with the consultation from the government and keeping the GNH philosophy of sustainable development under consideration, came out with certain policy that helped the country to have an environment which can nurture more and more financial assistance to the required areas. According to the annual report of Royal Monetary Authority of Bhutan, 2009, the government via RMA took certain measure to develop such a market which can nurture more and more financial institution so as to have a healthy competitive market for providing healthy loan and thereby decreasing the outlay to those area which are of concern.

This period of first democratically elected government also saw the commissioning of two of the power plants, Punatsangchhu I and Dagachhu projects (RMA, 2009), which helped the government to create the employment by being in the same lines of GNH's sustainable development goals and thus helping the economy also in terms of foreign exchange reserve.

With the mounting unemployment rate and being accused of "overusing" (Helena See, 2022). the philosophy of GNH the next government was formed under the prime ministership of Mr. Tshering Tobgay.

Mr. Tshering Tobgay wanted to be in line with the global trend and wanted to take the private sector as a catalyst for growth. He wanted to have an emphasis on private sector involvement in the country's economy to nurture the growth, but with the basic policy of GNH's sustainable development as a foundation (Helena See, 2022). Now the concern here was with the private sector coming in may dilute the basic structure of Bhutanese GNH policy. With the private sector and objective of keeping in lines with global trend of attracting FDI, Bhutan wanted to go ahead with these two as an important tool for development. This objective of keeping GNH first made certain hinderances to the attractiveness of Bhutanese economy. According to World Bank Publication "Investment Climate Assessment of Bhutan", authored by Santini, Tran and Beath in the year 2017, the idea of attracting FDIs and private sector in their economy had certain shortcomings, they were:

1) Limited access to key factors of productions like skilled labours and easy availability of finance thereby constraining the industries and enterprises to invest more and grow.
2) With low level of professional education and attractive packages by the government bodies has left the private organization to rely more on foreign skilled labours. But here also, the kind of restrictions and regulations which government of Bhutan has for expatriates makes it a unattractive market for skilled workforce to come to Bhutan. Instead, this has lead to the flow of unskilled labour to crowd the construction sites all across the Bhutan, thereby increasing the unemployment rate further for locals to work in these sites.
3) Logistics and transport infrastructure plays and major role in development of any economy and especially when economy is trying to come up at the world par. There is deficiency of transport infrastructure and logistical support all across Bhutan. This discourages the private sector to participate in the growth of economy. While over the years, road network have sufficiently improved in the country but the proper management and handling of the infrastructural back up support is still missing. In the border town of Phuentsholing there is no standard facility which can assist the smooth functioning of transport activity.

All these drawbacks were brought to the light by the respective authors in their annual report of World Bank. The major reason for all these points can be attributed to strict adherence to the GNH policy. It is not that GNH is full of issues pertaining to the hinderance of growth of economy, but GNH provides a yard stick for the development of sustainabnle development of economy.

Based on these shortcomings of economy, certain recommendations were also made for its smooth functioning. Some of them were:

1) Development of financial assistance tool to enable the private sector to have access of quick finances.
2) Reducing the state's too much involvement in developmental issues for private firms.
3) Encouraging the environment that can nurture and grow the public private partnership.
4) Development of logistical and transport support.

One the key factor identified by Helena See (2022) was, and still it holds true, the non-availability of skilled labour. This according to author this is because of their intolerant policy towards dilution of their culture and moreover it is one of the key pillars of GNH policy. Bhutanese government don't want to relax their culture with that of others and hence the immigration policy has made in a such way that it discourages more then encourage. It has one of the strictest immigration policies in the world (Santini, Tran, and Beath, 2017).

In the coming years the economy of Bhutan took a strange turn. The year of 2020 saw, world over, the arrival of pandemic which halted and showed a down turn to almost all the economies of the world. Same was the case with Bhutan. A country which has, one of the important source of foreign reserve and employment attributed to tourism, this pandemic hit the hardest. With almost 64000 dollar paying tourists (RMA, 2020) in the year ending 2019 to just 1 at the year ending 2021 (RMA, 2022), employment based on tourism was almost on the road. This was the testing times for the new government formed under Dr. Lotay Tshering in the year 2018.

TOURISM AND SUSTAINABLE DEVELOPMENT FEES

As we have discussed earlier, one of the major pillar for GNH policy of Bhutan is conservation of natural resources, which includes forest also, and culture. Keeping in lines with this, the thought for "high value, low volume" tourism policy which was later renamed as "high value, low impact" (Lhendup et al, 2020) was implemented in the strictest possible way. Bhutan Tourism industry continues to grow till 2019 making it one of the important contributors to the socio-economic development of the nation (BTM, 2019). It was the year 2019 which saw a significant rise of approximately 15% in the number of tourists as compared to the data of 2018 (BTM, 2019). Bhutan, being a preferred tourist destination for those who wants to have peace, won many prestigious awards for their effort in making the country of preference. The awards were like Green Destinations Gold Award by the ITB Berlin in Germany in March 2019 and also recognized as the number one travel destination for 2020 by the Lonely Planet in October 2019 (BTM, 2019). According to the Bhutan Tourism Monitor, 2019, tourism and related activity alone generated total revenue of approximately Rs. 2400 million. These activities were in the nature of tour operators, taxi services, hotels etc.

Preservation of nature and culture being one of the pillar for GNH policy for sustainable development, guided the tourism policy since the beginning of 1974 when Bhutan opened its gate for foreign tourists

Figure 3. Bhutan: Tourism financial status
Source: Bhutan Tourism Monitor 2019, Publication of the Tourism Council of Bhutan

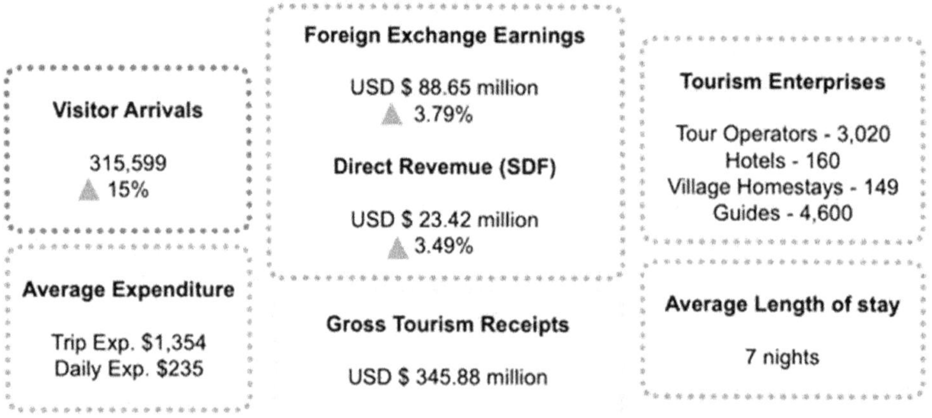

for the first time. To have all the things under control a Minimum Daily Package Rate of $ 130 per day was levied for the tourists. Somewhere in 2012 it was rose to $200 to $250 per day. The reason behind this increment was to reduce the number of tourists visiting the country and only to have the quality tourist who can understand the importance of nature and preservation of culture. Later in addition to that, $65 per person per night was introduced as Sustainable Development Fees (SDF) (The Hindu, 2022). But SDF was not applicable to those tourists who were coming from India. The amount to be received from this SDF was to be utilized for the purposeful uses like maintenance and construction of roads, giving medical benefits to the citizen of Bhutan and more importantly maintaining the tourism, and preservation of culture and nature. It was the advantageous position for Bhutan and India both as in the year 2018 saw almost 2.3 lakh visitors coming mainly from India. For Bhutan, it was helping in the various benefits like generation of employment and contribution to the economy of the country and for Indians, they had access to one of the most pristine and serene destination to visit and a very affordable cost.

According to the Tourism Council of Bhutan, it was getting difficult to maintain and preserve the culture and nature of Bhutan. With more and more number of tourist coming to the Bhutan there are issues pertaining to littering, unnecessary crowding was also increasing which according to the authority it was exactly opposite to the vision and objective of GNH. Till now tourists coming from Bangladesh, Maldives and India need not to pay the SDF charges. But with thc coming of Tourism Levy Bill of Bhutan 2022, now these tourists have also to be paid with the SDF of Rs. 1200 per person per day. Due to pandemic this was not implemented, but with reopening of border to international tourist in September, 2022 this scheme was levied. This scheme was implemented on the basis that it will be reviewed after two years.

According to the Prime Minister of Bhutan Lyonchhen Dr Lotay Tshering the new rules are exactly in lines with Bhutan's 'high value, low volume' policy (The Hindu, 2022).

This policy of government to apply SDF at the period when whole of the world and also Bhutan is facing shortages of tourists, impacted the tourism industry in Bhutan. Tourist now has to pay Rs. 1200 per day per person once they enter the border towards Bhutan. And also there was another rule which also impacted the small hotelier, that only, minimum, 3 star rated hotels are allowed to keep the tourists and non star rated hotels were debarred from keeping the international guest. Many of the small hotelier started investing in their hotel to upgrade it to star category. With that many small hotels who were not in position to upgrade their hotel, due to cash crunch during the pandemic, were left out of this opportunity and hence the impact was clearly seen in increase in number of unemployment. The unemployment rate rose to 5.9% which was increment of 1.1% as compared to the figure of 2021(Labour Force Survey Report, 2022). One the major reason attributed was COVID – 19 pandemic and its aftereffect.

The issue graved further in the border town of Phuentsholing, where out of 50 hotels there were 12 hotels which were star rated hotels (Kuensel, August 2022). Also, for them to upgrade it to star rated hotel is expensive. The third major issue was import of foreign workers to work in tourism industry of Bhutan, which was not allowed (Kuensel, August 2022). These factors effected Bhutan's economy, especially where major portion of country's GDP comes from tourism and its allied industry. Even after repeated request to the PMO of Bhutan, the effort were all in vain. PMO rejected the proposal to extend the imposition of SDF (Kuensel, July 2022).

The biggest beneficiary from all of this SDF impact was Indian side of border. Jaigaon, which is town bordering Bhutanese town of Phuentsholing. For an Indian trader or tourist who wishes to enter Bhutan via Phuentsholing terminal had to pay SDF. They stayed away from this payment, rather they asked the Bhutanese to come to Jaigaon and have the purchasing done from there. The traders of Phuentsholing were at the lost. Now the customer who would have made purchase from them, were purchasing from Jaigaon. Traders and hoteliers from Bhutanese side, because of low turnout of customer, either closed down or increased price. This increase in price had another impact on the economy of Bhutan. The real income of people of Bhutan fell down as the price of goods and hence the inflation rate increased. For example for food, inflation rate was 9.1% in 2021 which after easing the SDF regulations and other related restrictions fell down to 4.5% in 2022 (Kuensel, March, 2023).

The impact was seen on the tourist arrival before and after the SDF regime. There were approximately 350000 tourist arrival from January 2019 to February 2020 which left to only approximately 37000 tourist arrival since the opening of borders in September 2022 till 5 April 2023 (The Bhutanese, April 2023). This directly effects those who are related to this industry viz. taxi services etc.

But at the end, this should be taken into consideration that, the objective behind this step was to have 'high value, low volume' (The Hindu, 2022) tourist. There were the instances of tourist littering the places where they visit. "While this has been going for decades and contributes to the friendly people to people relations across the border littering by picnickers on the Bhutanese side is becoming an issue." (The Bhutanese, 2017). The problem as described by one of the members of parliament, that regional tourist (here India) come in their own vehicle, they drive their own vehicle, they move across the border, they bring their own utensils to prepare the food and once done they simply leave the waste behind.

According to Bhutan Tourism Monitor, 2016, in May 2016 alone there were record arrival of guest which was approximately 24000, out of which roughly 98% were Indian, 1.5% were from Bangladesh and 0.4% were Maldives. Further, according to this report, 52% of these tourists stayed in non-star rated hotels, 38% stayed in star rated hotels and whereas 7% of them stayed in lodges. These tourists who were staying up in lodges were those who were not under the ambit of 'high value, low volume' category, since they were not using any of guide facility nor they were paying any kind of SDF, rather their impact was huge.

With all these, government of Bhutan implemented the SDF to the tourists from neighboring countries (viz. India, Bangladesh and Maldives) which they called as regional tourists and increased the already existing SDF for non-regional tourist from $ 65 to $ 200. They proposed that the fund will be utilized for the development of tourist facilities all across the country. (The Kuensel, 2022). In this case, to revive the tourism industry, Thai government's effort is worth mentioning. They waived the visa fees. In turn they lost reserve in dollars, but for Thai government, it was not about losing in dollars, but in return, they gained in terms of tourists who casually visited the country and spent in dollars their which in turn helped the local shopkeepers, tourists service providers and others who are directly or indirectly related to tourism industry. At the end, even they lost huge amount but they gained more in terms of their tourism (The Kuensel, 2022).

Government of Bhutan wanted to have the sustainable development approach achieved through this tool of SDF which was a direct application of policy of GNH involving preservation of nature and conservation of culture. With all the protest and growing unrest from all the stakeholders, hence government of Bhutan waived SDF for border towns, to start with atleast, "Tourists visiting Samtse, Phuentsholing, Gelephu, and Samdrupjongkhar towns for 24 hours need not pay Sustainable Development Fee (SDF) beginning April 14" (The Kuensel, March, 2023).

SUSTAINABLE DEVELOPMENT GOALS AND BHUTAN

According to the Sustainable Development Goals I United Nations Development Programme (n.d.), the sustainable development goals are the "...universal call to action to end poverty, protect the planet, and ensure that by 2030 all people enjoy peace and prosperity."

The goals as defined by UN directly pertains to the development which can address the concerns of each and every part of society. It was developed in the year 2015.

It has a collection of 17 interconnected goals that guides a country or economy to develop their developmental agenda on the basis of these goals so identified.

These 17 sustainable development goals are:

- No Poverty: Eradicating poverty in all its forms
- Zero Hunger: Making sure all people–especially children–have sufficient and nutritious food all year.
- Good Health and Well-being: Good health is essential to sustainable development
- Quality Education: Achieving inclusive and quality education for all.
- Gender Equality: Ending all discrimination against women and girls.
- Clean Water and Sanitation: Safe and affordable drinking water for all
- Affordable and Clean Energy: Investing in solar, wind and thermal power, improving energy productivity, and ensuring energy for all.
- Decent Work and Economic Growth: To promote sustained economic growth, higher levels of productivity and technological innovation.
- Industry, Innovation, and Infrastructure: Investment in infrastructure and innovation
- Reduced Inequalities: Reduce inequality within and among countries.
- Sustainable Cities and Communities: Make cities and human settlements inclusive, safe, resilient, and sustainable.
- Responsible Consumption and Production: Ensure sustainable consumption and production patterns.
- Climate Action: Take urgent action to combat climate change and its impacts.

- Life Below Water: Conserve and sustainably use the oceans, seas, and marine resources for sustainable development.
- Life on Land: Protect, restore, and promote sustainable use of terrestrial ecosystems, sustainably manage forests, combat desertification, halt and reverse land degradation, and halt biodiversity loss.
- Peace, Justice, and Strong Institutions: Promote peaceful and inclusive societies for sustainable development, provide access to justice for all, and build effective, accountable, and inclusive institutions at all levels.
- Partnerships for the Goals: Strengthen the means of implementation and revitalize the global partnership for sustainable development.

UNDP play a major role in development and implementation of all these 17 sustainable development goals identified. The same case applies to Bhutan as well. Bhutan already had the groundwork done in the form of GNH policy. This policy itself has all the elements of sustainable development goals devised by UNDP. Though SDG was developed in the year 2015, Bhutan is already working on it when His Majesty, the fourth King announced the GNH to be based on all dimensions of policy making. Moving in lines with SDG and GNH, Bhutan at the end of its 12th Five year plan is all set to move up from the status of "Least Developed Country" (LDC) at the end of 2023. The 12th five year plan is based as the platform to transition up the ladder. While in the effort to achieve the target set, Bhutan has done commendable job in moving in the desired direction. According to "Sustainable Development Goals | United Nations in Bhutan", there has been improvement in income poverty which reduced from 23.2 percent to 8.2 percent in the year 2017 for the period of past 10 years. Income inequality has reduced and country has shown significant growth at an average of 7.5 percent. The share of industry and service sector has increased considerably during the last two decades. And more and more population are being urbanized, it can be clearly understood that the policy framed for achievement of GNH philosophies are in line with the framework set by SDG. With more and more indicators showing the "moderately improving" to "on the track maintaining the SDG achievement" on the dashboard of "Sustainable Development Report" pertaining to Bhutan, clearly indicates the seriousness of Bhutan towards the sustainable development of the society.

There are many doubts all across the world with respect to the achievement of SDGs goals by the end of 2030. World over countries is in the search of some examples which can be used as yardstick for the way they have achieved or yet to be achieved. Bhutan serves as one of the live examples for the same. This was made possible because of the three important interconnected core goals which Bhutan has identified for the achievement of SDG (Yangka, D, et al. 2018). These three goals identified are:

- Wealth
- Carbon Emissions
- Happiness

According to Bhutan government's national report on sustainable development, its GNH policy is exactly in lines with the goals set by UNDP as "Sustainable Development Goals".

Out of three integrated goals identified by Bhutanese government for the achievement of SDG, "wealth" pertains to GDP of the economy, "carbon emission" pertains to the emission of greenhouse gases and "happiness" pertains to the gross national happiness of the nation as identified by fourth king.

As far as gross domestic product is concerned, if we keep the effects of pandemic aside which effected the whole world in some or other way, overall Bhutan's GDP trend is impressive. It has shown clear movement from agriculture-based society to one which is shifting towards more of industrial society. According to the data provided by National Statistical Bureau, there is decline in contribution of agriculture to the GDP and there is increment in the sectoral contribution of industries in the GDP. Since the economic design of Bhutan is as such that it is based on the principles of GNH for sustainable development, the importance of emission is also needs to be taken under consideration. And hence the concept of "Carbon emission" was introduced. Here it is important to understand the concept of carbon neutrality. "Climate neutrality refers to the idea of achieving net zero greenhouse gas emissions by balancing those emissions so they are equal (or less than) the emissions that get removed through the planet's natural absorption; in basic terms it means we reduce our emissions through climate action." ("Sustainable Development Goals I United Nations in Bhutan"). With the target of maintaining the temperature at 2^0, Bhutan has ambitious target in its place of 1.5^0. The target becomes more challenging when Bhutan has agreed to enter WTO by the end of December 2023 (Kuensel Online, 2023). It is rather a dilemma whether to move forward with more and more companies coming in with the WTO agreement or remain at the level of industrialization where it is now. Though Bhutan has three years to prepare itself according to the requirement of WTO (Kuensel Online, 2023). During these three years Bhutan has sufficient time to modify / alter its industrial policy which should not interfere with the GNH principles of sustainable development and at the same time should be acting as a catalyst for rapid industrialization to take place. Government of Bhutan has already identified few areas to be developed as industrial park so that monitoring and regulations can be made while not harming the basic texture of GNH policy of sustainable development. These industrial parks are to be located at Pasakha, Jemina, Dhamdum, Jigmeling and Motanga. There is one more industrial park to be setup at Samtse. Government of Bhutan is also working towards the development of transport and other infrastructural requirements for the industrial sector to flourish. For this Government of Bhutan has entered an agreement with India and Bangladesh and soon they will have another with Nepal and Thailand too (Kuensel Online, 2023). In order to move from an import oriented approach to export oriented, there is need for faster clearance of industrial license, better banking and lending facility. With the current lending system of high rates the target seems to be bit difficult. But according to current PM Dr. Tshering, these hurdles will not come in the way for the country to move ahead with WTO strictly in lines with GNH philosophy of sustainable development.

These recent developments are highlighted especially when point 12 of SDG demands to ensure sustainable consumption and production pattern. If this has to be achieved, then this thing has to be taken under consideration that principles of GNH should not be hampered or altered. Hence GNH comes in immense pressure for balancing the aspiration of nation to prosper in the lines of global economies and at the same time with preservation and promotion of culture and environment. Thus, Bhutan can be seen as small nation with low-income level struggling to keep the status of carbon absorbing nation along with keeping the high index for gross national happiness (Yangka, D, et al. 2018).

As far as Bhutan's GNH policy is concerned, it already places a high weightage to sustainability to have the development in across all the domain. GNH is already an established tool to measure the societal development which is different from what is measured in GDP (Giannetti et al.). It already has points which is in common terms with principles of SDG. According to Centre for Bhutanese studies, the 9 domains of GNH are interconnected with each other and are to be taken all as mutually exclusive to each other and not different from each other. Further more if we have detailed look onto what 17 points of SDG has to offer, we will find that both SDG and GNH has unique and multidimensional approach towards the definition of growth of country along with society.

GNH and SDG can be considered as two different policies with same approach. Yangka, D, et al. has established that most of the points of GNH are same as that of one which is in SDG policies. Hence the application of GNH policy in nation building can directly lead to the application of SDG. Further with this Yangka, D, et al. has suggested that with GNH policies in place it is easier for Bhutan to implement SDG. Hence it serves as a model for other nations which seeks for SDG.

CONCLUSION

The points discussed above highlighted the approach of Bhutan towards its willingness to have sustainable development of its economy. It's been in the priority of the policymakers since the beginning, to develop a plan that can inculcate all the stakeholders of the society so that sustainable development can be achieved. Though some of literatures highlighted the wrong timing of implementation of SDF to regional tourists which effected the tourism industry, but the policymakers' strong belief towards GNH and SDG made the Bhutan stand strong even in the difficult times of pandemic. With the commitment towards sustainable development, the SDG points have already been included in GNH and hence by following its principle the SDG is being achieved thereby serving as an example to the world.

The chapter will form the basis for the future researchers who can study further the impact of adaptation of these policies into the governance pertaining specifically to migration of worker from agriculture to industry and other non-agricultural domain. It can serve as the base for future researcher to study into the use and impact of technology in the implementation of the same.

REFERENCES

1st Five Year Plan (1961-1966). (n.d.). Gross National Happiness Commission, Royal Government of Bhutan.

Bhutan Tourism Monitor. (2019). Publication of the Tourism Council of Bhutan.

Denyer, S. (2007). *Bhutan Wonders If TV Really Brings Happiness.* Reuters. www.reuters.com/article/us-bhutan-television-idUSSP29261020070513

Drechsler, W. (2019). The Reality and Diversity of Buddhist Economics. *American Journal of Economics and Sociology, 78*(2), 523–560. doi:10.1111/ajes.12271

Dubey. (1996). Buddhism in Bhutan and Sri Lanka: A Comparative Study. In *Bhutan: Society and Polity.* New Delhi: Indus Publishing Company.

Dunne, J. (2015). First Thoughts Toward a Buddhist Economics. In M. Ricard & T. Singer (Eds.), *Caring Economics: Conversations on Altruism and Compassion Between Scientists, Economists and the Dalai Lama.* Picador.

Giannetti, B. F., Agostinho, F., Almeida, C. M. V. B., & Huisingh, D. (2015, January). A Review of Limitations of GDP and Alternative Indices to Monitor Human Wellbeing and to Manage Eco-System Functionality. *Journal of Cleaner Production*, *87*(1), 11–25. doi:10.1016/j.jclepro.2014.10.051

Givel, M., & Figueroa, L. (2015). *Early Happiness Policy as a Government Mission of Bhutan-A Survey of the Bhutanese Unwritten Constitution From 1619 to 1729*. Academic Press.

Helliwell, J. F., Layard, R., Sachs, J. D., De Neve, J.-E., Aknin, L. B., & Wang, S. (Eds.). (2022). *World Happiness Report 2022*. Sustainable Development Solutions Network.

History of GNH. (n.d.). GNH Centre Bhutan. www.gnhcentrebhutan.org/history-of-gnh/

International Monetary Fund. (1995). Bhutan: Background Paper. doi:10.5089/9781451806212.002

Jha, M. K. (2023, March 29). *The Financial Express*. https://www.financialexpress.com/business/defence-doklam-crisis-bhutan-china-talks-raise-concerns-for-india-3026076/

Kershaw, R. (2001). *Monarchy in South-East Asia: The Faces of Tradition in Transition*. Routledge.

Khan, S., Magd, H., Ansari, M., Sharma, P., Tirwa, I., & Bhuyan, U. (2022). *Analysis of Regional Tourists in Bhutan before Covid-19*. .2022.v4i1.39-58 doi:10.56411/anusandhan

Kose & Ohnsorge. (Eds.). (2020). A Decade after the Global Recession: Lessons and Challenges for Emerging and Developing Economies. World Bank. doi:10.1596/978-1-4648-1527-0

Kuensel, Bhutan's Daily. (2023a). *Govt. Waives SDF for Border Towns*. Kuensel Online. kuenselonline.com/govt-waives-sdf-for-border-towns/

Kuensel, Bhutan's Daily. (2023b). *Inflation Drops to 4.33 Percent in January*. Kuensel Online. kuenselonline.com/inflation-drops-to-4-33-percent-in-january/

Kuensel, Bhutan's Daily. (2022). *Phuentsholing Hoteliers Say SDF Could Affect Their Businesses*. Kuensel Online. kuenselonline.com/phuentsholing-hoteliers-say-sdf-could-affect-their-businesses/

Labour Force Survey Report. (2022). National Statistics Bureau, Bhutan.

Lhendup, K., & Panda, B. (2020). The Fable of Regional Tourists' Loyalty to Bhutan. *Journal of Tourism and Hospitality Management*, *8*(2). Advance online publication. doi:10.15640/jthm.v8n2a6

Luechauer. (n.d.). *Gross National Happiness of Bhutan and its False Promises*. https://www.gsdmagazine.org/the-false-promises-of-bhutans-gross-national-happiness/

Masaki, K., & Tshering, J. (2021). Exploring the origins of Bhutan's gross national happiness. *Journal of South Asian Development*, *16*(2), 273–292. doi:10.1177/09731741211039049

Nantichas. (2017). *Philosophy of Sufficiency Economy*. www.chaipat.or.th/eng/concepts-theories/sufficiency-economy-new-theory.html#:~:text=Sufficiency%20Economy%20is%20a%20philosophy

National Statistical Bureau. (2004). *Bhutan National Accounts Statistics*. National Statistical Bureau.

Newspaper, Bhutan's Daily. (2023). *Bhutan to Join WTO*. Kuensel Online, kuenselonline.com/bhutan-to-join-wto/

Newspaper, Bhutan's Daily. (2022). *Tourism Development: Looking beyond SDF*. Kuensel Online. kuenselonline.com/tourism-development-looking-beyond-sdf/

Rai, M. (2023). 15,354 Tourists Arrive in 2023 so Far. *The Bhutanese*. thebhutanese.bt/15354-tourists-arrive-in-2023-so-far/#:~:text=There%20has%20been%20a%20total

Royal Government of Bhutan (RGoB). (2012). *Bhutan: In Pursuit of Sustainable Development—National Report for the United Nations Conference on Sustainable Development 2012*. Royal Government of Bhutan.

Royal Monetary Authority of Bhutan, Annual Report 2008/09

Royal Monetary Authority of Bhutan, Annual Report 2020

Royal Monetary Authority of Bhutan, Annual Report 2022

SachsJ. (n.d.). https://www.aljazeera.com/opinions/2011/9/10/the-economics-of-happiness

Sangay. (2016). *Gross National Happiness – Royal Bhutanese Embassy, New Delhi*. Royal Bhutanese Embassy. www.mfa.gov.bt/rbedelhi/bhutan-at-glance/gross-national-happiness/

Santini, M., Tran, T., & Beath, A. (2017). *Investment Climate Assessment of Bhutan: Removing Constraints to Private Sector Development to Enable the Creation of More and Better Jobs*. World Bank Group. http://documents.worldbank.org/curated/en/673381506689928071/pdf/120144-ICA-PUBLIC-P153221-Bhutan- Competitiveness-Policy-Note-Web.pdf

Sarracino, F., Greyling, T., Peroni, C., O'Connor, K., & Rossouw, S. (2021). *A year of pandemic: levels, changes and validity of well-being data from Twitter. Evidence from ten countries*. Academic Press.

See, H. (2022). The Two Faces of Gross National Happiness: Can Bhutan's Nation-Building Strategy Also Be a Sustainable Alternative Development Paradigm? *Journal of Contemporary Asia*, *52*(3), 452–470. doi:10.1080/00472336.2021.1933139

Seldon, P. (2017). Regional Tourists Turn into Litterbugs in Bhutan. *The Bhutanese*. thebhutanese.bt/regional-tourists-turn-into-litterbugs-in-bhutan/.

Sumeda. (2022). Explained: Bhutan's Strategy to Levy Sustainable Development Fee on Tourists. *The Hindu*. www.thehindu.com/news/international/bhutan-strategy-daily-tax-tourists-sustainable-development-fee-tax/article65943766.ece

Sustainable Development Goals | United Nations Development Programme. (n.d.). *UNDP*. Retrieved May 11, 2023, from https://www.undp.org/bhutan/sustainable-development-goals

Sustainable Development Goals | United Nations in Bhutan. (2023). bhutan.un.org/en/sdgs

Sustainable Development Goals | United Nations in Bhutan. (2023). bhutan.un.org/en/sdgs

The Economist Newspaper. (n.d.). Why is Finland so happy? *The Economist*. https://www.economist.com/the-economist-explains/2018/03/26/why-is-finland-so-happy

Ura, K., Alkire, S., Zangmo, T., & Wangdi, K. (2012). *An extensive analysis of GNH index*. Centre for Bhutan Studies. doi:10.35648/20.500.12413/11781/ii036

Walcott, S. M. (2011, Spring). Making the Most of Geographic Disadvantage: Modernizing Bhutan, Education about. *Asia*, *16*, 1.

Wangmo, T. (2016). *Starting Strong on the SDGs in Asia: Readiness in Bhutan*. IGES Discussion Paper, Institute for Global Environmental Strategies.

Yangka, D., Newman, P., Rauland, V., & Devereux, P. (2018). Sustainability in an emerging nation: The Bhutan case study. *Sustainability (Basel)*, *10*(5), 1622. Advance online publication. doi:10.3390u10051622

Yangka, D., Rauland, V., & Newman, P. (2019). Carbon neutral policy in action: The case of Bhutan. *Climate Policy*, *19*(6), 672–687. doi:10.1080/14693062.2018.1551187

Chapter 13
Green Cement:
A Step Towards Sustainable Development

Sushma Singh
Bhilai Institute of Technology, Durg, India

Urvashi Shrivastava
Bhilai Institute of Technology, Durg, India

ABSTRACT

With the focus on sustainable development of the global arena in terms of construction and infrastructure, it is necessary to evaluate the impact of one of the ingredient "cements" on the environment. Therefore, this chapter investigates the literature available in the manufacturing process of cements, alternatives to it, and in particular, the emergence of "green cement," which assumes to be environmentally friendly. The action taken by the cement companies is clean and green, with profitability and sustainability as the prime focus by lesser CO2 emission. By innovating new technologies and adopting modernizations, the companies are focusing on developing such concrete, which is an environmentally friendly and economically sustainable. The future of green cement holds the promise of reduced carbon, enhanced durability, and improved performance. As environmental regulations become more stringent, the demand for green cement is expected to rise significantly. This presents a significant opportunity for the construction industry to embrace greener practices and contribute to a more sustainable environment.

INTRODUCTION

Our planet, where human and nature have lived for millions of years in harmony, portrayed a relationship of mother-child. Dwelling like nomads, it took years for human to realize the need of permanent shelter. Owing to his demanding quality thrust, there has been evolution from stone to wood and getting civilized from bricks to cement. The frail humanity of not sticking to quality assurance during demand peaks have made it more harmful for the environment (Chen, Habert, Bouzidi, & Jullien, 2010a). The present age is all construction with cement at core in industrialization and competition. Due to the depletion of natural fuel supplies, the shortage of raw materials, the exponentially rising demand for cement, and

DOI: 10.4018/978-1-6684-9503-2.ch013

climate-related environmental issues, the cement industry is facing numerous challenges in the twenty-first century (Imbabi, 2012). For managing climate, there are few things to be done. Firstly, the plants which are installed now will be having waste heat recovery system, secondly, they all will be having the system to utilize all type of non-fossil fuel, thirdly, as soon as possible, they will have access to renewable energy, and on top of that, cement manufacturing companies will produce blended cement, which is low carbon cement that uses waste material from other manufacturing processes, so it's a win-win situation and a great example of circular economy. (Anwar, Miyagawa, & Gaweesh, 2000). Innovation is needed for sustainability and Indian Cement Industry is looking forward in creating history by developing such technologies which will bring modernization and development along with sustainability (Naqi, 2019).

Nearly 8% of the environmental problems facing the planet are caused by the cement industry. In developing nations such as India, the concrete jungle is the new wave. India is the second largest producer of cement in the world, used to build bridges, dams, roads, sidewalks, and skyscrapers (Shah, 2004). However, it has also strained our environment and releases enormous amount of CO_2 into the atmosphere (Habert, 2014).

Every year, more than 4 billion tons of cement are produced, accounting for around 8% of global carbon dioxide emissions. To rescue, there comes something that will change the world forever – the Green Cement. Imagine roadways that don't absorb water in the monsoon, houses that don't wear and tear overtime, water resistant, heat resistant and damage proof (P-C, 2000). Learning from the past to build a better future, the new technologies have come up with a contemporary product inspired by the power of nature. It makes use of industrial waste that would otherwise be sent to landfill and it doesn't require as much heat during its production instead locking CO_2 inside the material and releasing up to 80% less of the pollutant (Zhang, 2021). Companies across the globe are working towards replacing the traditional method with this new technology and with the rising awareness among consumers surrounding the negative impact of carbon emissions combined with the various environmental policies introduced by government bodies, the industry is changing. Manufacturing and usage of green cement can significantly boost the cement industry's efforts towards a greener, cleaner, better and sustainable future (Damtoft, Lukasik, Herfort, Sorrentino, & Gartner, 2008).

HISTORY

Cement the world's most important building material with 4 gigatons being produced every single year; that is equivalent to 8 million international space stations or 12 million Boeing 747 jets. That's a lot of cement what you may not know, however is, that it is actually a mass producer of carbon dioxide as its production requires limestone and clay to be heated to around 1500 degrees Celsius in a chemical process which releases enormous amounts of carbon dioxide into the atmosphere. Reports state that currently it contributes towards 8% of the world's emissions and with escalating demand for homes, that number is only going to rise unless the industry adapts. India is the world's second-largest cement producer, with 329 million tonnes (MT) produced in FY20 and 381 MT projected by FY22. However, cement production has a variety of side effects that contribute to the already-increasing environmental degradation. The cement industry accounts for about 5% of global man-made CO_2 emissions, with the chemical process accounting for 50% and fuel combustion accounting for 40%. Due to the growing global population and the need for shelter and housing, cement production is expected to exceed 4.5 million tonnes by 2050. As a relatively cheap, extremely strong and incredibly versatile material, cement or more accurately

Figure 1. Specific thermal energy consumption in a clinker manufacturing process (Habert, 2014)

Kiln process	Thermal energy consumption (GJ / $t_{clinker}$)
Wet process	5.85–6.28
Long dry process	4.60
Shaft kiln	3.70–6.60
1-stage cyclone preheater	4.18
2-stage cyclone preheater	3.77
4-stage cyclone preheater	3.55
4-stage cyclone preheater plus calciner	3.14
6-stage cyclone preheater plus calciner	<2.93

concrete, that it's made from, is ubiquitous all over the world. In fact, it's the second most consumed product on earth beaten only by water. In total, concrete accounts for 8% of all global carbon dioxide emissions. So, it is a problem, and to be fair, the cement industry is well aware of it. However, there is room for innovation in this field, and smarter and cleaner technologies are required to control the current situation and look ahead to a more sustainable and clean future.

The cement industry creates happiness by providing cement for home building, having said that the energy consumption by this industry is more in specific requirement of energy to burn limestone which emit two things one is CO_2 which is inherent in the limestone and secondly it creates a product called clinker which is used to produce cement. In this process 50 to 60% of emission is due to burning of lime stone and CO_2 emission and another contributor is the energy consumption (Figure 1) through coal or any other fossil fuel or the power used through electricity (Wang, 2016).

These two are the major contributors and hence the challenge with the industry is how to bring down the CO_2 emission and reduce the consumption of energy and continue to give right quality of cement (Abdel-Ghani, 2018). At the global level the CO_2 emission pattern of cement is on an average 800 to 850 kg per ton of cement whereas if we look at the cement companies in India it may be ranging from say 600 to 750 kg per ton of cement. The reason for this difference being Indian Cement Companies are very conscious of energy cost and at the same time tries to be profitable. The proportion of polluting gases released in production of Portland cement is described in (Figure 2) which includes oxides of carbon, nitrogen and sulfur. Thus, the companies have been innovating and modernizing cement plants and cement technology in continuous way.

For sustainability of construction and cement industry, the companies are reducing the consumption and requirement of limestone by replacing it with fly ash of power plants, slags of steel plants and other waste products. Thus, increasing the consumption of waste products including slag and fly ash and at the same time producing better quality of cement. New investments in machinery and plant are made using the finest energy-efficient technologies, focused on helping the energy process in a way that substantially reduces its consumption (Shrivastava, 2017). For example, Dalmia cement is one of the lowest power consumption companies not only in India but also across the globe in terms of per ton power consumption (Mishra, 2013). The cement companies are focusing at first by transforming the power consumption from fossil fuel to 100% non-fossil fuel, thus companies are progressing in a big way, secondly with the

Figure 2. Portland cement production expressed in g/kg cement or g/kg clinker** (Habert, 2014)*

CO_2		SO_2		NO_x		CKD / PM-10	
Ref	Values	Ref	Values	Ref	Values	Ref	Values
[1]	870 g/kg*	[3]	0.40–0.60 g/kg*	[3]	2.4 g/kg*	[3]	0.1–10 g/kg*
[2]	810 g/kg*	[5]	0.82 g/kg*	[5]	1.2 g/kg*	[5]	0.49 g/kg*
[3]	800 g/kg*	[6]	0.58 g/kg*	[6]	1.5 g/kg*	[6]	0.04 g/kg*
[4]	820 g/kg*	[7]	0.27 g/kg**	[7]	1–4 g/kg**	[7]	200 g/kg**
[5]	690 g/kg*	[8]	0.54 g/kg**	[9]	2.5 g/kg**	[9]	0.1–0.3 g/kg**
[6]	810 g/kg*					[10]	150–200 g/kg**
[7]	900 g/kg*						
[10]	895 g/kg*						
Mean	814 g/kg*	Mean	0.5 g/kg*	Mean	2.5 g/kg*	Mean	25 g/kg*

use of waste hot gases of the cement plants by converting it into power and thirdly transforming thermal energy transition to renewable energy.

Between 2011 and 2013, China used more cement than the USA used in the entire 20th century and they are still going gangbusters now as they have got through their pandemic lockdowns. In 2021, they produced 2.5 billion tons of the stuff. That's more than the rest of the world put together. And that's a problem because with current production methods every tonne of cement results in a ton of carbon dioxide emissions. It's not just China though, obviously. The challenge at India level as well as across the globe is how to handle 55% to 60% of CO_2 emission which comes from the process of manufacturing cement. The Eastern Development Bank (EDB) has been studying the utilization of carbon capturing projects that many companies are creating to handle carbon dioxide emissions. And they have identified that carbon capturing technique is technically and physically viable and at the same time it needs collaborators and partners like fertilizer companies or soda sectors which can use CO_2 in their products which otherwise they generate by burning fossil fuels. So, the cement companies are tying up with those companies which require CO_2 thereby further reducing CO_2 emissions in atmosphere to 470 kg per ton of cement manufactured that's almost 50% of current emission by most of the western cement manufacturing companies. The construction sector's innovation of green cement is a game-changer. The carbon footprint of cement production around the world can be reduced, thanks to this environmentally friendly, sustainable cement. The usage of green cement can drastically reduce the 8% of carbon dioxide emissions produced by conventional cement manufacture, according to reports, and this percentage can be increased even further. Here's everything you need to know about green cement - the future of Sustainable Construction.

GREEN CEMENT: THE FUTURE OF CONSTRUCTION

The world is focusing on eco-friendly designs that provide a balance between nature and building structures, known as green architecture. This concept is being adopted in civil structure engineering, and many countries are switching to the use of green concrete as their basic component for the concrete mix.

Portland cement is the traditional component of the concrete mix, and reinforced concrete structures are mostly composed of reinforced concrete structures. The need is obvious for the structural engineering world shifting in this so-called green concrete which carry many advantages over Portland cement, as increase in environment damages such as global warming and its effect on human health such as kidney failure and respiratory diseases such as asthma, pulmonary ossification and lung cancer. Green cement is produced when the percentage of Portland cement will be substituted by the eco-friendly cement materials and concrete mix (Chen H. Q., 2023). These eco-friendly cement materials should be mixed with basic cement material to be considered as green cement i.e. green cements are produced from eco-friendly waste materials and in most of the cases the manufacturing process is same as the manufacturing process of Portland cement, but the production of this concrete causes the least harm to the environment (Chen, Lin, & Huang, 2011). The manufacture of cement involves in the emission of large amounts CO_2 which is harmful to the environment. Thus, green cement is being used to reduce the emission of CO_2. Green Cement is considered as a revolutionary topic in the history of the concrete industry. This was first invented in Denmark in the year 1998 by Dr. WG. The design and manufacture of green cement or we can say green concrete should follow reduce, reuse and recycle technique (Boesch & Hellweg, 2010). The common materials used to produce green concrete are recycled demolition waste aggregate, recycled concrete aggregate, blast furnace slag, recycled glass aggregate and fly ash. Apart from reduction in CO_2 emission, it is also being used for reducing the use of natural resources like lime stone, natural river sand, natural rocks, etc. and to utilize wastes formed during production of other materials. It is currently used in the construction of bridges, columns, dams and much more (Phair, 2006).

Green cement is the building material of the future due to its high tensile strength and corrosion resistance. It is expected to grow to 56.6 billion US dollars by 2027, and is a low-carbon product that reduces the carbon footprint of cement production. According to Mark Group, the green cement market was worth 27.02 billion US dollars in 2021 and is expected to grow to 56.6 billion by 2027. Due to significant growth in the building, construction, and infrastructure industries, the India Green Cement Market is expected to grow rapidly by 2023. This is due to the need to reduce carbon emissions caused by the heating of clay, sand, and pulverised limestone in the cement manufacturing process. Green cement requires little heating during production, reducing its carbon footprint. Furthermore, rising urbanisation and industrialization, as well as the increasing use of Green Cement in residential and commercial applications, will drive market growth. Due to rising constructions and demand for green and sustainable buildings, the residential segment is expected to have the largest market share by 2023.

WHAT IS GREEN CEMENT?

Green Cement is an eco-friendly cement that uses a carbon-negative process of manufacturing. The majority of the industry's leftover trash is one of the main raw materials utilized to make green cement. The primary ingredients used in the production of green cement are fly ash and slag from the blast furnace (Jansson, 2015). Green cement is cement that is designed to be ecologically friendly and reduce the carbon footprint of cement production. The word "green" is a prefix that refers to something that is environmentally friendly or helps to preserve the environment. Thanks to international symposiums boosting awareness of sustainability and climate change, the importance of green cement is continuously being realized by all countries. Different industrial organizations have each patented a variety of processes and technology for making green cement (Kırgız, 2018).

Because of the utilization of latest innovative technologies, production is incredibly energy-efficient. This procedure emits a very little amount of carbon dioxide. According to reports, this technique reduces the carbon footprint by 40%. The cost of green cement is lower because the majority of the raw materials used in its production are industrial waste (Elinwa & Mahmood, 2002). Cement replacement materials that have been documented and validated in laboratory tests and practice include fly ash, blast furnace slag and silica fumes. They are a byproduct of coal combustion, iron smelting, and the production of elemental silicon or ferro silicon alloys in electric arc furnaces.

GREEN CEMENT ADVANTAGES

Green cement, according to experts, is the future of sustainable construction due to its ability to reduce carbon emissions.

- Green cement is created from industrial wastes such as slag and fly ash, which can contribute to land pollution. It uses industrial waste such as fly ash, silica fume, and blast furnace, which may require the disposal of several acres of land. It also uses a variety of industrial wastes and by-products, making the process more cost-effective and reducing the amount of land required for disposal (Zhang C. Y., 2021).
- Waste products are used as raw materials, lowering manufacturing costs and making it both economical and environmentally friendly.
- Additionally, cement produced with fossil fuels necessitates a significant amount of energy. However, the production of green cement is technology-driven and requires very little energy.
- Green cement is an excellent choice for bridge and road construction due to its high tensile strength and corrosion resistance, making it more durable than traditional cement.
- Green cement is a strong concrete that can withstand acid rain and temperature changes while lowering carbon dioxide emissions by up to 80%. It requires less heat during production and emits up to 80% less carbon dioxide.
- Green cement production is energy efficient, requiring less energy and reducing the consumption of non-renewable resources such as coal. It contains industrial by-products and is resistant to temperature fluctuations, lowering heating and cooling costs.
- Green cement is advantageous to contractors because it does not require additional tools or equipment, dries faster, and can be applied within 72-96 hours, saving time. It also allows for faster curing and setting times.
- Green cement has a longer life than traditional cement because it is more fire resistant and resistant to deteriorating agents such as acid rain.

THE SUSTAINABILITY OF GREEN CEMENT

Green cement is the ideal solution for stronger, longer-lasting concrete. It is environmentally friendly, reduces carbon dioxide emissions, and requires less energy to produce. It is extremely long-lasting and durable, with high early strength, resilience, durability, crack resistance, and low chloride permeability. It has three to four times the corrosion resistance of ordinary cement.

Green concrete has a higher tangibility than ordinary cement due to its use of calcined clay and limestone, which reduce porosity and improve mechanical strength. It is also easily hybridisable, which encourages more durable and sustainable building. This results in a higher tensile strength than conventional concrete. Green cement is a novel product or manufacturing method that is frequently regarded as a risky financial investment. To meet the needs of the construction industry, national governments must promote the production of green cement and provide incentives to investors and manufacturers. Home inspectors should be hired to keep records of green cement structures, allowing for certification of such structures and payment of builders.

The production of ordinary cement requires a large amount of heat, resulting in the emission of carbon dioxide. This is a major source of air pollution, so the concept of green cement was developed to reduce emissions and protect the environment. It burns limestone and clayey materials at a temperature of 1200 degrees Celsius to produce clinker. Green cement is a viable option for lowering carbon dioxide and other harmful emissions. It is made using a technologically advanced process that produces less harmful gases during the unit production, i.e. clinker production. It is also known as "cement produced with the help of a carbon-reducing process."

BENEFITS OF USING GREEN CEMENT

The word Sustainable Construction came into existence due the claim given by experts that by using Green Cement sustainability can be achieved. The reasons behind their claim are as follows:

- Industrial Wastes like fly ash and slag are used to make green cement. This can contribute to resolving a major issue of industrial waste disposal, which contributes significantly to land pollution. As a result, it keeps land from becoming a dumping ground and, eventually, from being destroyed. Green cement uses a variety of industrial wastes and by-products such as fly ash (a by-product of the coal combustion process) and blast furnace slag (a by-product of the iron manufacturing process), making the process more cost-effective and reducing the amount of land required for disposal of such products.
- Because the raw materials used are mostly waste products, manufacturing costs of green cement are significantly less as compared to normal cement. Moreover, production of green cement is both economic and environmental friendly as it does not use any fossil fuel in its production process. Additionally, cement produced with fossil fuels necessitates a significant amount of energy. However, the production of green cement is technology-driven and requires very less energy.
- Because of its high tensile strength and corrosion resistance, green cement is an excellent choice for bridge and road construction. It has a lower shrinkage rate than conventional cement, making it more durable. Green cement is a strong concrete that can withstand the negative effects of acid rain and temperature shifts.
- The carbon footprint of green cement production is approximately 80% lower than that of traditional cement production as it does not require as much heat production as needed by traditional cement production. Hence it minimises the carbon di oxide emissions by 80%.
- As compared to traditional types of cement, green cement has proven to be more beneficial to contractors, due to the fact that green cement does not requires the use of any additional tools or

equipment. Due to its faster drying properties, the time can be saved and the coating can be applied within 72-96 hours. This type of cement also has a faster curing and setting rate.

- Green Cement is more durable as compared to traditional cement as it has better resistance to deteriorating agents such as acid rain and has better fire resistance properties.
- In this competitive era providing quality product at low price is the key for sustainability of any business and an industry where huge investments are required; profitability is equally necessary and producing a product which is cost effective serves the purpose.

VARIETIES OF GREEN CEMENT

The common types of green cement used in the construction industry can be briefly explained as follows:

1. Magnesium Oxychloride Cement (MOC)
    - This type of green cement contains magnesium oxide (MgO) powder and a concentrated solution of magnesium chloride ($MgCl_2$) as its major components.
    - Magnesium oxide (MgO) powder and Magnesium chloride ($MgCl_2$) are the residual products obtained from mining of magnesium.
    - This type of green cement has a quick setting time and a high compressive strength. Additionally, Magnesium Oxide absorbs CO_2 from atmosphere but its strength can be reduced significantly by water. Hence, additives such as fly ash (15%), silica fume, phosphoric acid, and soluble phosphates can be used to improve its durability and strength. However, it can cause steel corrosion and is not suited for reinforced concrete structures unless proper precautions are followed.
2. Ekkomaxx Cement
    - Ceratech Company in the United States has developed and manufactured Ekkomaxx Cement, a type of environmentally friendly cement.
    - Ekkomaxx Cement consists primarily of fly ash (approximately 95%) and other additives, including renewable liquid additives (about 5%).
    - This green cement, manufactured by Ceratech Company, adheres to the United States Green Building Code and the International Code Council, resulting in a carbon footprint of zero.
    - Ekkomaxx Cement exhibits exceptional qualities such as high resistance to cracking, sulphate attack, and corrosion. It also offers increased strength, resilience, and durability.
    - Ceratech Company produces Ekkomaxx Cement in the United States, using a composition of 95% fly ash and 5% renewable liquid additives. This cement aligns with standards set by the International Code Council and the United States Green Building Council, ensuring an almost nonexistent carbon footprint.
    - The production process of this cement reduces the use of virgin materials by 95% and decreases water requirements by 50%.
    - Ekkomaxx Cement boasts several key characteristics, including high early strength, resilience, crack resistance, low chloride permeability, resistance to sulphate attacks, durability, corrosion resistance, which exceeds that of conventional cement by over three times, and superior resistance to freezing and thawing compared to normal cement.

3. Geopolymer cement

   Geopolymer cement, also known as alkali-activated cement, is a type of green cement that is made from aluminosilicates. Aluminosilicates are obtained from industrial by-products like fly ash. Geopolymer cement is a more environmentally friendly alternative to ordinary Portland cement because it emits 95% less carbon dioxide. Geopolymer cement is competitive with ordinary Portland cement in performance and cost.

   Here are some additional details about geopolymer cement:

   Geopolymer cement is made by reacting aluminosilicates with an alkaline solution. This reaction creates a solid material that has the properties of cement.

   Geopolymer cement is strong, durable, and resistant to chemicals. It is also fire-resistant and has good thermal insulation properties.

   Geopolymer cement is a sustainable material because it is made from waste materials. It also produces significantly less greenhouse gas emissions than ordinary Portland cement.

   Geopolymer cement is a promising new material that has the potential to revolutionize the construction industry. It is a more environmentally friendly and sustainable alternative to ordinary Portland cement.

4. Ferrocrete
   - Ferrocrete is a form of environmentally friendly cement created by combining two primary industrial by-products: silica from the glass industry and iron from the steel industry.
   - When this mixture undergoes curing in the presence of carbon dioxide, it has the potential to produce carbon-negative cement.
   - The University of Arizona is credited with the initial invention of Ferrocrete Cement.
   - Ferrocrete Cement is produced by blending waste by-products, silica and iron, derived from the steel and glass industries. This mixture is then cured using carbon dioxide, resulting in a material that has the potential to be carbon-negative. The innovation of ferrocrete was pioneered by scientists at the University of Arizona.

5. Sequestrated Carbon Cement

   Sequestered carbon cement is a type of green cement that is made from seawater or brine that is mixed with carbon dioxide. It is a substitute or alternative for Portland cement. The production process of this cement involves filtering carbon dioxide-containing gases through seawater.

   The Calera Corporation in California developed this cement. In this process, calcium and magnesium are stripped from seawater and react with carbon dioxide to produce high-quality cement. This cement is white, air-permeable, and stronger than regular Portland cement.

   Here are some additional details about sequestered carbon cement:

   Sequestered carbon cement is a more environmentally friendly alternative to Portland cement because it produces significantly less greenhouse gas emissions.

   Sequestered carbon cement is also a sustainable material because it uses seawater as a raw material.

   Sequestered carbon cement is a promising new material that has the potential to revolutionize the construction industry. It is a more environmentally friendly and sustainable alternative to Portland cement.

6. Calcium Sulfoaluminate Cement

   Calcium Sulfoaluminate Cement is a form of environmentally friendly cement that decreases energy consumption by 25% and reduces carbon emissions by 20%.

- It exhibits an exceptionally fast setting rate, allowing it to reach the strength equivalent to conventional cement within 24 hours instead of the usual 28 days.
- This type of cement is widely utilized for concrete projects involving bridge decks and airport runways. The production process of calcium sulfoaluminate cement involves a kiln that operates at a temperature of 1232.2°C (2250°F), significantly lower than the 1426.6°C (2500°F) required for conventional cement. This results in lower CO_2 emissions into the atmosphere. The rapid setting properties of calcium sulfoaluminate cement enable it to achieve the 28-day strength of conventional concrete in just 24 hours.
- Due to its rapid setting characteristics, it is commonly employed in construction projects where quick concrete setting is vital, such as bridge decks and airport runways. Additionally, when a larger quantity of gypsum is added, calcium sulfoaluminate cement can function as a shrinkage compensating cement.
- This type of cement can deliver energy savings of up to 25% and provides environmental benefits by reducing CO2 emissions by approximately 20% when compared to Portland cement.

7. Cement Produced Using Superheated Steam

   Cement produced using superheated steam refers to a cement production method known as steam curing or autoclave curing. In this process, the cementitious materials, such as cement, aggregates, and water, are mixed together to form a paste or mortar. The mixture is then placed in a chamber or autoclave where it is exposed to high-pressure steam at elevated temperatures.

   The superheated steam accelerates the hydration process of the cement, leading to rapid strength development and curing. This method is often used for the production of precast concrete products, such as blocks, pipes, and panels, where a shorter curing time is desired to increase productivity.

   By using superheated steam curing, the cement gains early strength and achieves the desired properties in a shorter period compared to traditional curing methods. This allows for faster demolding, handling, and transportation of the precast concrete products. Additionally, the controlled environment inside the autoclave ensures uniform curing and minimizes the risk of surface cracking or defects.

   It's important to note that the production of cement itself does not involve the use of superheated steam. The curing process of cementitious materials using superheated steam is a technique employed in specific applications within the construction industry.

8. Cement Produced with Reactive Hydrothermal Liquid-phase Densification

   Cement produced with reactive hydrothermal liquid-phase densification refers to a cement production method that involves a specialized densification process using a reactive hydrothermal liquid. This technique is typically employed to enhance the properties and performance of cementitious materials.

   In this process, cementitious materials, such as cement, aggregates, and water are mixed together to form a paste or slurry. The reactive hydrothermal liquid, which typically contains specific additives or chemicals, is then introduced into the mixture.

   Under controlled temperature and pressure conditions, the reactive hydrothermal liquid facilitates a chemical reaction within the cementitious mixture. This reaction promotes the densification of the cement matrix, resulting in improved strength, durability, and other desirable characteristics.

   The exact mechanisms and specific additives used in reactive hydrothermal liquid-phase densification can vary depending on the desired properties and application requirements. This technique is often utilized in the production of high-performance cementitious materials for applications such as construction, infrastructure, and specialized industrial uses.

Overall, the use of reactive hydrothermal liquid-phase densification in cement production aims to optimize the performance and quality of cement-based materials by enhancing their microstructure and properties through a carefully controlled chemical process.

INDUSTRY ANALYSIS

In 2022, the worldwide green cement market is expected to be worth $31.1 billion USD. The leading green cement businesses are progressively focused on R&D expenditures to modernize their manufacturing technologies in order to increase green cements' quality and performance.

To address the price-sensitive demography in emerging nations, prominent market participants are producing cost-effective product variations with increased durability compared to conventional cement. Furthermore, producers are concentrating on the development of cement formulas that allow for a lower carbon footprint by utilizing eco-friendly and alternative raw materials. They are also utilizing social media as well as traditional advertising initiatives to promote public knowledge of green cement.

To expand their geographical footprint, the leading players are also engaging in smart mergers and acquisitions (M&As), as well as developing partnerships and collaborations with construction businesses, architects, and engineers. Other factors influencing the market include rapid urbanization, rising environmental concerns, significant growth in global construction and infrastructure development projects, favorable government initiatives encouraging the use of green construction materials and green buildings, and rising disposable income levels. The market value is predicted to reach US$ 63.3 billion by 2028, increasing at a CAGR of 12.3% between 2023 and 2028.

FEW TOP COMPANIES PRODUCING GREEN CEMENT WORLDWIDE

1. **CarbonCure Technologies Inc.**: It is a company based in Canada. It is a multinational corporation that creates cutting-edge carbon capture and utilisation solutions for the concrete and cement industries. Their method is pumping CO2 into the concrete mixture, which permanently stores the carbon, resulting in a stronger and more ecologically friendly end product. They work with cement manufacturers, construction businesses, and industry players in North America, Europe, and Asia-Pacific, and they spend extensively in sustainable building practises and the implementation of cutting-edge sustainable technology in collaboration with concrete producers.
2. **CEMEX S.A.B. de C.V.:** Established in 1906, this company is based in Monterrey, Mexico. It manufactures and sells cement, ready-mix concrete, and aggregates, and is known for pioneering sustainable cement manufacturing methods. It also performs research and development to minimise greenhouse gas emissions. CEMEX has a strong presence in over 50 countries throughout the world, with large operations in the United States, Europe, Asia, Africa, and the Middle East.
3. **CRH plc:** Situated in Dublin, Ireland in 1970, CRH plc is a prominent producer of carbon-neutral cementitious materials. They make significant investments in green technology, new production methods, and initiatives such as the use of alternative fuels and raw materials. They are present

in over 30 countries and have a substantial business presence in North America, South America, Europe, and Asia-Pacific.

4. **LafargeHolcim:** LafargeHolcim is a worldwide company that is a leading manufacturer and supplier of construction materials such as cement, aggregates, and ready-mix concrete since 2015, operated from Zug, Switzerland. It replaces clinker in its green cement formulation with supplemental cementitious materials such as fly ash, slag, and limestone calcined clay, and undertakes R&D to support sustainable development practises. The corporation has offices in more than 70 countries, including Europe, North America, Asia-Pacific, Latin America, and Africa.
5. **Fortera Corporation:** Based in California, USA since 2007, this company specialises in the development of new carbon capture and utilisation solutions for the cement industry. Their patented process produces green cement, which may be used as a supplemental cementitious material or combined with Portland cement. North America, Europe, Asia-Pacific, and Latin America are among the nations and areas in which the corporation works.
6. **Votorantim cimentos S.A.:** Votorantim Cimentos is a Brazilian firm that has engaged in R&D to improve the environmental performance of its cement products. It has a global presence and activities in several countries, and it works with academic institutes and industry organisations to promote sustainable development in cement industry.
7. **Anhui Conch Cement:** Anhui Conch Cement is a Chinese cement producer that was established in 1997. It manufactures and distributes cement and other construction products, has green supply chains in place, and adheres to high environmental regulations. It has made global investments in cement manufacturing facilities and projects, notably in Southeast Asia, South Asia, Central Asia, and Africa.

SOME OF THE INDIAN PLAYERS IN THIS INDUSTRY

1. **Kiran Global Chem Limited:** Kiran Global Chem Limited, established in 1979, operated from Pondicherry, India is a renowned maker of green cement under brand Geocement, made from the polymerization of Silicon (Si) and Aluminium (Al) ions that is exceptionally energy-efficient and sustainable. Geoactivator is a proprietary liquid made up of several industrial byproducts that are combined to provide an energy-efficient binding solution. The firm performs considerable R&D operations concentrating on carbon emission reduction, and their product is widely used in several green building projects.
2. **UltraTech Cement Ltd.:** UltraTech Cement Ltd., from 1983 in Maharashtra, is a multinational corporation with place of effective management in India and is the country's largest producer of ready-mix concrete, white cement, and grey cement. To improve sustainability, they have incorporated measures such as alternate fuels and waste heat recovery systems. Through their green cement options, they have earned a reputation for applying sustainable production practises, minimising environmental effect, and lowering carbon footprint. Manufacturing factories, grinding units, and distribution networks are strategically placed throughout India, and they also have a significant business presence in the Middle East, Africa, Europe, and Asia-Pacific.
3. **ACC Ltd.:** ACC Ltd., an Ambuja Cements subsidiary and member of the Adani Group, primarily operates in India and has a significant local market presence. It is involved in the introduction of alternative fuels and raw materials into green cement manufacturing, such as biomass, waste-derived

fuels, and agricultural wastes. To lower clinker consumption, it has also built waste heat recovery systems and incorporated fly ash into cement manufacturing. It also performs substantial research and development to focus on novel cement processing methods. ACC Ltd. is leveraging global experience and resources to increase its market reach outside India.

4. **Dalmia Cement**: Dalmia Cement, one of the leading cement manufacturers in India, has made significant strides in promoting green cement. They have developed and introduced various eco-friendly cement products, such as Dalmia DSP and Dalmia Cement One, which have lower carbon footprints compared to traditional cement.
5. **Shree Cement:** Shree Cement has been at the forefront of adopting green cement practices in India. They have implemented innovative measures to reduce energy consumption and carbon emissions in their manufacturing process. Shree Cement's products, such as Shree Jung Rodhak and Shree Ultra, focus on sustainability and environmental conservation.
6. **Ambuja Cements:** Ambuja Cements, a part of the Lafarge Holcim group, is committed to sustainable cement production. They have introduced Ambuja Compocem, a green cement product that utilizes fly ash and slag, reducing the carbon footprint and promoting the efficient use of waste materials.
7. **JSW Cement:** JSW Cement is India's foremost green cement firm, having a current capacity of 17 MTPA across production sites in Vijayanagar, Andhra Pradesh, Salboni, West Bengal, Jajpur, Odisha, Dolvi, Maharashtra, and Fujairah, UAE. Shiva Cement, our subsidiary, is now investing over 1,500 crores in a 1.36 MTPA clinker facility project in Sundergarh, Odisha. The project entails the installation of a 1 MTPA grinding plant and accompanying infrastructure.

The table 1 tabulate the specific strength of cement companies in India, these Indian companies are actively investing in research and development, collaborating with technology providers, and adopting advanced manufacturing processes to produce greener cement. Their initiatives contribute to India's sustainable development goals and demonstrate a commitment to environmental stewardship in the construction industry. Moreover, ACC Ltd, Shree Cement Ltd, and Ultratech have inked a collabora-

Table 1. Distinctive strength of cement companies

Types of Cement	Distinctive Strength of Cement Companies
Kiran Global Chem Limited	Geocement, made from the polymerization of Silicon (Si) and Aluminium (Al) ions that is exceptionally energy-efficient and sustainable.
UltraTech Cement Ltd	Is country's largest producer of ready-mix concrete, white cement, and grey cement.
ACC Ltd	Involved in the introduction of alternative fuels and raw materials into green cement manufacturing, such as biomass, waste-derived fuels, and agricultural wastes.
Dalmia Cement	Have developed and introduced various eco-friendly cement products, such as Dalmia DSP and Dalmia Cement One, which have lower carbon footprints compared to traditional cement.
Shree Cement	Shree Cement's products, such as Shree Jung Rodhak and Shree Ultra, focus on sustainability and environmental conservation.
Ambuja Cements	They have introduced Ambuja Compocem, a green cement product that utilizes fly ash and slag, reducing the carbon footprint and promoting the efficient use of waste materials.
JSW Cement	Investing over 1,500 crores in a 1.36 MTPA clinker facility project in Sundergarh, Odisha. The project entails the installation of a 1 MTPA grinding plant and accompanying infrastructure.

tion agreement to encourage low-carbon developments in India. The agreement was signed in Geneva by the World Business Council for Sustainable Development's Cement Sustainability Initiative and the International Finance Corporation (IFC). The plan will serve as a potential transition path for the Indian cement sector in order to cut direct emissions by 18% by 2050.

CHALLENGES AND OBSTACLES IN CURRENT SCENARIO

The few challenges encountered in green concrete applications in construction include difficulties in meeting regulatory standards such as minimum clinker concrete levels and chemical composition of cement, a paucity or insufficient durability data spanning up to 20 years or more, differentiation of green concrete for various applications, and moving towards additional research and development to foster a deeper knowledge of green concrete chemistry. This involves the rewriting of different building regulatory norms in order to make them more ecologically friendly and to stimulate the use of green concrete.

Reasonable and affordable technologies and guidelines are required for systematic and structured processing and production of green concrete. Along with this sufficient performance data is also required to justify and inform changes in construction codes and standards. Data from the field on application of green concrete are also very insufficient. Green concretes' field applications in various structural forms, as well as standardization, are required to promote the generation of long-term data and guide their implementations. In addition, further durability data on shrinkage, creep, abrasion, and ASR are necessary.

Furthermore, to support the long-term development and deployment of green concrete in field applications, novel and cost-effective activators are required. Cheap and inexpensive characterization methodologies are also required, especially in developing countries where the cost of research and development is excessively high. Construction companies, universities, and research organisations should be recognised for pioneering the development and use of green concrete in infrastructure projects (Baykara, 2017).

To raise knowledge about the benefits of green concrete, appropriate and sufficient hands-on training and re-training should be provided to build-environment specialists. This will help to spread green concrete practises in the building industry. Similarly, the difficulties encountered in its adoption by construction and consulting firms should be addressed (Shwekat, 2018).

Green concrete research clusters should also be formed to promote the continuous innovation of green concrete products and construction practises. Indigenous manufacturing practises should be encouraged in order to provide low-cost green concrete products while reducing reliance on costly foreign technologies (Benhelal, 2012).

Furthermore, efforts to promote green concrete in construction should be integrated to reduce research duplication, speed green concrete applications, and generate best practises to permanently embed it in the building industry (Liu, 2021).

CONCLUSION

The future scope of green cement holds tremendous potential for addressing the environmental challenges posed by traditional cement production. As sustainability and carbon reduction become increasingly important global concerns, green cement offers a promising solution. The ongoing research and development in green cement technology are paving the way for innovative and eco-friendly cement formulations. These advancements aim to further reduce carbon emissions, minimize resource consumption, and enhance the overall sustainability of the construction industry.

With the introduction of alternative materials, such as fly ash, slag, and silica fume, as well as the utilization of renewable additives, green cement is continuously evolving to meet the demands of a greener future. The development of new production methods, like steam curing and reactive hydrothermal liquid-phase densification, further expands the possibilities for environmentally friendly cement production.

Moreover, collaborations between industry leaders, researchers, and government entities are fostering a supportive environment for the commercialization and widespread adoption of green cement technologies. Initiatives such as partnerships and regulatory incentives play a crucial role in promoting the use of green cement in construction projects globally. The future of green cement holds the promise of reduced carbon footprints, enhanced durability, and improved performance. As awareness of sustainability grows and environmental regulations become more stringent, the demand for green cement is expected to rise significantly. This presents a significant opportunity for the construction industry to embrace greener practices and contribute to a more sustainable and resilient built environment.

By embracing the potential of green cement and investing in its development and implementation, we can work towards a future where construction materials are not only strong and durable but also environmentally friendly, helping to mitigate climate change and preserve our planet for future generations.

REFERENCES

Abdel-Ghani, N. T.-S.-H. (2018). Utilization of by-pass cement kiln dust and air-cooled blast-furnace steel slag in the production of some "green" cement products. *HBRC Journal, 14*(3), 408-414.

Aïtcin, P.-C. (2000). Cements of yesterday and today: Concrete of tomorrow. *Cement and Concrete Research, 30*(9), 1349–1359. doi:10.1016/S0008-8846(00)00365-3

Anwar, M., Miyagawa, T., & Gaweesh, M. (2000). Using rice husk ash as a cement replacement material in concrete. *Waste Management (New York, N.Y.), 1*, 671–684.

Baykara, H. C., Cornejo, M. H., Murillo, R., Gavilanes, A., Paredes, C., & Elsen, J. (2017). Preparation, characterization and reaction kinetics of green cement: Ecuadorian natural mordenite-based geopolymers. *Materials and Structures, 50*(3), 1–12. doi:10.161711527-017-1057-z

Benhelal, E. Z., Zahedi, G., & Hashim, H. (2012). A novel design for green and economical cement manufacturing. *Journal of Cleaner Production, 22*(1), 60–66. doi:10.1016/j.jclepro.2011.09.019

Boesch, M., & Hellweg, S. (2010). Identifying improvement potentials in cement production with life cycle assessment. *Environmental Science & Technology, 44*(23), 9143–9149. doi:10.1021/es100771k PMID:21047057

Chen, A., Lin, W.-T., & Huang, R. (2011). Article. *Materials & Design, 32*, 635–642.

Chen, C., Habert, G., Bouzidi, Y., & Jullien, A. (2010a). Environmental impact of cement production: Detail of the different processes and cement plant variability evaluation. *Journal of Cleaner Production, 18*(5), 478–485. doi:10.1016/j.jclepro.2009.12.014

Chen, H. Q., Qin, R., Chow, C. L., & Lau, D. (2023). Recycling thermoset plastic waste for manufacturing green cement mortar. *Cement and Concrete Composites*, *137*, 104922. doi:10.1016/j.cemconcomp.2022.104922

Damtoft, J., Lukasik, J., Herfort, D., Sorrentino, D., & Gartner, E. (2008). Sustainable development and climate change initiatives. *Cement and Concrete Research*, *38*(2), 115–127. doi:10.1016/j.cemconres.2007.09.008

Elinwa, A., & Mahmood, Y. (2002). Ash from timber waste as cement replacement material. *Cement and Concrete Composites*, *24*(2), 219–222. doi:10.1016/S0958-9465(01)00039-7

Habert, G. (2014). Assessing the environmental impact of conventional and 'green' cement production. Woodhead Publishing.

Imbabi, M. S., Carrigan, C., & McKenna, S. (2012). Trends and developments in green cement and concrete technology. *International Journal of Sustainable Built Environment*, *1*(2), 194–216. doi:10.1016/j.ijsbe.2013.05.001

Jansson, H. B., Bernin, D., & Ramser, K. (2015). Silicate species of water glass and insights for alkali-activated green cement. *AIP Advances*, *5*(6), 067167. doi:10.1063/1.4923371

Kırgız, M. S. (2018). Green cement composite concept reinforced by graphite nano-engineered particle suspension for infrastructure renewal material. *Composites. Part B, Engineering*, *154*, 423–429. doi:10.1016/j.compositesb.2018.09.012

Liu, Y. Z., Zhang, Z., Hou, G., & Yan, P. (2021). Preparation of sustainable and green cement-based composite binders with high-volume steel slag powder and ultrafine blast furnace slag powder. *Journal of Cleaner Production*, *289*, 125–133. doi:10.1016/j.jclepro.2020.125133

Mishra, A. P. (2013). Green cement for sustainable concrete using marble dust. *International Journal of Chemtech Research*, *5*(2), 616–622.

Naqi, A. J., & Jang, J. (2019). Recent progress in green cement technology utilizing low-carbon emission fuels and raw materials: A review. *Sustainability (Basel)*, *11*(2), 537. doi:10.3390u11020537

Phair, J. W. (2006). Green chemistry for sustainable cement production and use. *Green Chemistry*, *8*(9), 763–780. doi:10.1039/b603997a

Shah, S. P. (2004). Development of 'green'cement for sustainable concrete using cement kiln dust and fly ash. *Proceedings of the International Workshop on Sustainable Development and Concrete Technology*, 15-23.

Shrivastava, S. R. (2017). A systematic literature review on green manufacturing concepts in cement industries. *International Journal of Quality & Reliability Management*, *34*(1), 68–90. doi:10.1108/IJQRM-02-2014-0028

Shwekat, K. W., & Wu, H.-C. (2018). Benefit-cost analysis model of using class fly ash-based green cement in masonry units. *Journal of Cleaner Production*, *198*, 443–451. doi:10.1016/j.jclepro.2018.06.229

Wang, L. C., Chen, S. S., Tsang, D. C. W., Poon, C.-S., & Shih, K. (2016). Recycling contaminated wood into eco-friendly particleboard using green cement and carbon dioxide curing. *Journal of Cleaner Production*, *137*, 861–870. doi:10.1016/j.jclepro.2016.07.180

Zhang, C. Y., Yu, B., Chen, J.-M., & Wei, Y.-M. (2021). Green transition pathways for cement industry in China. *Resources, Conservation and Recycling*, *166*, 105355. doi:10.1016/j.resconrec.2020.105355

Zhang, K. S., Shen, P., Yang, L., Rao, M., Nie, S., & Wang, F. (2021). Development of high-ferrite cement: Toward green cement production. *Journal of Cleaner Production*, *327*, 129487. doi:10.1016/j.jclepro.2021.129487

Chapter 14
Characterising a Green Supply Chain Using Multi-Dimensional Construct:
Evidence From an Emerging Economy

Chetan V. Hiremath
Kirloskar Institute of Management, India

Chandan Chavadi
https://orcid.org/0000-0002-7214-5888
Presidency Business School, India

Sudhansu Sekhar Nanda
https://orcid.org/0000-0003-0974-181X
Kirloskar Institute of Management, India

Shankargouda C. Patil
Rani Channamma University, India

ABSTRACT

The study proposed a construct for a green supply chain, which aims to analyze the different elements that contribute to a network being considered "green." The study found that there are many types of research done on green products, but very few on green corporations and supply chains. Also, there is a lack of consensus in definitions and previous studies have not examined the inter-relations among the variables that contribute to a supply chain being considered green. The review of the literature reveals the range of factors, from the procurement of raw materials to end customer perceptions which make the supply chain green. In a simple random sampling, 125 responses were collected from Karnataka to explore the interrelations between the factors. The analysis evaluates whether the underlying structure of a green supply chain is uni-dimensional or multi-dimensional. Whether the factors have a hierarchical relationship and are factors independent or correlated, the analysis found that there are four correlated factors without any higher order.

DOI: 10.4018/978-1-6684-9503-2.ch014

INTRODUCTION

Going green is no more an option but a compulsion to organizations across the globe. Recently, there has been unprecedented economic growth catalysed by technology and globalization which have further fuelled excessive consumption and undue exploitation of natural resources. The environment has been exploited beyond what it can sustain. Off late consumers, activists, governmental and non-governmental agencies are concerned about the devastating effects of unsustainable consumption and production patterns. In the past few years, there has been an outcry from scientists, environmental activists about the irreversible impact on the environment owing to unnecessary human activity (Vachon, S. and Klassen, R.D., 2006; Susana G *et al.*, 2011; Miroshnychenko, I, 2017; World Economic Forum, 2021; Jili *et al.*, 2023).

The present environmental hazards are compelling customers and companies to shift towards sustainable living by adopting and promoting products that are sustainable and pro-environment, commonly referred as "Green Products'' (Wasik, 1996; Elkington and Makower, 1988; Unruh and Ettenson, 2010; Chen & Chai, 2010). Despite all the uproar for sustainable living via sustainable products, there is a lack of consensus on what actually constitutes a sustainability, both from the manufacturer's perspective as well as from the consumers perspective (de Freitas Netto, S.V *et al*,2020). Marketers are challenged to produce and promote green products.

The widespread phenomenon of consuming green products is a manifestation of customers' concern for the environment and his/her own wellbeing. To that extent, higher prices are no more a concern for consumers, who exhibit pro-environment behaviour. To encash the emerging opportunities, more and more companies are weaving their organizational strategies around pro-environmental practices (Unruh and Ettenson, 2010; Nguyen-Viet 2023).

Many international frameworks have been set up to monitor the green practices of the firms (Shetty and Bhat,2022). Studies have established a relationship between sustainable practices and firm's performance. Developing countries are in no way behind developed countries in terms of ESG reporting (Bose and Khan,2022). But studies suggest that developing counties face many challenges in adopting suitable green practices (Rupa and Saif, 2022; Alzubi and Akkerman, 2022; Yadav and Kumar, 2023). Also, the divergent priority of Nations and differed level of commitments across the globe make it difficult for the companies to adopt standard green practices and reporting them (Shetty and Bhat,2022; Bose and Khan,2022).

REVIEW OF LITERATURE

There are myriads of concepts about green products put forth by several studies. The extant review of literature affirms lack of universal and coherent definition in terms of what constitutes green product and green practices (Orsato, 2006; Albino et al., 2009; L´opez-Gamero et al., 2009; Ritter *et al.*, 2015, Han *et al.*, 2023). Moreover, the literature is also scattered over the concept of green (Rivera-Camino, 2006). There are many perception challenges with the definition of Green. Also, few studies address how different approaches towards sustainability yield different results (Figg and Thorpe, 2023). The following items (Table No. 1) were extracted based on the comprehensive review to understand what made firms green.

Table 1. Characteristics of green

Efficient products (V1)	Shamdasami, 1993; Mangun & Thurston, 2002.; Bearse et al., 2009.;Chen & Chang 2013
Fewer resources to produce (V2)	Zappelli *et al* 2016; Khandelwal & Yadav, 2014
Better quality products (V3)	Mangun & Thurston, 2002.; Bearse et al., 2009.; Chen & Chang 2013
Natural ingredients (V4)	Pedro Pereira Luzio 2013
Recycled raw material (V5)	Chen, T, B.& Chai, L.T, 2010; Khandelwal & Yadav, 2014
Nontoxic (V6)	Khandelwal & Yadav, 2014; Chen, 2001Chen, 2001
Biodegradable Raw material (V7)	Chen, T, B.& Chai, L.T, 2010; Khandelwal & Yadav, 2014
Biodegradable product (V8)	Chen, T, B.& Chai, L.T, 2010; Khandelwal & Yadav, 2014
Organic (V9)	Mangun & Thurston, 2002., Bearse et al., 2009.; Parker, Segev, & Pinto, 2009
Green Packing (V10)	Chen, T, B.& Chai, L.T, 2010; Jeevan P, et al., 2017; Nguyen-Viet, 2023
Positioning as green product (V11)	Jeevan P, et al., 2017; Leonidou et. al. 2011; Ahmed et al 2023
Ethical Commitment by Corporate (V12)	Mazar & Zhong 2010; Bratanova et al.2012
Local procurement (V13)	Bezenc & Blili, 2010
Sustainable practices (V14)	Chen, 2001; Zappelli et al 2016;
Social conscience (V15)	Pedro Pereira Luzio 2013
Green manufacturing (V16)	Zappelli *et al* 2016; Yenipazarli, A 2015
Low carbon footprint (V17)	Chen, 2001; Chen, T, B.& Chai, L.T, 2010

Source: compiled by authors based on Review of Literature

PRODUCT

Products which are contributing to a sustainable world and meet the consumer needs without harming the environment are termed as green products (Shamdasami, 1993). Attributes such as energy efficiency, environmentally friendly packaging, biodegradability, organic products are generally chosen dimensions to define a green product (Mangun & Thurston, 2002., Bearse et al., 2009., Chen & Chang 2013). Products which are produced employing procedures and processes which are energy efficient are also referred to as a dimension of a green product (Massawe & Geiser, 2012). Thus, customers views become critical in deciding which companies are producing green products (Shevchenko et al, 2023).

INPUTS

Some studies tag products whose inputs are safe to the environment, recyclable and are less intensively packaged as green products (Chen, T, B.& Chai, L.T, 2010). The terms recyclable, environmentally safe, reusable, devoid of phosphate and ozone friendly usually mean green products (Khandelwal & Yadav, 2014). Another term which is synonymously used to describe a green product is the word 'organic product' (Parker, Segev, & Pinto, 2009)

Green products can also be defined as products made out of non-renewable resources and nontoxic chemicals (Chen, 2001). Green products are endowed with several advantages, such as: low water and energy consumption leading to lower pollution. Also, their packaging can be recycled (Eyraud *et al.*, 2013; Zappelli *et al* 2016).

COMMUNICATION

Sustainability can also be deciphered from the way where economic agents communicate their commitment to green products. One of the dimensions of sustainability is packaging that should not be overlooked (Jeevan P, et al., 2017). There is a dire need among business houses to weave their communication in accordance to the changing expectations among green consumers. The heightened concern among businesses and customers, for the environmental friendliness has led to increased green advertising (Leonidou *et. al.*, 2011; Ahmed *et al.*, 2023).

COMMITMENT

People tend to associate an element of morality in consuming and promoting green products (Mazar & Zhong 2010; Bratanova *et al.*2012; Deci & Ryan, 2000). Green consumer segmentation is a very compounded exercise, as consumers are scattered in their perception of green products. A wide range of environmental concerns exhibited by consumers spanning from energy efficient products to products with minimum hazards. Few past studies have demonstrated that virtuousness and the act of being virtuous are very significant elements in green consumerism. Over and above the intrinsic benefits, there is a holistic benefit associated with green consumerism (Spielmann, N, 2020). A strong and significant association has been found between undue consumption and its negative repercussions on the environment (Svensson & Wagner, 2012). A simple and a minor tweaking of the existing brown product and projecting it as a green product may not suffice the manufacturers, as consumers are viewing the corporated commitment to continuous innovation and sustainable practices (Yenipazarli, A 2015).

Consumers tend to adopt those products from the sources which are trustworthy, genuine and exhibit friendly relationships (Bezenc & Blili, 2010; Centobelli *et al*, 2023). By doing so, the satisfaction of aligning the behaviour with the beliefs is derived.

Green consumers when consuming green products tend to forego some of the expectations such as appealing packaging, fragrance, variety, design, style, etc (Pedro Pereira Luzio 2013).

METHODOLOGY

This section discusses the philosophical stand taken, extraction methods adopted and the reason for selecting them. In the later part of the section whether shortlisted items have any structure, if so the nature of their interrelations are explored.

PHILOSOPHICAL VIEW

The philosophical stand that depends on what she thinks reality is. Either it is realism or relativism. In realism the researcher assumes that there is only one truth and in relativism many. This stand influences the way one enquires, measures, creates and applies the knowledge. If the researcher believes in independent truth not influenced by the observer it is called objectivism. On the other hand, if an observer creates the narrative and it is accepted as truth it is called subjectivism. The present study takes a structural realism approach which is a type of realism. Where it is believed that theory explains truth, but incompletely (James Ladyman, 1998). Also, the study assumes constructionism, that the theory is an interplay between the belief of an observer and the object under study (Tom Andrews, 2012).

This assumption is essential in social sciences as most of the concepts like intelligence, loyalty, fairness etc are human abstracts of what exists. The concept is an abstraction and reality is a nexus of concepts. When the relationships between many such concepts are measured, it is known as grand theory (C. Wright Mills 1959).

The study assumes that there is an underlying construct called *green* and it could be measured. Also, it is assumed that the construct is influenced by the opinion of the observers and it is possible that what observers presently think as green could be a partial truth about what actually constitutes green. As a result, it will have true component, error component and random variation (Lawley, D. & Maxwell, A., 1973; Grace *et al*, 2008). If all the observer's belief about green is the same, then their views about green should also be the same. As a result, when their beliefs are measured, they all converge (Spearman, 1904).

SAMPLING DESIGN

The study used simple random sampling. The respondents were from tier two cities. In total 125 responses were collected, of which ten were dropped due to incompleteness. This sample size is sufficient to undertake the study (MacCallum et al 1999,2001).

The responses were collected from tier two cities. In India Tier two cities are the ones with the population between 10000 to 100000. The geographical areas were selected with the assumption that if the trend has reached these cities it has become a part of daily practices (Singh, P. and Siddiqui, A.A. 2021), they represent (Ministry of Finance, Govt. of India, 2015). As a result, findings would represent the population.

MEASUREMENT

The present study explores the construct called *"Green"*. The Green construct was defined by rationally selected items, which were well bounded by the rationality. These items were measured using a practical seven-point scale. (De Vellis, R. F. 2016). The summaries are as follows (Table 2).

The above table explains the summary statistics of all the 17 variables considered for the study along with, univariate normality, Shapiro-Wilk's test values (Shapiro and Wilk, 1965). All the variables considered for the study do not come from the normally distributed populations. The selection of the principal component extraction method depends on whether the underlying data follows normal distribution.

Table 2. Data summary

Variable	n	Mean	Std.Dev	Median	Min	Max	25th	75th	Skew	Kurtosis	W-S Test Value	P Value	Result
V1	106	5.8	1.2	6.0	2.0	7.0	5.0	7.0	-0.7	-0.26	0.87	*	NO
V2	106	5.3	1.5	5.5	1.0	7.0	4.0	7.0	-0.6	-0.38	0.89	*	NO
V3	106	5.7	1.3	6.0	1.0	7.0	5.0	7.0	-0.7	-0.07	0.86	*	NO
V4	106	5.9	1.2	6.0	1.0	7.0	5.0	7.0	-1.2	1.69	0.83	*	NO
V5	106	5.6	1.4	6.0	2.0	7.0	5.0	7.0	-0.9	-0.26	0.85	*	NO
V6	106	5.7	1.4	6.0	1.0	7.0	5.0	7.0	-1.4	1.93	0.81	*	NO
V7	106	6.1	1.1	6.0	1.0	7.0	5.3	7.0	-1.4	2.59	0.77	*	NO
V8	106	6.0	1.3	6.0	1.0	7.0	5.0	7.0	-1.7	3.17	0.75	*	NO
V9	106	5.9	1.2	6.0	2.0	7.0	5.0	7.0	-1.0	0.37	0.82	*	NO
V10	106	4.7	2.0	5.0	1.0	7.0	4.0	6.0	-0.5	-0.82	0.89	*	NO
V11	106	4.6	1.8	5.0	1.0	7.0	3.3	6.0	-0.5	-0.73	0.91	*	NO
V12	106	5.8	1.2	6.0	1.0	7.0	5.0	7.0	-1.0	1.33	0.85	*	NO
V13	106	4.7	1.8	5.0	1.0	7.0	4.0	6.0	-0.5	-0.64	0.91	*	NO
V14	106	5.7	1.3	6.0	1.0	7.0	5.0	7.0	-1.0	0.87	0.86	*	NO
V15	106	5.6	1.3	6.0	1.0	7.0	5.0	7.0	-0.8	0.57	0.88	*	NO
V16	106	5.2	1.5	5.0	1.0	7.0	4.0	6.0	-0.6	-0.41	0.91	*	NO
V17	106	5.8	1.4	6.0	1.0	7.0	5.0	7.0	-1.1	0.96	0.83	*	NO
* <0.001													

Source: Output of MVN package (Korkmaz S *et al*, 2014)

PRINCIPAL COMPONENT EXTRACTION METHODS

The sample size of *115* was sufficient to undertake this study (MacCallum et al 1999,2001). The data adequacy of sample (MSA 0.88) and sphericity of data (K=101.41, df=16, p=0) were adequate to continue with the analysis (Bartlett, M. S. 1951; Kaiser, H. F 1974, 1992). Also, it was observed that the data was not normally distributed (Madria Skewness =2281, Madria Kurtosis =20.3, p<0). Based on these findings the study used "Principal Axis (PA) Analysis" (Kim, J.-O *et al* 1978,1987; Fabrigar *et al* 1999). Throughout the study oblique rotation was used as the underlying structure was unknown and "Promax rotation" was used to get "simple structure" (Bryant and Yarnold 1995, Revelle and Rocklin 1979; Thurstone 1947).

As the underlying structure of *Green* is unknown the following questions were answered,

1. Whether these shortlisted items have any underlying structure?
2. Whether the structure is unidimensional or multidimensional?
3. Are these dimensions single order or have higher order?

WHETHER THE SHORTLISTED ITEMS HAVE ANY UNDERLYING STRUCTURE

Figure 1. Response details of the respondents
Source: Primary data

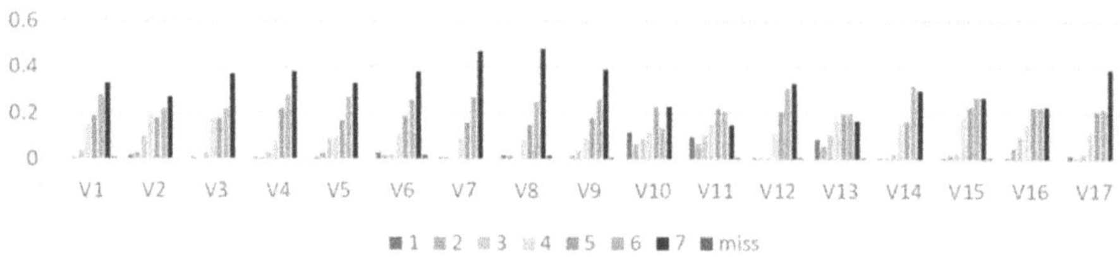

The average Inter -rater reliability was more than 90%. This indicated that respondent's views about what they believed about *Green* converged (Shrout 1979; MacGraw & Wong,1996). Also, the internal consistency among the items was measured using *alpha* (Cronbach, L. 1951). For the data collected, alpha values were (0.9, 0.92, 0.94) at 95% confidence interval. The high value of alpha indicated high internal consistency among items. Based on these results further analysis was carried out.

WHETHER THE STRUCTURE IS UNIDIMENSIONAL OR MULTIDIMENSIONAL

The knowledge of dimensionality of a construct helps understand it better (Law *et al*, 1998). The factor analysis was used to understand the contextual structure of the data (McGrath R. E., 2005; Spearman1904; Hair *et al* 2014: Morrison, 1990; Lawley, D., & Maxwell, A. 1973).

Since, the structure of the data collected decides the underlying construct (M. Sarstedt *et al*, 2017). It would be improper to assign dimensionality to a construct as it is context dependent (Bollen, 2011). Thus, tests were conducted to ascertain its dimensionality.

TEST FOR UNI-DIMENSIONALITY

To abide by the rule of parsimony, we started with a single factor (unidimensional) construct using the "Principal Axis Method". The factor loadings are shown in the figure No. 2. The total variance explained was 43%. Which is less than recommended 60%. We decided to explore multidimensional constructs to explain more variance (Nunnally and Bernstein, 1994; Hair et al. 2014).

Figure 2. Single factor construct for green
Source: Output of psych Package (Revelle W, 2021)

Table 3. Number of dimensions to extract

#	n_Factors	Method	Family
1	1	Acceleration factor	Scree
2	1	R2	Scree_SE
3	1	TLI	Fit
4	1	RMSEA	Fit
5	2	Bentler	Bentler
6	3	CNG	CNG
7	4	beta	Multiple_regression
8	4	Optimal coordinates	Scree
9	4	Parallel analysis	Scree
10	4	Kaiser criterion	Scree
11	4	SE Scree	Scree_SE
12	4	BIC	Fit
13	6	CRMS	Fit
14	8	t	Multiple_regression
15	8	p	Multiple_regression
16	15	Bartlett	Barlett
17	15	Anderson	Barlett
18	15	Lawley	Barlett

Source: Output of a Parameters package (Lüdecke D *et al*, 2020)

TEST FOR MULTIDIMENSIONALITY

Though the decision with respect to number of factors to extract has been discussed at length, there is no agreement among scholars with respect to exact number to extract (Kaiser HF, 1992, Horn, 1965). Also, many methods of factor extraction have been used (Lawley and Maxwell, 1963; MacCallum and Tucker, 1991, MacCallum et al., 2007). The above table (No.3) lists some popular methods of factor extraction and numbers of factors to extract. About 33.33% methods (i. e six out of 18) suggested four factors for the given data. The further analysis was performed using four factors.

Figure 3. Exploratory factor analysis with four correlated factors
Source: Output of psych package (Revelle W, 2021).

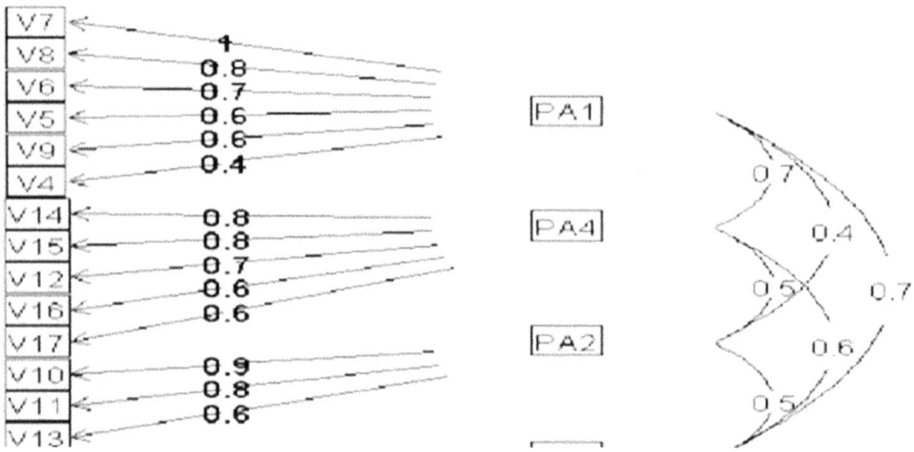

The factor loadings are shown in the above Figure 3. However, factors share correlation between them, e.g. correlation between PA1 and PA4 is 0.7. There is a possibility of higher dimension (Zinbarg, R. E *et al* 1997). The study proceeded to check whether higher dimensions exist.

ARE THESE DIMENSIONS SINGLE ORDER OR DO THEY HAVE HIGHER ORDERS?

The bi factor model captures the underlying common factor for correlated latent variables (LV) The higher dimensional models can be either hierarchical or may have a common factor (Zinbarg, R. E et al 1997). Both the higher order models are shown in the figure No4 a and b.

In hierarchical models LV create their own LVs (Fig No 4a). The higher order LV partially captures the correlation among the lower LVs (Holzinger & Swineford, 1937). On the other hand, the common factor model explains the underlying common structure among all the items and the interactions among LVs (Fig No 4 b). The remaining variance is explained by the orthogonal group factors and the error terms (Schmid & Leiman, 1957).

Figure 4a. Hierarchical model

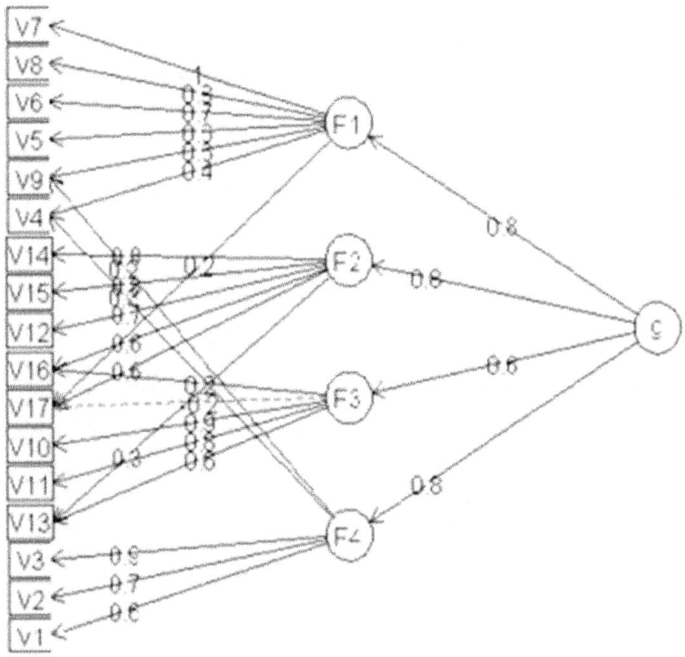

Figure 4b. Common factor model
Source: Output of psych package (Revelle W, 2021).

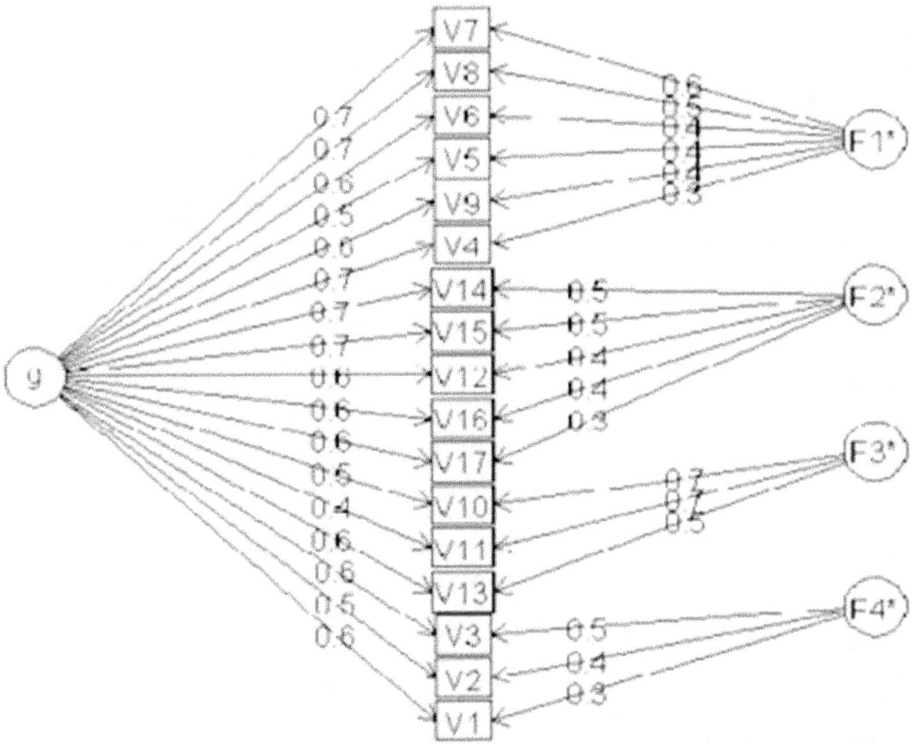

WHICH IS A BETTER MODEL?

The fit indices indicate that the correlated four factor model is better than the other two (χ^2 =129.79, df= 114, RMSE = 0.08, BIC = -221.33). as it has a smaller chi square, value lower RMSE index and BIC. The single factor model (χ^2=431.02, df= 119, RMSE = 0.151, BIC = -133.63) is slightly better than higher order model (χ^2=447.9, df= 119, RMSE = 0.155, BIC = -116.75) However, Chi square difference test (χ^2=2.642,df=5,p=.978) indicates that there is no significance difference between single

Table 4. Four factor correlated model

Variable	item	PA1	PA4	PA2	PA3	h2	u2	com
V15	12	0.9				0.8	0.2	1
V14	11	0.9				0.7	0.3	1
V12	9	0.7				0.6	0.4	1.3
V16	13	0.6				0.6	0.4	1.3
V17	14	0.5				0.6	0.4	1.9
V8	6		1			0.9	0.1	1
V7	5		0.9			0.8	0.2	1
V6	4		0.6			0.5	0.5	1.1
V10	7			0.9		0.7	0.3	1.1
V11	8			0.8		0.7	0.3	1
V13	10			0.5		0.6	0.4	1.6
V3	3				0.9	0.7	0.3	1.1
V2	2				0.8	0.6	0.4	1.3
V1	1				0.6	0.5	0.5	1.1

	PA1	PA4	PA2	PA3
SS loadings	3.05	2.41	1.97	1.79
Proportion Var	0.22	0.17	0.14	0.13
Cumulative Var	0.22	0.39	0.53	0.66
Proportion Explained	0.33	0.26	0.21	0.19
Cumulative Proportion	0.33	0.59	0.81	1

With factor correlations of

	PA1	PA4	PA2	PA3
PA1	1	0.7	0.5	0.6
PA4	0.67	1	0.3	0.6
PA2	0.54	0.3	1	0.5
PA3	0.61	0.6	0.5	1

Source: Output of psych package (Revelle W, 2021)

factor model and correlated four factor model. But, RMSE index and BIC are in favour of the latter (Hu & Bentler, 1998, Kline, R. B. 1998; Bollen et al, 2014,) . The four-factor model with correlation was adopted with following changes.

The variables V4 and V5 were removed from the list as first one had < 0.5 loading on PA1 . and second had communality <0.5. After this iteration variable V9 was removed for both of the above reasons. The revised model was obtained as shown below,

All the loadings and communalities of the revised model were above 0.5. In total 66% of the variance was explained by these four factors. The model fit statistics were (χ^2= 52.23, df=41, p=.11, RMSE =0.048, BIC=-142.31). The Tucker Lewis index was 0.971. The overall fit of the model was good.

CONFIRMATORY FACTOR ANALYSIS

The dual approach of EFA followed by CFA is strongly recommended for structural analysis. In CFA, we try to *confirm* the item variables obtained from literature review and EFA. The above exploratory model was further subjected to rigorous testing using Confirmatory Factor Modelling. In CFA, model fit is measured. A measurement model fit should be assessed to ensure structural model fit. Model fit is a measure of fit between theoretical model and observed data model. The measurement model is assessed for goodness of fit. The goodness of fit is ascertained by CMIN/df ratio. The CMIN/df ratio should be between 3.0- 5.0 (Hair, *et al* 2014). Additionally, model fit is ascertained by values of CFI, TLI. The values for CFI and TLI of 0.95 were considered ideal. But, now any values greater than 0.9 are accepted. However, there is no change in acceptable values of. RMSEA (0.08) (Schumacker & Lomax, 2010; Hu and Bentler 1999). The present measurement model yielded an acceptable model fit (CFI=.948, TLI =0.933 and RMSE = 0.077).

Figure 5. Confirmatory factor analysis
Source: Output of sem Plot (Epskamp, S., 2015)

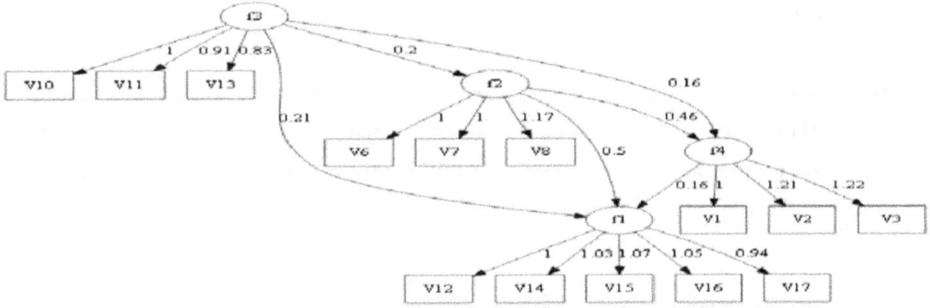

DISCUSSION

Table 5. Factors and items

Factors	Items
Commitment (PA1)	Ethical Commitment by Corporate (V12)
	Sustainable practices (V14)
	Social conscience (V15)
	Green manufacturing (V16)
	Low carbon footprint (V17)
Green Practice (PA2)	Green packing (V10)
	Positioning as green product (V11)
	Local procurement (V13)
Product (PA3)	Efficient products (V1)
	Fewer resources to produce (V2)
	Better quality products(V3)
Inputs (PA4)	Non-toxic (V6)
	Biodegradable Raw material (V7)
	Biodegradable (V8)

Source: compiled by authors

The four-factor correlated model sheds light on what makes products green. The four factors are commitment, sustainable practice, product and inputs. The number of items reduced from 17 to 14. This indicates that the initial selection of items based on review of literature was good.

COMMITMENT (PA1)

Social conscience (V15) and sustainable practices (V14) of the management outweigh their Ethical commitment (V12). The respondents believe that both the parties exceeded ethical commitment by the firms. Whereas green manufacturing (V16) and low carbon footprints (V17) did not matter much as they are already a part of sustainable practice.

GREEN PROMOTION (PA2)

The green packaging (V10) helps companies to position (V11) themselves as committed. Customers are not much concerned about the local procurement (V13). But consider it important.

PRODUCT (PA3)

All believed that the green products essentially mean better quality products (V3) and consume less resources for manufacturing (V2) and perform better (V1).

INPUTS (PA4)

The green products are biodegradable (V8) as they make use of biodegradable raw materials (V7). Also, green products use non-toxic (V6) materials.

MANAGERIAL IMPLICATIONS

The green companies claim that their ethical is not enough, but include sustainable practices with social consciousness along with existing practices like green manufacturing.

The companies should show their commitment for green practices by the way of local procurement and green packaging.

The customers expect green companies to produce performing, better quality products with fewer resources.

The produce of a green company should produce bio-degradable products using non-toxic and bio degradable raw materials only.

At factor level Company's commitment (PA1) is most important followed by which type of inputs (PA4) they use. Later it is important for the firm to promote itself as a green company (PA2) and deliver quality products (PA3). Since all these factors are correlated, companies cannot neglect any one of them.

The company should not only walk its talk but should also talk its way.

REFERENCES

Ahmed, R. R., Streimikiene, D., Qadir, H., & Streimikis, J. (2023). Effect of green marketing mix, green customer value, and attitude on green purchase intention: Evidence from the USA. *Environmental Science and Pollution Research International*, *30*(5), 11473–11495. doi:10.100711356-022-22944-7 PMID:36094715

Alzubi, E., & Akkerman, R. (2022). Sustainable supply chain management practices in developing countries: An empirical study of Jordanian manufacturing companies. *Cleaner Production Letters*, *2*, 100005. doi:10.1016/j.clpl.2022.100005

Andrews, T. (2012). What is Social Constructionism? *The Grounded Theory Review*, *11*(1), 39–46.

Azevedo, S. G., Carvalho, H., & Machado, V. C. (2011). The influence of green practices on supply chain performance: A case study approach. *Transportation Research Part E, Logistics and Transportation Review*, *47*(6), 850–871. doi:10.1016/j.tre.2011.05.017

Bartlett, M. S. (1951). The effect of standardization on a Chi-square approximation in factor analysis. *Biometrika*, *38*(3/4), 337–344. doi:10.2307/2332580

Bezencon, V. & Blili, S. (2010). Ethical products and consumer involvement: what's new? *European Journal of Marketing, 44*(9/10), 1305-1321.

Bollen, K. A. (2011). Evaluating Effect, Composite, and Causal Indicators in Structural Equation Models. *Management Information Systems Quarterly*, *35*(2), 359–372. doi:10.2307/23044047

Bollen, K. A., Harden, J. J., Ray, S., & Zavisca, J. (2014). BIC and Alternative Bayesian Information Criteria in the Selection of Structural Equation Models. *Structural Equation Modeling*, *21*(1), 1–19. doi:10.1080/10705511.2014.856691 PMID:31360054

Bose, S., & Khan, H. Z. (2022). Sustainable development goals (SDGs) reporting and the role of country-level institutional factors: An international evidence. *Journal of Cleaner Production*, *335*, 130290. doi:10.1016/j.jclepro.2021.130290

Bratanova, B., Loughnan, S., & Gatersleben, B. (2012). The moral circle as a common motivational cause of cross-situational pro-environmentalism. *European Journal of Social Psychology*, *42*(5), 539–545. doi:10.1002/ejsp.1871

Bryant, F. B., & Yarnold, P. R. (1995). Principal-components analysis and confirmatory factor analysis. In L. G. Grimm & P. R. Yarnold (Eds.), *Reading and understanding multivariate statistics* (pp. 99–136). American Psychological Association.

Centobelli, P., Cerchione, R., Chiaroni, D., del Vecchio, P., & Urbinati, A. (2020). Designing business models in circular economy: A systematic literature review and research agenda. *Business Strategy and the Environment*, *29*(4), 1734–1749. doi:10.1002/bse.2466

Chen, C. (2001). Design for the environment: A quality-based model for green product development. *Management Science*, *47*(2), 250–263. doi:10.1287/mnsc.47.2.250.9841

Chen, T. B., & Chai, L. T. (2010). Attitude towards the environment and green products: Consumers perspective. *Management Science and Engineering, 4*(2), 27-39.

Chen, Y. S., & Chang, C. H. (2013). Greenwash and green trust: The mediation effects of green consumer confusion and green perceived risk. *Journal of Business Ethics*, *114*(3), 489–500. doi:10.100710551-012-1360-0

Cronbach, L. (1951). Coefficient alpha and the internal structure of tests. *Psychometrika*, *16*(3), 297–334. doi:10.1007/BF02310555

de Freitas Netto, S. V., Sobral, M. F. F., Ribeiro, A. R. B., & Soares, G. R. L. (2020). Concepts and forms of greenwashing: A systematic review. *Environmental Sciences Europe*, *32*(1), 19. doi:10.118612302-020-0300-3

Deci, E. L., & Ryan, R. M. (2000). The "what" and "why" of goal pursuits: Human needs and the self-determination of behavior. *Psychological Inquiry*, *11*(4), 227–268. doi:10.1207/S15327965PLI1104_01

DeVellis, R. F. (2016). *Scale development: Theory and applications* (Vol. 26). Sage Publications.

Elkington, H., & Makower. (1988). *The green consumers*. New York: Penguin Books.

Epskamp, S. (2015). semPlot: Unified visualizations of structural equation models. *Structural Equation Modeling*, *22*(3), 474–483. doi:10.1080/10705511.2014.937847

Eyraud, L., Clements, B., & Wane, A. (2013). Green investment: Trends and determinants. *Energy Policy*, *60*, 852–865. doi:10.1016/j.enpol.2013.04.039

Fabrigar, L. R., Wegener, D. T., MacCallum, R. C., & Strahan, E. J. (1999). Evaluating the use of exploratory factor analysis in psychological research. *Psychological Methods*, *4*(3), 272–299. doi:10.1037/1082-989X.4.3.272

Figge & Thorpe. (2023). Circular economy, operational eco-efficiency, and sufficiency. An integrated view. *Ecological Economics, 204*(B), 107692.

Grace, J., & Bollen, K. (2007). Representing general theoretical concepts in structural equation models: The role of composite variables. *Environmental and Ecological Statistics*, *15*(2), 191–213. doi:10.100710651-007-0047-7

Hair Jr, J., Sarstedt, M., Hopkins, L., & Kuppelwieser, V. (2014). Partial least squares structural equation modeling (PLS-SEM): An emerging tool in business research. *European Business Review*, *26*(2), 106–121. doi:10.1108/EBR-10-2013-0128

Holzinger, K. J., & Swineford, F. (1937). The Bi-factor method. *Psychometrika*, *2*(1), 41–54. doi:10.1007/BF02287965

Horn, J. L. (1965). A rationale and test for the number of factors in factor analysis. *Psychometrika*, *30*(2), 179–185. doi:10.1007/BF02289447 PMID:14306381

Hu, L., & Bentler, P. M. (1998). Fit indices in covariance structure modeling: Sensitivity to under parameterized model misspecification. *Psychological Methods*, *3*(4), 424–453. doi:10.1037/1082-989X.3.4.424

Hu, L., & Bentler, P. M. (1999). Cut off criteria for fit indexes in covariance structure analysis: Conventional criteria versus new alternatives. *Structural Equation Modeling*, *6*(1), 1–55. doi:10.1080/10705519909540118

Huang, H., Long, R., Chen, H., Sun, K., Sun, Q., & Li, Q. (2023). Why don't more people engage in green practices in China? A policy-oriented approach to promoting green transformation in five consumption areas. *Environmental Impact Assessment Review*, *101*, 107099. doi:10.1016/j.eiar.2023.107099

Jeevan, P., & Bhargav, V. (2016). A Study on Green Packaging - A Case Study Approach with Reference to Dell Inc. *International Educational Scientific Research Journal, 2*(7).

Kaiser, H. F. (1974). An index of factorial simplicity. *Psychometrika*, *39*(1), 31–36. doi:10.1007/BF02291575

Kaiser, H. F. (1992). On Cliff's Formula, the Kaiser-Guttman Rule, and the Number of Factors. *Perceptual and Motor Skills*, *74*(2), 595–598. doi:10.2466/pms.1992.74.2.595 PMID:1408626

Khandelwal, U., & Yadav, S. K. (2014). Green marketing and sustainable development: Marketing professionals' attitude towards green marketing. *International Journal of Marketing and Business Communication, 3*(1).

Kim, J.-O., & Mueller, C. W. (1978). *Introduction to factor analysis. What it is and how to do it.* Sage.

Kim, J.-O., & Mueller, C. W. (1987). *Factor analysis. Statistical methods and practical issues.* Sage.

Kline, R. B. (1998). *Principles and practice of structural equation modeling.* Guilford Press.

Korkmaz, S., Goksuluk, D., & Zararsiz, G. (2014). MVN: An R Package for Assessing Multivariate Normality. *The R Journal, 6*(2), 151–162. doi:10.32614/RJ-2014-031

Lopez-Gamero, M., Molina-Azorın, J., & Claver-Cortes, E. (2009). The whole relationship between environmental variables and firm performance: competitive advantage and firm resources as media to variables. *Journal of Environmental Management, 90*(10), 3110–3121.

Ladyman, J. (1998). What is structural realism? *Studies in History and Philosophy of Science, 29*(3), 409–424. doi:10.1016/S0039-3681(98)80129-5

Law, K., Wong, C., & Mobley, W. (1998). Toward a Taxonomy of Multidimensional Constructs. *Academy of Management Review, 23*(4), 741–755. doi:10.2307/259060

Lawley, D., & Maxwell, A. (1973). Regression and Factor Analysis. *Biometrika, 60*(2), 331–338.

Leonidou, L. C., Leonidou, C. N., Palihawadana, D., & Hultman, M. (2011). Evaluating the green advertising practices of international firms: A trend analysis. *International Marketing Review, 28*(1), 6–33. doi:10.1108/02651331111107080

Liu, J. (2023). Sustainability and natural resources management in developed countries: The role of financial inclusion and human development. *Resources Policy, 80*, 103143.

Lüdecke, D., Ben-Shachar, M., Patil, I., & Makowski, D. (2020). parameters: Extracting, Computing and Exploring the Parameters of Statistical Models using R. *Journal of Open Source Software, 5*(53), 2445. doi:10.21105/joss.02445

MacCallum, R., Widaman, K., Preacher, K., & Hong, S. (2001). Sample Size in Factor Analysis: The Role of Model Error. *Multivariate Behavioral Research, 36*(4), 36. doi:10.1207/S15327906MBR3604_06 PMID:26822184

MacCallum, R. C., Widaman, K. F., Zhang, S., & Hong, S. (1999). Sample size in factor analysis. *Psychological Methods, 4*(1), 84–99. doi:10.1037/1082-989X.4.1.84

Mangun, D., & Thurston, D. L. (2002). Incorporating component reuse, remanufacture, and recycle into product portfolio design. *Engineering Management, IEEE Transactions on, 49*(4), 479-490.

Massawe, E., & Geiser, K. (2012). The dilemma of promoting green products: What we know and don't know about biobased metalworking fluids. *Journal of Environmental Health, 74*(8), 8–16. PMID:22533119

Mazar, N., & Zhong, C. B. (2010). Do green products make us better people? *Psychological Science, 21*(4), 494–498. doi:10.1177/0956797610363538 PMID:20424089

McGrath, R. E. (2005). Conceptual complexity and construct validity. *Journal of Personality Assessment, 85*(2), 112–124. doi:10.120715327752jpa8502_02 PMID:16171410

McGraw, K. O., & Wong, S. P. (1996). Forming inferences about some intraclass correlation coefficients. *Psychological Methods, 1*(1), 30–46. doi:10.1037/1082-989X.1.1.30

Mills, C. W. (1959). *The sociological imagination*. Oxford University Press.

Miroshnychenko, I., Barontini, R., & Testa, F. (2017). Green practices and financial performance: A global outlook. *Journal of Cleaner Production, 147*, 340–351. doi:10.1016/j.jclepro.2017.01.058

Morrison, D. F. (1990). *Multivariate Statistical Methods*. McGraw-Hill.

Nguyen-Viet, B. (2023). The impact of green marketing mix elements on green customer based brand equity in an emerging market. *Asia-Pacific Journal of Business Administration, 15*(1), 96–116. doi:10.1108/APJBA-08-2021-0398

Nunnally, J., & Bernstein, I. (1994). *Psychometric Theory New York* (3rd ed.). McGraw Hill.

Orsato, R. J. (2006). Competitive environmental strategies: When does it pay to be green? *California Management Review, 48*(2), 127–143. doi:10.2307/41166341

Parker, B., Segev, S., & Pinto, J. (2009). *What it means to go green: Consumer perception of green brands and dimensions of greenness*. Florida International University.

Pedro Pereira Luzio, J., & Lemke, F. (2013). Exploring green consumers' product demands and consumption processes. *European Business Review, 25*(3), 281–300. doi:10.1108/09555341311314825

Revelle, W. (2020). psych: Procedures for Psychological, Psychometric, and Personality Research. Northwestern University, Evanston, Illinois. R package version 2.0.9

Revelle, W. (2021). *psych: Procedures for Psychological, Psychometric, and Personality Research*. Northwestern University, Evanston, Illinois. R package version 2.1.3. https://CRAN.R-project.org/package=psych

Revelle, W., & Rocklin, T. (1979). Very Simple Structure: An Alternative Procedure For Estimating The Optimal Number Of Interpretable Factors. *Multivariate Behavioral Research, 14*(4), 403–414. doi:10.120715327906mbr1404_2 PMID:26804437

Ritter, A., Borchardt, M., Vaccaro, G., Pereira, G., & Almeida, F. (2015). Motivations for promoting the consumption of green products in an emerging country: exploring attitudes of Brazilian consumers. *Journal of Cleaner Production, 106*, 507–520.

Rivera-Camino. (2006). Re-evaluating green marketing strategy: A stakeholder perspective. *European Journal of Marketing, 41*(11/12), 1328–1358.

Rupa, R. A., & Saif, A. N. M. (2022). Impact of Green Supply Chain Management (GSCM) on Business Performance and Environmental Sustainability: Case of a Developing Country. *Business Perspectives and Research, 10*(1), 140–163. doi:10.1177/2278533720983089

Sarstedt, M., Ringle, C., & Hair, J. (2017). Partial Least Squares. *Structural Equation Modeling*, 1–40. Advance online publication. doi:10.1007/978-3-319-05542-8_15-1

Schmid, J., & Leiman, J. M. (1957). The development of hierarchical factor solutions. *Psychometrika*, *22*(1), 53–61. doi:10.1007/BF02289209

Schumacker, R. E., & Lomax, R. G. (2010). *A beginner's guide to structural equation modeling* (3rd ed.). Routledge Academic.

Shamdasani, P., Chon-Lin, G. O., & Richmond, D. (1993). Exploring green consumers in an oriental culture: Role of personal and marketing mix factors. *Advances in Consumer Research. Association for Consumer Research (U. S.)*, *20*(1), 488–493.

Shapiro, S. S., & Wilk, M. B. (1965). An analysis of variance test for normality (complete samples). *Biometrika*, *52*(3–4), 591–611. doi:10.1093/biomet/52.3-4.591

Shetty, S. K., & Bhat, K. S. (2022). Green supply chain management practices implementation and sustainability–A review. *Materials Today: Proceedings*, *52*(3), 735–740. doi:10.1016/j.matpr.2021.10.135

Shevchenko, T., Saidani, M., Ranjbari, M., Kronenberg, J., Danko, Y., & Laitala, K. (2023). Consumer behavior in the circular economy: Developing a product-centric framework. *Journal of Cleaner Production*, *384*, 135568. doi:10.1016/j.jclepro.2022.135568

Shrout, P. E., & Fleiss, J. L. (1979). Intraclass correlations: Uses in assessing rater reliability. *Psychological Bulletin*, *86*(2), 420–3428. doi:10.1037/0033-2909.86.2.420 PMID:18839484

Singh, P., & Siddiqui, A. A. (2021). Innovation, ICT penetration, trade and economic growth in developing and developed countries: a VECM approach. Competitiveness Review. doi:10.1108/CR-05-2021-0074

Spearman, C. (1904). 'General intelligence,' objectively determined and measured. *The American Journal of Psychology*, *15*(2), 201–293. doi:10.2307/1412107

Spielmann, N. (2020). Green is the New White: How Virtue Motivates Green Product Purchase. *Journal of Business Ethics*. Advance online publication. doi:10.100710551-020-04493-6

Svensson, G., & Wagner, B. (2012). Business sustainability and E-footprints on Earth's life and ecosystems: Generic models. *European Business Review*, *24*(6), 543–552. doi:10.1108/09555341211270555

Thurstone, L. L. (1947). *Multiple factor analysis*. University of Chicago Press.

Unruh, G., & Ettenson, R. (2010). Growing green. *Harvard Business Review*, *88*(6), 94–100.

Vachon, S., & Klassen, R. D. (2006). Extending green practices across the supply chain: The impact of upstream and downstream integration. *International Journal of Operations & Production Management*, *26*(7), 795–821. doi:10.1108/01443570610672248

Wasik, J. F. (1996). *Green marketing and management: A global perspective*. Blackwell Publishers Inc.

World Economic Forum. (2021). *The Global eco-weakening: how consumers are driving sustainability, climate breakthroughs: The Road to COP26 and Beyond*. Author.

Yadav, H., Soni, U., & Kumar, G. (2023). Analysing challenges to smart waste management for a sustainable circular economy in developing countries: A fuzzy DEMATEL study. *Smart and Sustainable Built Environment*, *12*(2), 361–384. doi:10.1108/SASBE-06-2021-0097

Yenipazarli, A., & Vakharia, A. J. (2015). Green, greener or brown: Choosing the right color of the product. *Annals of Operations Research*, *250*(2), 537–567. doi:10.100710479-014-1781-5

Zappelli, C., Barbulova, A., Apone, F., & Colucci, G. (2016). Effective Active Ingredients Obtained through Biotechnology. *Cosmetics*, *3*(4), 39. doi:10.3390/cosmetics3040039

Zinbarg, R. E., Barlow, D. H., & Brown, T. A. (1997). Hierarchical structure and general factor saturation of the Anxiety Sensitivity Index: Evidence and implications. *Psychological Assessment*, *9*(3), 277–284. doi:10.1037/1040-3590.9.3.277

Chapter 15
Impact of Digital Culture Integration on Performance of Infosys Ltd.
A Pre- and Post-Merger Analysis

Abhay Kant
Agra Public Group of Institutions, India

Renu Bharti
Dayalbagh Educational Instiution, India

Shobhit Sagar
Mewar Institute of Management, Ghaziabad, India

ABSTRACT

The customer is an important resource for achieving the digitalisation capabilities. Organisations mainly emphasise the customer taste and preference because the customer is the king of the market as well as a help in assessing the internal as well as external process that can simplify the various points that are useful to make a balance between customers and life cycle of products and services. The main aim of the chapter is to know the digital culture transformation and its impact on financial performance of Infosys. The present study utilizes secondary data sources. The secondary data sources primarily consist of published studies in various international and national journals, magazines, and conference proceedings. The statistical tools such as the ratio analysis and sample test were used to test impact of digital culture transformation. The study found that electronic payment growth (EPG) did not significantly impact four important financial metrics, namely net interest margin (NIM), current ratio, advance growth, and capital adequacy ratio (CAR).

DOI: 10.4018/978-1-6684-9503-2.ch015

INTRODUCTION

The financial industry has seen a gradual shift in the usefulness of banks as a result of the implementation of data innovation. Currently, banks offer IT-based products and services to their customers. Bank clients are ending up being significantly mentioning and curious about the new advancement based monetary things and organizations. Banks are now able to meet the needs of their customers in a satisfactory manner thanks to innovation, which has altered the entire banking process. Although IT is not limited to exchange handling and the executives' data framework, it has created a serious environment for banks to keep their customers. Indian banking refers to these data mechanical changes as portable banking. It is one of the emerging trends in Indian banking and is playing a significant role in strengthening the financial sector and improving service quality. Any business that is worthy of at least some respect has tried to adapt its practices, innovations, and procedures to meet the new needs of its customers and representatives. A recent Deloitte study reveals that 86% of leaders believe that representatives will have more freedom and influence over their managers than they did in the past. Nevertheless, innovation is fundamentally linked to change: Only 56% of noting relationship in PwC's 2022 Mechanized degree of knowledge concentrate on broadened their readiness on contraptions and new cycles. By focusing on hierarchical culture and making coordinated efforts to keep up with it during change, this gap between associations and representatives' acceptance of new developments can be bridged. Reluctance to surrender existing methodologies keeps them from spreading out and achieving new goals that add to the advancement of their laborers and affiliation.

The foundation of an association's way of life is based on the Advanced Culture Initiative, such as how representatives communicate with one another, the processes that define the activity, and so on. A painstakingly changed culture takes on different techniques to these qualities, and a report by MIT and Capgemini perceives the going with as key credits in embracing an electronic first culture: nimbleness and adaptability, client-centricity, development, information-driven navigation, coordinated effort, open culture, a computerized first approach, and From my point of view, industry professionals have disengaged from these key credits for a variety of complex reasons. Zeroing in on clients is a huge sign of painstakingly changed affiliations. When such associations' forerunners consider the client's perspective, they adjust internal and external processes, focusing, most importantly, on addressing problem areas at every touchpoint in the client life cycle.

Growing up in a computerized society necessitates taking calculated risks. Because their employees are always on the lookout for novel ideas, organizations that embrace controversial advancements will typically be the leaders in this group. Information investigation provides organizations with the opportunity to further develop their independent direction as well as to identify barricades well in advance of their potential to disrupt operations. Another indication of a computerized culture is the promotion of coordinated effort. It encourages representatives from across divisions and capabilities to meet and improve processes, resulting in increased efficiency. Supporting straightforwardness allows a specialist in a painstakingly drawn in culture to associate for help when they hit a deterrent rather than worrying about the reactions of their seniors and partners. In addition, it builds trust and helps representatives freely share their thoughts, ideas, and opinions.

REVIEW OF LITERATURE

Khaira, N. Triyonggo, Y, and Sukmawati, A. (2023) this study is to identify the characteristics of digital talent and analyze the influence of HR management, digital literacy, HR risk, digital leadership, and digital culture on PT Telkom Indonesia's digital talent. This study uses primary data obtained from 140 respondents of PT Telkom Indonesia employees from 14 work locations of digital talent employees. The method used in this research is Structural Equation Modeling - Partial Least Square (SEM-PLS) analysis. The results of this study indicate that HR management, digital literacy, HR risk, digital leadership, and digital culture affect digital talent.

Lyu Y, Zhang L and Wang D (2023), examines the impact of the digital revolution on low-carbon development. To survey the level of computerized change and low-carbon improvement in 17 assembling enterprises across 30 Chinese territories from 2012 to 2018, this study utilizes the Very Slack Based Measure (SBM) model, a multi-provincial information yield model, a high-layered fixed impact model, and an intercession model. It additionally looks at the impacts of advanced change on low-carbon improvement. China's manufacturing sector showed an upward trend in terms of low-carbon development level and digital transformation throughout the study period, despite significant regional and industrial disparities. The robustness test still holds true, indicating that digital transformation can significantly contribute to the manufacturing sector's low-carbon development.

Ionaşcu, I., Ionaşcu, M., Nechita, E., Săcărin, M. furthermore, Minu, M., (2022) examines the digitalization efforts of companies listed on the primary EU stock exchanges in light of the European Green Arrangement's goals, which call for extremely ambitious targets for natural protection. based on a large number of businesses listed on the European Association's financial markets. The study's findings have significant implications for EU administrative bodies, which can implement digitalization-related policies with positive effects on native habitat through the promotion of sustainable strategies. The results moreover attest the leaders procedures of large European recorded associations, considering digitalization and viable development, which can be contacted various kinds of components.

Hautala-Kankaanpää, T. (2022) examines the impact of digital platforms and supply chain capability on operational performance and tests the mediation effect of supply chain capability. The data were harvested from 194 Finnish manufacturing companies, and structural equation modeling was used to test the hypotheses. The findings show that digital platforms positively and significantly affect supply chain capability. Moreover, supply chain capability mediates the relation between digital platforms and operational performance. Further, this study confirms that digital culture is a contextual factor that explains the differences in the effects of digital platforms on firm performance.

Camodeca, R., Almici, A. (2021) overviews the hypothesis that exceptional advancement adds to the achievement of Down to earth Improvement Targets inside the UN 2030 Arrangement. A text-based examination of the digitalization efforts of Italian businesses was carried out to test this hypothesis; A relapse examination was used to connect the obtained results to the ESG scores of the selected companies. The investigation concentrated on Italian FTSE MIB-registered businesses for the years 2016 to 2019. The revelations show a positive association among digitalization and Legitimate Improvement Targets, highlighting the relevance of modernized development in doing the sensibility plan.

Bakon, Kinn Abass and Elias, Nur Fazidah and Abusamhadana, Ghassan A. O. (2021) outlines the accessible exploration on the computerized separation, culture, and e-government achievement. A writing survey was led covering observational investigations on the computerized gap, culture, and e-government achievement. Result shows that observational investigations of e-government outcome in agricultural nations are uncommon, most exploration on advanced partition were led in created nations and zeroed in on ICT access, rather than complex methodology and furthermore concentrates on that examine various components of the computerized partition impact on e-government triumphs specifically are nearly non-existent in IS culture studies, the effect of social aspects on e-government progress specifically is missing; and the investigation of individual level social aspects effect on e-government achievement gets little consideration from analysts.

Santos Nonato (2021) addresses how digital culture and digital teaching resources have been articulated in the Emergency Remote Teaching. It is a Survey applied to 502 public and private school and higher education teachers in Bahia State, Brazil. This research reveals that the experience during the pandemic allows a glimpse of a new dynamics of the digital culture in the post-pandemic educational processes due to the teachers' learnings related to the use of Digital Information and Communication Technologies in their pedagogical practice, pointing to a new scenario for digital pedagogical resources in Education.

Ricci, F., Scafarto, V., Ferri, S., and Tron, A. (2020) examined whether there is a correlation between the degree to which Italian listed companies uncover data on their digitalization drives and the valuation of securities on the stock In addition, it is anticipated that the relationship between securities exchange valuation and digitalization exposure will be influenced by the standing for corporate maintainability. The study's results show that members of financial exchanges incorporate digitalization-related data into their business valuation cycles and that companies with better corporate maintainability achieve higher valuations from announcing their digitalization initiatives.

Sehgal, Supriya and Garg, Pratibha (2020) analyzes the perception of and expectations of other bank employees from the role of their bank's HR personnel in BPR programs and to compare the selected banks vis-à-vis the gap between perception and expectations from the role played by HR personnel in BPR programs. A questionnaire was designed and mailed to the Operations and Front-line executives of the selected Indian Private Sector banks. Suitable hypothesis were developed to assess and evaluate the perceptions and expectations of bank employees. The data obtained from the responses was used to perform the empirical analysis. Various parametric, non-parametric tests and descriptive statistics formed the parts of the extensive empirical analysis. The empirical analysis reveals that even though there is a strong perception of HR playing an important role in BPR programs, the expectations from HR remain very high.

M. F. Mubarak, F. A. Shaikh, M. Mubarik, K. A. Samo, and S. Mastoi (2019) concentrates on has perceived the impact of industry 4.0 headways including huge data, computerized genuine systems, web of things and interoperability, on the presentation of Little and Medium-sized Attempts (SMEs) in Pakistan. In the urban areas of Karachi, Lahore, Peshawar, Islamabad, Gujrat, and Sialkot, an important survey was created and distributed haphazardly. After employing a variety of regression techniques with SPSS, it was discovered that the immaterial impact of the internet of things and huge amounts of information have a significant positive impact on future business execution.

Omari, Khodor (2019) analyzes the effects of E-Commerce and E-Business on the DT enterprise in Lebanon using a qualitative research method with 10 small business owners in a private sector. This study concludes that it is difficult to realign the company because there needs to be restructuring of the company's culture and business process of organization (Veselovsky, M.Y., Izamailova, M.A., Yunusov, L.A., Yunusov, I. A., 2019).

Elisabetta R. and Claudio V. (2018) examines the kinds of business esteem that businesses can gain from large-scale information investigation speculations, the immediate effects it has on a company's monetary performance, and the interceding effects of market performance and customer loyalty. The outcomes demonstrate the absence of a market execution intervention impact and the presence of a consumer loyalty intervention impact.

NEED OF THE STUDY

The study is needed because digitalization is the most recent trend or manifestation across all industries. It refers to the use of digital technologies to alter a business model and provide new revenue and value-adding options. Over recent many years, innovation on the planet has progressed rapidly. Anything that humans are capable of comprehending has been forever altered by technology. The banking sector was one of the first to adopt information technology (IT). This boom in technology has affected all industries. To keep up with the competition and provide their customers with the best services, global banks are rapidly moving toward digitalization.

OBJECTIVES

The main aim of paper is to know the digital culture transformation and its impact on financial performance of Infosys.

RESEARCH METHODOLOGY

The present study utilizes secondary data sources. The secondary data sources primarily consisting published studies in various international and national journals, magazines and conference proceedings. The statistical tools such as the ratio analysis and sample test were used to test impact of digital culture transformation.

DATA ANALYSIS AND INTERPRETATION

Table 1. Showing the various key ratios and digital culture growth

Year	Net Interest Margin	Return on Assets	Current Ratio	Earning per Share	Digital Culture Growth
2019	20.12	18.62	3.01	33.67	40.28
2020	19.67	19.19	2.89	36.35	54.58
2021	21.00	19.24	2.76	42.39	56.74
2022	20.44	21.38	2.14	5029	47.81
2023	18.78	22.97	1.94	55.49	65.85

Source: Author calculation

Table 2. Regression between organizational performance and digital culture growth

	R Square	Constant	Digital Culture Growth	Sig.
Net Interest Margin	25.50	22.33	-.044	.386
R.O.A	39.40	13.92	.120	.257
Advance growth	31.10	4.015	-.028	.329
Current ratio	41.6	10.92	.617	.240

FINDINGS

- The statement suggests that there is a positive correlation between Net Interest Margin (NIM) and DIGITA. This means that as EPG increases, NIM is expected to increase as well. However, when a regression analysis was conducted to test the relationship between these two variables, the results were not statistically significant. This means that the null hypothesis - which states that there is no significant relationship between NIM and EPG - cannot be rejected. In other words, the evidence does not support the idea that changes in EPG lead to changes in NIM.
- This lack of statistical significance could be due to a number of reasons, such as insufficient sample size or inadequate variable selection. It could also be possible that the relationship between NIM and EPG is complex and cannot be fully captured by a linear regression analysis.
- Based on these findings, we cannot conclude that changes in EPG will necessarily lead to changes in NIM. It is important to note, however, that the lack of statistical significance does not necessarily mean that there is no relationship between these variables. It simply means that we cannot be confident in the existence or strength of such a relationship based on the available data and the analysis conducted. Further research and analysis may be needed to better understand the relationship between NIM and EPG, and to identify any potential causal factors that may be at play.
- The findings suggest that there is a negative correlation between Return on Assets (ROA) and Electronic Payments Growth (EPG). This implies that as EPG increases, we would expect ROA to decrease. However, when a regression analysis was conducted to test the relationship between these two variables, the results were not statistically significant. This means that we cannot reject the null hypothesis, which states that there is no significant relationship between ROA and EPG. Therefore, we cannot conclude that changes in EPG lead to changes in ROA.
- It is important to understand that a lack of statistical significance in a regression analysis does not necessarily mean that there is no relationship between the variables being analyzed. It simply means that the observed data is not sufficient to confirm the presence or strength of any relationship.
- The absence of a statistically significant relationship between ROA and EPG could be due to a variety of factors. For example, there may be other variables that influence ROA, such as macroeconomic conditions or company-specific factors, that were not included in the analysis. Additionally, the relationship between ROA and EPG could be non-linear, meaning that a simple linear regression model may not accurately capture the relationship.
- Therefore, while the current analysis suggests that there is no significant relationship between ROA and EPG, further research may be needed to explore the relationship in more detail. It is also

important to note that correlation does not necessarily imply causation, and that other factors may be influencing both ROA and EPG independently.

- The statement suggests that there is a positive correlation between Advance Growth and Electronic Payments Growth (EPG). This indicates that as EPG increases, we would expect Advance Growth to also increase. However, when a regression analysis was conducted to examine the relationship between these two variables, the results were not statistically significant. This means that we cannot reject the null hypothesis, which states that there is no significant relationship between Advance Growth and EPG. Therefore, we cannot conclude that changes in EPG lead to changes in Advance Growth.
- It is important to note that a lack of statistical significance in a regression analysis does not necessarily mean that there is no relationship between the variables being analyzed. There may be other factors that influence Advance Growth, such as market conditions or company-specific factors, that were not captured in the analysis. Additionally, the relationship between Advance Growth and EPG may be non-linear, meaning that a simple linear regression model may not accurately capture the relationship.
- Therefore, while the current analysis suggests that there is no significant relationship between Advance Growth and EPG, further research may be necessary to better understand the relationship between these variables. It is also important to consider that correlation does not necessarily imply causation, and that other factors may be influencing both Advance Growth and EPG independently.
- The findings suggests that there is a positive correlation between Current Ratio and Electronic Payments Growth (EPG). This indicates that as EPG increases, we would expect Current Ratio to increase as well. However, when a regression analysis was conducted to test the relationship between these two variables, the results were not statistically significant. Therefore, we cannot reject the null hypothesis, which states that there is no significant relationship between Current Ratio and EPG.
- The statement suggests that there is a negative correlation between Capital Adequacy Ratio (CAR) and Electronic Payments Growth (EPG). This means that as EPG increases, we would expect CAR to decrease. However, when a regression analysis was conducted to examine the relationship between these two variables, the results were not statistically significant. Therefore, we cannot reject the null hypothesis, which states that there is no significant relationship between CAR and EPG.
- While the current analysis did not find a significant relationship between CAR and EPG, it is important to consider that there may be other factors at play. For example, there may be external economic or regulatory factors that impact both CAR and EPG independently.
- Moreover, it is worth noting that correlation does not necessarily imply causation. While there may be a negative correlation between CAR and EPG, it is not clear whether changes in EPG are causing changes in CAR or if there are other factors involved. For instance, companies with lower CAR may be more likely to adopt electronic payment systems as a means of reducing costs or increasing efficiency.
- Overall, while the current analysis did not identify a significant relationship between CAR and EPG, further research may be necessary to fully understand the potential causal factors underlying this relationship. Nonetheless, the findings do suggest that changes in EPG may not have a direct impact on CAR, at least in the short term.

CONCLUSION

The study found that Electronic Payment Growth (EPG) did not significantly impact four important financial metrics, namely Net Interest Margin (NIM), Current Ratio, Advance Growth, and Capital Adequacy Ratio (CAR). This suggests that despite the significant growth and adoption of digital payment systems, they may not have a direct impact on the financial performance of organizations in the short term. This conclusion may be surprising given the widely held belief that digitalization has a positive impact on the operating performance of organizations. One possible explanation for this discrepancy is that the adoption of digital payment systems may not be widespread or mature enough to have a significant impact on the financial metrics being measured. Another explanation could be that there are other factors at play that are more influential in determining the financial performance of organizations. Moreover, while digital payment systems may enhance operational efficiency, their impact on financial metrics may be indirect and take longer to materialize. For example, the cost savings from digital payments may be reinvested in other areas of the organization, leading to improvements in other financial metrics over time. It is also important to consider that the impact of digitalization may vary depending on the specific industry or organization being studied. The study may have focused on a sample that was not representative of the wider industry, or may have overlooked other important factors that can affect financial performance. Overall, the conclusion that EPG does not significantly impact NIM, Current Ratio, Advance Growth, and CAR highlights the need for further research and analysis. Future studies could explore the relationship between digitalization and financial performance in greater depth, examining factors such as the maturity of digital payment systems, the impact of cost savings on other areas of the organization, and the industry-specific implications of digitalization.

REFERENCES

BakonK. A.EliasN. F.AbusamhadanaG. A. O. (n.d.). Culture and Digital Divide Influence on E-Government Success of Developing Countries: A Literature Review. *Journal of Theoretical and Applied Information Technology, 98*(9), 1362-1378. https://ssrn.com/abstract=3762293

Camodeca, R., & Almici, A. (2021). Digital Transformation and Convergence toward the 2030 Agenda's Sustainability Development Goals: Evidence from Italian Listed Firms. *Sustainability, 13*(21), 11831. doi:10.3390/su132111831

Elisabetta, R., & Claudio, V. (2018). Investments in big data analytics and firm performance: An empirical investigation of direct and mediating effects. *International Journal of Production Research, 56*(15), 5206–5221. doi:10.1080/00207543.2018.1427900

Hautala-Kankaanpää, T. (2022). The impact of digitalization on firm performance: Examining the role of digital culture and the effect of supply chain capability. *Business Process Management Journal, 28*(8), 90–109. doi:10.1108/BPMJ-03-2022-0122

Hossnofsky, V., & Junge, S. (2019). Does the market reward digitalization efforts? Evidence from securities analysts' investment recommendations. *Journal of Business Economics, 89*(8/9), 965-994. doi:10.1007/s11573-019-00949-y

Ionaşcu, I., Ionaşcu, M., Nechita, E., Săcărin, M., & Minu, M. (2022). Digital Transformation, Financial Performance and Sustainability: Evidence for European Union Listed Companies. *Amfiteatru Economic, 24*(59), 94–109. doi:10.24818/EA/2022/59/94

Khaira, N., Triyonggo, Y., & Sukmawati, A. (2023). Examining the Mediating Role of Digital Culture on Digital Talent in the Telecommunication Sector. Jurnal Aplikasi Manajemen, 21(2), 439–454. doi:10.21776/ub.jam.2023.021.02.13

Loebbecke, C., & Picot, A. (2015). Reflections on societal and business model transformation arising from digitization and big data analytics: A research agenda. *The Journal of Strategic Information Systems, 24*(3), 149-157. . doi:10.1016/j.jsis.2015.08.002

Mubarak, M. F., Shaikh, F. A., Mubarik, M., Samo, K. A., & Mastoi, S. (2019). Impact of digital transformation on business performance: A study of Pakistani SMEs. *Engineering, Technology & Applied Scientific Research, 9*(6), 5056–5061.

Nonato, S. (2021). Digital Culture And Digital Teaching Resources: An Overview Of Teaching During The Covid-19 Pandemics. *Revista Práxis Educacional, 17*(45), 1–25.

Nwankpa, J. K., & Roumani, Y. (2016). IT Capability and Digital Transformation: A Firm Performance Perspective. In *Association for Information Systems (AIS), 37 International Conference on Information Systems. 11-14 December 2016, Dublin, Ireland.* Association for Information Systems (AIS).

OmariK. (2019). The Effects of E-Commerce and Digital Transformation in Today's Business/Lebanon Case Studies. *Proceedings of the Ninth International Conference on Engaged Management Scholarship.* doi:10.2139/ssrn.3454124

Ricci, F., Scafarto, V., Ferri, S., & Tron, A. (2020). Value relevance of digitalization: The moderating role of corporate sustainability. An empirical study of Italian listed companies. *Journal of Cleaner Production, 276*, 1-8. . doi:10.1016/j.jclepro.2020.123282

Scott, S. V., Van Reenen, J., & Zachariadis, M. (2017). The long-term effect of digital innovation on bank performance: An empirical study of SWIFT adoption in financial services. *Research Policy, 46*(5), 984–1004. doi:10.1016/j.respol.2017.03.010

Sehgal, S., & Garg, P. (2020). The Human Challenges of the Digital World: The Paradigm of Employee Expectations in the Banking Sector. *First Pan IIT International Management Conference – 2018.* doi:10.2139/ssrn.3743389

Chapter 16
The Mediating Role of Job Satisfaction Between Work-Life Balance and Employee Retention

Soni Rathi
https://orcid.org/0000-0002-2997-8622
Chandigarh University, India

Praveen Kumar
Chandigarh University, India

ABSTRACT

Work-life balance is a well-researched area, and it does not only escalate the effectiveness and retention of employees, but also reforms satisfaction in both the professional as well as personal lives. The current study highlights the impact of WLB on employee retention and JS and the influence of job satisfaction on employee retention. It further explores the sub-factors of work-life balance and job satisfaction of distinct sectors. The chapter deals with surveying the past literature reviews on work-life balance, job satisfaction, and employee retention and discovering the association among them in various sectors of industries in India and other nations. It is found via this study that work-life balance has an influence on the employee retention and it also influences job satisfaction of employees. Further, there is a relationship also depicted between job satisfaction and retention of employees at the workplace. The limitations and future directions are also discussed.

1. INTRODUCTION

Work-life balance is an exclusive and comprehensive concept with various connotations and distinct phenomena. The harmony between professional and nonprofessional domains has been the subject of much attention in public discourse. Despite its simplicity of understanding, this concept is a term that

DOI: 10.4018/978-1-6684-9503-2.ch016

is still difficult to conceptualize and measure. There are two crucial domains of people's lives: personal and work. Many researchers are heeding the boundary between the personal and professional lives of employees at the workplace on account of workplace modifications and the demographic background of employees (Saeed & Farooqi, 2014). Work-life balance has recently become notable for balancing professional time and spare or leisure time. This is because there has been great importance on work-life balance worldwide. Furthermore, people seeking jobs decide to maintain an equilibrium between work life and private life at the workplace (Newaz & Zaman, 2012).

The first concern of the management is work-life balance, which means aiding workers in accomplishing their household lives and fulfilling professional life commitments. As a result, it reforms employees' personal life and work-life satisfaction. Moreover, employees stay loyal to the organization if they are content. The employees' performance is connected with the success of the organization. The association between both domains represents the main point, which is workplace contentment. Therefore, work-life balance is considered a basis of job satisfaction and a need for reconciliation among both domains (Hafeez & Akbar, 2015). Job satisfaction is the primary source of efficiency and effectiveness of an organization. Apart from the salary, productive activity is aimed at the rewards or results of an employee. Thus, to keep the best employees at the workplace, an organization must consider this factor to satisfy employees (Mustapha & Zakaria, 2013). An organization's competitive advantage is determined by employee retention, as it is crucial and associated with the organization's efficiency, productivity, and sustainability. Human resources are the assets of any organization. Therefore, retaining the workforce is achieved by encouraging them to stay loyal to the company for as long as possible or until the project is finished. It is beneficial for both the organization and the employee to retain employees despite it now has been a critical job (Biason, 2020). Considering the above facts, this analysis aims to probe the effect of WLB on JS and employee retention in organizations.

1.1 Statement of Problem

Nowadays, in this cut-throat world, organizations have been spending a plethora of time & money on keeping the employees in the organizations and rendering them satisfied at a professional level to reform productivity and aid them in fulfilling the organizational requirement. The literature review depicted that many surveys have mentioned the organizations' work-life balance, employee retention, and job satisfaction lately. The current investigation holds the key to identifying the association between work-life balance, job satisfaction, and employee retention. It further examines the effect of work-life balance on employee retention and job satisfaction. This review paper is embarked with the following research questions:

1. What is the effect of work-life balance on employee retention and job satisfaction?
2. What is the relationship between job satisfaction and employee retention?
3. What are the sub-dimensions that influence work-life balance and job satisfaction?

2. LITERATURE REVIEW

2.1 Work-Life Balance

The WLB was first introduced in 1986, which refers to effectively handling several duties in the workplace, household, and in all facets of life. This issue is crucial to all industries, organizations, and employees (Naithani, 2009). According to previous studies of work-life balance, more than half of the employees' found no balance between their family life and professional life. In order to maintain a balance between both lives, rigorous alternatives must be created (Subha, 2013). An effective equilibrium between personal and professional lives keeps people satisfied, jovial, and fulfilled. Work-life balance is a level at which people are content and participate in both personal and professional lives efficiently and equitably (Saikia, 2011). Greenhaus et al. (2003) represented that work-life balance has three constructs such as time balance (gives equal time to work life and personal life), involvement balance (mental involvement with personal and professional life adversities), and satisfaction balance (renders equitable satisfaction among personal and professional life). Arif & Farooqi (2014) introduced various policies of work-life balance for organizations that aid and support employees to be efficient in personal and professional domains, aid to increase motivation & productivity, boost the morale of employees, minimize conflict among work life & personal life and accept employees as people with priorities. Harim et al. (2023) revealed that the presence of a WLB and its supportive culture leads to job satisfaction, organizational commitment, dwindled work–family conflict, and escalates employees' intent to stay. Lazar et al. (2010) indicated that supervisors and workers could benefit when there is a prosperous and successful equilibrium between professional and personal life. Therefore, this equilibrium among both domains reforms the quality of personal life and work-related results.

WLB is an essential tool for providing contentment to employees at the workplace (Bhalerao, 2013). There is a need for work-life balance in organizations which aids them to retain their essential employees, better life balance, minimize stress and increase work satisfaction. Wise et al. (2003) mentioned that employees must know work-life practices. Nowadays, organizations put more emphasis on flexi-time and child-caring, but they need to focus on other duties such as family or personal life. As a result, these practices aid in reducing absenteeism and boosting employee morale. Magnus & Viswesvaran (2006) depicted that supervisor's support, coworkers' support, family support, flexitime, and adequate maternity leaves, etc., aid in managing work life and personal life as well as try to minimize the negative side of it, i.e., work-life conflict among employees. Poelmans et al. (2008) found numerous factors to keep an equilibrium between work life and personal life, such as multiple roles, fairness or equality in various roles, satisfaction and fulfillment of role salience of multiple roles, a link among conflict & facilitation, and perceived control among multiple roles. Other factors include increasing the recruitment or involvement of female employees and the expected international need for more efficient workers. Kelliher & Anderson (2010) identified that at the workplace, those employees who spend less time at work and work from home can enjoy more satisfaction than other employees, and due to that, they are more committed to organizations. Kulkarni & Kulkarni (2016) depicted that for a successful business organization, there is a need for positive work-life balance results such as satisfaction of critical work and career development among employees. Glass & Estes (1997) the policies of work-life balance mean to support flexible work options, child care, elder care, and personal, parental & family leaves. Employers provide work-life balance support at the workplace, such as flextime, job sharing, moving from a full-time to a part-time job, flexible working hours, and working from home. Zheng et al. (2016) indicated that work-life balance policies indirectly aid the workforce in reforming their well-being, attitudes, and life coping strategies.

2.1.1 Antecedents of WLB

Work-life balance is influenced by many factors, such as individual, social, family, and organizational factors (Burke & Greenglass, 1999). Work-life balance has taken support from various areas and people, such as family support, friends support, coworkers support, and supervisor support (Kreiner et al., 2009). Organizational strategies and familial factors both carry the highest impact on the web. In the professional domain of personnel, organizational strategies have played a key role indirectly in family life. The personal life of personnel is hampered when they need more time to spend with their family members. Further, on account of that, the repercussion of personnel will be hampered, too (Majumder et al., 2019). Various factors affect individual-related, family-related, and work-related work-life balance, such as societal support, organizational problems, stress issues, information technology, professional problems, personal problems, work overload, individual issues, and lack of knowledge (Vyas & Shrivastava, 2017).

In previous studies, some variables have depicted a significant association with WLB, such as work contentment (Saif et al., 2010) and remote work. While some antecedents have a negative influence on work-life balance, likely employees' burnout (Zeng & Chen, 2020) and overload at work (Poulose & Dhal, 2020). A positive association was identified among the antecedents of WLB, like family service, infant care, office hours, and boosted workforce morale (Uddin et al., 2020; White & Maniam, 2020). Individual, organizational, societal, and other factors are the four categories of work-life balance variables. Firstly, individual factors such as personality, well-being, and emotional intelligence. Secondly, organizational factors are work arrangements, work-life balance policies, organization support, superior support, colleague support, job stress, and role conflict. Next, societal factors are likely childcare facilities, spouse support, family support, social support, and personal or family support. Eventually, other factors are age, gender, marital status, parental status, and experience (Shobitha & Sudarshan, 2014; White & Maniam, 2020).

WLB of employees is enhanced by professionalism, job engagement, and organizational support (Jaharuddin & Zainol, 2019; Lamprinou et al., 2021). Moreover, overstrain at work and automation development are crucial in organizations, whereas excessive strain at work and machinery usage could harm employees' work-life balance. Kundnani & Mehta (2016) found six factors that are highly related to the WLB of personnel in the banking sector, such as working conditions, amenities, performance, support, freedom, and contentment. Moreover, it is also found that there is a significant relation amid work from home and household duties, as well as a positive relation between working hours and family duties. Moreover, a positive association is found between the working environment and the pessimistic behavior of supervisors & colleagues. On the contrary, there is a negative relation between deadlines/schedules/targets and amicable work in a bank that satisfaction will be escalated by targets or schedules, and working in a bank will be reduced. Further, there is a negative relation between overtime and amicable work in a bank: an increment in overtime and depreciation in satisfaction (see Figure 1).

Figure 1. Work-Life Balance and its Antecedents

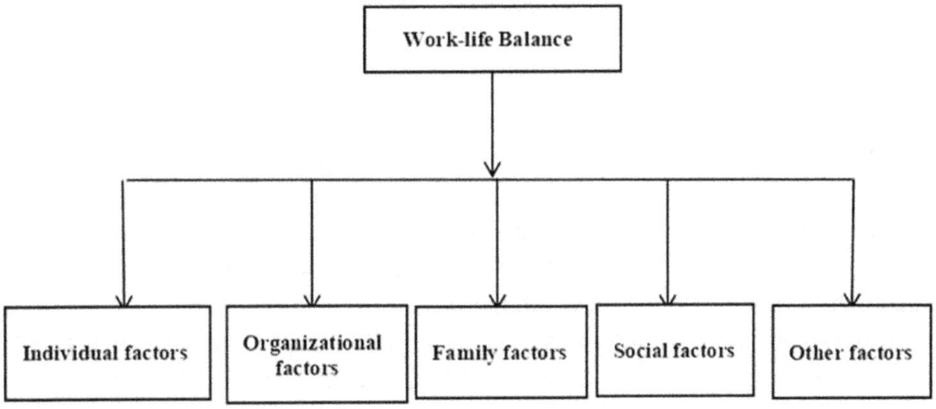

2.1.2 Theories of Work-Life Balance

2.1.2.1 Segmentation Theory

Based on segmentation theory, work and family domains operate separately and do not interact (Edwards & Rothbard, 2000; Zedeck, 1992). Bruke & Greenglass (1987) first showed the link between family and work. Since the industrial revolution, space, function, and time have inherently divided life and work (Gragnano et al., 2020). Segmentation of work and life domains occurs due to the suppression of work-related emotions, behaviors, thoughts, and pleasures in daily life and similarly restraining them at work. (Piotrkowski, 1979). This makes employees handle their life skilfully. The segmentation theory has, however, the weakest empirical support for linking employee social lives to work. Therefore, it is viewed as a theoretical tool only. As the segmentation model asserts, work and non-work are distinct and do not influence one another (Guest, 2002). According to this theory, work-life balance studies explain how distinct domains of employee lives influence strain arising from distinct roles in an organization (Zedeck, 1992).

2.1.2.2 Spill-Over Theory

The spill-over theory (Kumar & Janakiram, 2017) was cited as the most prominent theory in personal and professional lives (Staines, 1980). Employees have difficulty balancing work and life, often leading to workplace spillover (Bell et al., 2012). Generally speaking, this approach makes roles similar because one role's skills influence the other's. Work-family relationships, whether vertical or horizontal (Sirgy et al., 2001) and positive or negative spillover, are analyzed using this theory (Morris & Madsen, 2007; Xu, 2009). It is believed that positive spill-over occurs when work and life satisfaction and achievement can influence similar emotions in both. On the other hand, positive and negative spill-overs can occur in both directions, from work to life and life to work (Balmforth & Gardner, 2006). According to Young & Kleiner (1992), family and work are interconnected in spill-over theory primarily due to the effects of what happens at home on what happens at work. The skills, emotions, attitudes, and behaviors acquired during a family role are carried over to a professional role and vice versa (Kelly & Voydanoff, 1985). In work-life balance studies, this theory is most commonly used to analyze the kinship between work and non-work domains (Rincy & Panchanatham, 2014).

2.2 Job Satisfaction

The term job satisfaction is very complex and perceived by different people differently. Sometimes it is considered a part of encouraging people but quite varied from it. It is people's emotions, opinions, and feelings connected with achieving goals at work (Mullins, 2005). It also refers to the employee's general attitude toward his job. It is accustomed to explaining how a person is satisfied with her/his work. Organizations have developed many training programs and benefits packages to promulgate and encourage a reliable workforce at the workplace. Employees can be valuable assets if they stay at the workplace longer (Dhir et al., 2020; Sahito & Vaisanen, 2020). Employees' satisfaction with work used to be described if they were cheerful and fulfilling their needs and desires (Deshpande et al., 2012). In this study, various factors aid in examining or measuring the employees' satisfaction level at work, for instance, financial benefits, employee relations, work environment, and role clarity. Organizational success is based on employees' contentment and excellent workplace performance, and financial success will be reformed (Naseem et al., 2011). Satisfied employees increase the quality of work. It is shown that human resource contentment is the most crucial variable of organizational success. There are various factors in job satisfaction, such as social relationships, workers' reliability, compensation, opportunity, promotions, etc. (Aman et al., 2023; Abdulla et al., 2011; Rojikinnor et al., 2022). However, pay or salary is one of the most important factors of it among employees. Employees believe that salaries will escalate job satisfaction. It will encourage employees and aid them to perform better. It is also represented that when other job characteristics are positive, the salary can affect job satisfaction (Al-Zoubi, 2012; Hwang & Jang, 2020; Hermanto et al., 2018).

Kelliher & Anderson (2010) unveiled that employees are more satisfied and committed to the organization than others when they spend less time at the workplace and work from home. Therefore, due to this, employees are more committed to the organization. Andrade & Westover (2022) discovered that compared to the private sector, employees in the public sector enjoy more job satisfaction than public ones. Furthermore, in the public sector, performance is recognized as a factor of job satisfaction. Girisha (2019) represented that content employees are punctual and productive and become the cause of lesser accidents or mistakes at the workplace.

On the other hand, employees tend to leave their jobs when they are not content with their work. Employees' achievement and success at the workplace are depicted as job satisfaction. It is related to personal well-being and productivity directly. Employee satisfaction leads to excitement, happiness, recognition, income, promotion, and a feeling of fulfillment (Kaliski, 2007). Employees believe in equal rewards and salaries (Uka & Prendi, 2021). However, intrinsic and extrinsic rewards are not linked with employee satisfaction directly. Employees observe work satisfaction as a repercussion which tells how the employees handle the work significantly. Job experience appraisal is meant for an employee's enjoyment and optimism (Locke, 1976). Employees' productivity, performance, and morale played a key role in job satisfaction. This study has found various job satisfaction factors, such as working conditions, colleagues, challenging jobs, and rewards (Yadav & Dabhade, 2013). Healthy organizational cultures provide an environment that may inspire, motivate, meet, and exceed strategic goals and deliver team results. People working in toxic organizational cultures will find it arduous to sustain organizational success and strategic goals (Burgess et al., 2021). According to Statista, in 2022, 38% of employees are looking for better opportunities, 45% are comfortable with their careers, and the other 45% are content with their current occupations. In 2019, reported by Statista, 89% of employees were highly content with their jobs in India, followed by Mexico, the U.S., China, and so on.

2.2.1 Antecedents of Job Satisfaction

Herzberg et al. (1957) indicated via their theory that either motivators are (achievement and responsibility) available or unavailable; they cause job contentment and discontentment, known as intrinsic factors. In contrast, hygiene factors mean job context characteristics such as organizational policies, supervisors, and pay, which are extrinsic factors. Due to inadequacy, this may cause job dissatisfaction; on the other hand, job satisfaction may be caused when they are available. Hemmasi et al. (1992) unveiled that contentment with pay or salary was associated with job satisfaction positively and significantly. In order to promulgate job satisfaction among employees, there are six factors likely opportunity, strong leadership, enhanced relative strength, high work standards, equitable compensation & rewards, and adequate authority (Sahito & Vaisanen, 2017; Furso, 2020). An employee's job satisfaction is influenced by two crucial factors, such as rewards and recognition (Jun et al., 2006; Chi et al., 2023; Wijayathunga & Rathish). For job satisfaction, professional recognition comes in second, followed by wages, salary, compensation, interpersonal relationships, job-for-life, career advancement, favorable working conditions, supervision, task achievement, professional advancement, and job security and promotion (Munir & Tobi, 2020; Osakwe, 2003; Rice et al., 2022; Rayos & Borbon, 2022).

Salary, opportunities for advancement, degree of challenge of the job, autonomy, working conditions, and interaction with colleagues and students are the significant variables of work satisfaction among teachers (Sahito & Vaisanen, 2017; Bashir & Gani, 2020; Toropova et al., 2021). Wangai (2012) considered various satisfaction factors in this study, like gender, age, marital status, professional qualification, job experience, pay, working conditions, job security, leadership, and organizational culture (see Figure 2). Sahito & Vaisanen (2017) unearthed numerous factors of job satisfaction, likely organizational administration, organizational associations & practices, communication system, selection, recruitment, promotion process and system, compensation package, staff relationships, relationship with students, evaluation and feedback system, working atmosphere, appreciation & rewards, opportunities for career growth, exposure & advancement, politics groups, commitment to duties and responsibilities, consideration for personal matters and support. Kapur (2018) represented various factors influencing the job satisfaction of employees, such as knowledge & information, technology usage, seminars and workshops, rendering ideas and suggestions, inculcation of morality and ethics of people, working environmental conditions, helpfulness and cooperation, conflict resolution methods, salary and reimbursements, reward and compensation, promotional opportunities, activities and events, autonomy and training. Abuhashesh et al. (2019) identified other factors that impact job satisfaction, such as wages, office hours, schedule, organizational style, organizational benefits, strain, and flexibility. Therefore, it leads to productivity, motivation, work performance, and life satisfaction.

Figure 2. Job Satisfaction and its Antecedents

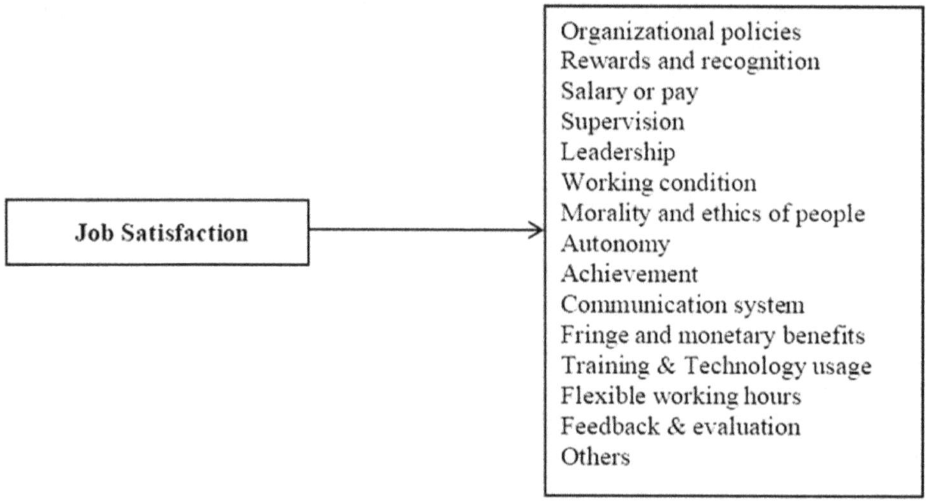

2.2.2 Theories of Job Satisfaction

2.2.2.1 Herzberg Two-Factor Theory

This theory was obtained from (Chu & Kuo, 2015). Herzberg et al. (1959) proposed a two-factor theory suggesting that particular elements of working life contribute to job contentment and inspiration, while others lead to dissatisfaction. Additionally, the antecedents that result in contentment or discontentment are discrete. As a result, he believes "the alternate of job satisfaction is not job dissatisfaction. However, it is no satisfaction." Whereas "the alternate of job dissatisfaction is not job satisfaction, however, it is satisfaction" (Herzberg, 2008). Motivation and hygiene are considered to be different factors that contribute to job satisfaction or dissatisfaction. Rewards, recognition, responsibilities, a sense of importance, and belonging to the organization are more intrinsic motivation factors. Further, a high workplace status, perks, and insurance coverage are more extrinsic hygiene factors (Hackman & Oldham, 1976). According to Syptak et al. (1999), job satisfaction was strongly influenced by the following variables: salary, recognition, supervision, working condition, achievement, corporate policies (administration & procedures), work itself, interpersonal relationships, responsibility, opportunity, promotion, and status.

2.2.2.2 Equity Theory

Equity theory depicts how a person perceives equity concerning societal kinship. A societal exchange is supposed to reveal the total number of inputs derived from kinship with the output and how much effort is put in by others (Walster et al., 1973). Equity relates to the relationship between efforts and rewards given to coworkers within a company. It is obvious that equity means treating workers equally, and if some workers feel others are receiving higher rewards for similar work, they will naturally restrain. Seeing a colleague who performs inefficiently and unproductively receive the same salary or rewards or perks as the employee who is putting in hard work may probably demotivate him or her to maintain the same level of productivity and efficiency consistently (Thangaswamy & Thiyagaraj, 2017). Some factors have been extracted after reviewing this theory, such as time, effort, hard work, commitment, ability, adaptability, flexibility, tolerance, determination, enthusiasm, personal sacrifice, trust in superiors, support from coworkers & colleagues, skills, job security, esteem, salary, employee benefits, expenses, recognition, reputation, responsibilities, and sense of achievement, praise, appreciation, stimuli and so on.

2.3 Employee Retention

Retaining and maintaining adequate talent is the priority among corporate executives and human resource specialists (Saufi et al., 2023). Retention is beneficial and refers to the long-term commitment or association that employees have towards the firms, defined as being connected to workers that the firm requires more than other firms. Employee retention is a prominent tool for better performance. Cut-throat competition can be seen in organizations to keep employees long-term. These organizations' main concerns are finding or attracting skilled workers and choosing, recruiting, and retaining top talents. Employee retention requires employers to implement work-life balance policies to benefit both parties and motivate employees to achieve goals (Johnson et al., 2000). Traditionally, keeping or retaining workers in the organization had been represented as a consistent obligation to work with a particular organization. Cheng et al. (2019) depicted that higher perceived job quality positively relates to the intention to stay. Organizations striving to reform their performance have made employee retention a key strategy component.

Additionally, organizations that are struggling to retain the right employees have turned out to be one of their most imperative antecedents to maintain success over the long run (Mandhanya, 2015). The most vital quality of work life contributes significantly to the intention to stay among employees. This means if the employees are content with their quality of work life, work world, work design, and work life/home life, the stronger will the employee commitment be to the organization and, ultimately, their intention to stay (Agus & Selvaraj, 2020).

Employee retention is a management initiative that focuses on creating high levels of employee satisfaction, ultimately retaining employees. Beynon et al. (2015) identified that training is crucial for keeping employees and boosting contentment at a professional level. Ventures must retain professionals because it eradicates the cost of recruitment and replacement; it also ensures that organizations remain knowledgeable and expert throughout (Tymon et al., 2011). Kaye & Evans (2000) deduced that money and perks are important to employees. However, they are also interested in meaningful & challenging tasks or work, competent bosses, and opportunities to learn and grow. Attracting and retaining talented employees plays a pivotal role for any organization because the workforce's knowledge and skills make the organization financially successful. However, in a cut-throat business environment, retaining skilled workers is perilous (Bartlett & Ghoshal, 2013). There is a win-win situation for superiors and subordinates when workers work in the organizations for a long time (Dutta & Banerjee, 2014).Moreover, recruitment and training costs are not the only ones to consider when reducing turnover costs. To prevent talent from being poached, retaining employees is crucial. Various antecedents represent the value of employee retention (Biason, 2020).

2.4 Work-life Balance and Employee Retention

WLB has a significant impact on staff retention (Garg, 2016). It is also stated that organizations should provide flexibility in working hours, division of work, and time off from work. Hence, organizations must support employees in maintaining harmony between work and family life. Specific investigations have been unsuccessful in identifying a theoretical connection between WLB and intent to quit (Riley, 2006), while others have found that WLB directly impacts staff turnover rates and intention to leave (Moore, 2007). Saufi et al., (2023) depicted that sustaining WLB practices were positively associated with job satisfaction and negatively associated with employees' intention to leave.

Further, job satisfaction significantly mediates the relationship between WLB practices and employee intention to leave. Too & Kwasira (2017) showed that rewards and compensations influence organizational performance more than work-life balance. They have also depicted that work-life balance is positively related to employee retention. A policy that promotes work-life balance aids workers in managing both their personal and professional lives more effectively, and reforms attributes such as organizational attachment and turnover intention (Wood et al., 2020; Saufi et al., 2023) and retention (Samroodh et al., 2022). Work-life balance influences employee retention positively in the organization. This result was supported by some researchers such as Hashim et al., (2016); Silaban & Margaretha (2021).

2.5 Job Satisfaction and Employee Retention

A key element of employee retention is job satisfaction, which happens only when employers make employees feel physically and mentally cozy and comfortable. The inflationary trends in the market economy are forcing the job market to become more cut-throat. Employers increasingly find keeping employees in their organizations challenging because the talent pool needs to grow. To be a good employer, one must know how to own top talents for extended periods (Alkandari & Hammad, 2009). Employees' feelings, whether they are positive or negative, about their work are reflected in their job satisfaction. In addition, it indicates how well expectations are met and how well actual awards match expectations. A person's performance in the workplace will significantly influence how satisfied they are at work (Davis & Nestrom, 1985). A positive association is depicted between job satisfaction and employee retention. Job satisfaction is a reliable and related predictor of employee retention, and it also reforms retention techniques (Biason, 2020). Employee satisfaction can positively impact organizational productivity, performance, quality of work, profits, commitment to the organization, and turnover and absenteeism (Chaturvedi & Sangwan, 2016). It has been identified that job satisfaction influences employee retention (Karodia et al., 2016; Bharadwaj et al., 2022). A theoretical model has been created after reviewing previous literature reviews and theories, which are given below in Figure 3.

2.6 Work-Life Balance and Job Satisfaction

WLB policies have depicted a significant association with job satisfaction (Saif et al., 2011). Rani et al. (2011) illustrated that work contentment is positively related to the web and negatively related to acknowledgment, activities or tasks, and superiors & subordinates. On the other hand, most of the employees seem dissatisfied with their job, so they cannot have a proper work-life balance (Arunashantha, 2019). Yadav & Dabhade (2013), Arif & Farooqi (2014), Syuzairi et al. (2023) found a favorable and positive association between work-family balance and job satisfaction. Rania et al. (2011) unveiled that work-life balance is attached to recognition, work task, and benefits, eventually leading to job satisfaction. Furthermore, pay is an intervening factor between the work-life balance and employee satisfaction. Guest (2002) identified various work-life balance policies which aid employees in fulfilling their family

Figure 3. Theoretical Framework of the Study

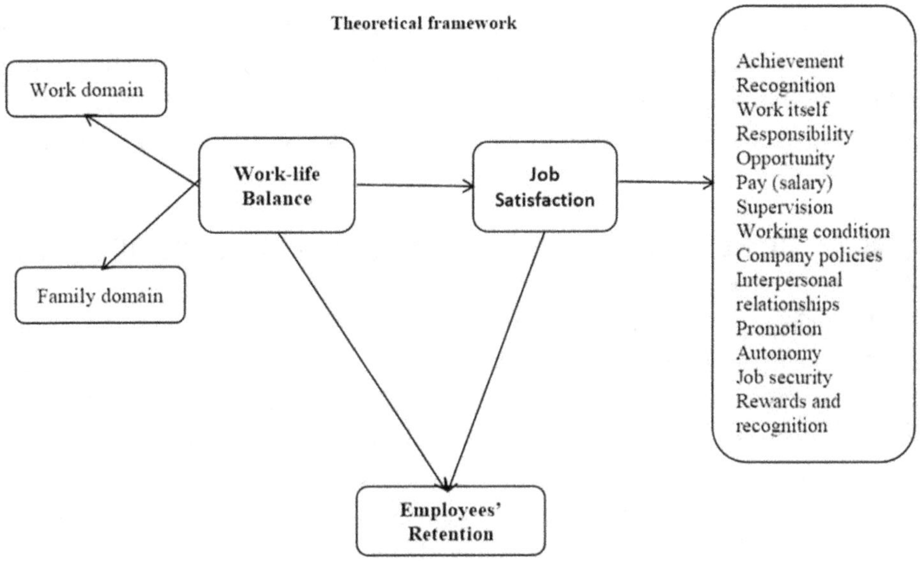

Source: Author(s) own (Fig 3)

duties and getting satisfied such as flexi-time, healthcare, sports centers, and entertainment programs. Implementing these policies provides employees satisfaction and encourages them to perform better.

Work-life balance policies render numerous guidelines to employees who enjoy the benefits of these policies. Further, they take advantage of them to get satisfaction at the workplace. These policies influence employees with higher salaries compared to those with lower wages (Ueda, 2012). The outcome of a positive work-life balance aids the staff in the organizations to achieve work-life balance, satisfaction, and career development, which are vital for booming business schemes or strategies (Kulkarni & Kulkarni, 2016). Gayathiri & Ramakrishnan (2013), Bocean et al. (2023), Kim et al. (2022) represented a positive association between web and job contentment. It is an onus of the organizations to provide work arrangements that will aid workers to enjoy time flexibility and devote some time to their families. Thus, employees get satisfied at work due to these work arrangements. It is probable that millennial employees held a high work-life balance compared to multi-job holders. Contentment in personnel's personal and professional lives appeared to be moderate yet significant predictors of work-life balance. In contrast, supervisor satisfaction, growth opportunity, and social life are crucial for female employees to provide joy in their professional lives. (Kumar & Chaturvedi, 2017). The key objectives of the investigation based on the previous review are as follows: to determine the effect of work-life balance on employee retention, to identify the impact of work-life balance on job satisfaction, and to assess the influence of job satisfaction on employee retention.

3. METHODOLOGY OF THE STUDY

The literature on work-life balance, employee retention, and job satisfaction has been reviewed from 1970 to 2023. This is indicated that the current investigation has reviewed previous literature on work-life balance, employee retention, and satisfaction at a professional level. Further, this study has also adopted two work-life balance theories and two job satisfaction theories to create a theoretical modal.

4. DISCUSSION, CONCLUSION, LIMITATIONS, AND FUTURE DIRECTIONS

In this study, the main focus is on the influence of work-life balance on employee retention and job satisfaction. Additionally, it explores factors associated with work-life balance and job satisfaction. It is identified via this paper that work-life balance has a robust impact on employee retention and job satisfaction at the workplace. Silaban & Margaretha (2021) stated that WLB and employee retention are significantly related. Arif & Farooqi (2014) represented a positive association between work-life balance and job satisfaction. A positive association is depicted between job satisfaction and employee retention. Job satisfaction is a reliable and related predictor of employee retention, and it also reforms retention techniques (Biason, 2020). According to the OECD report, the top 10 nations ranked highest for the quality of work-life balance in the year 2020. There are a plethora of factors on the web, such as job domain, household domain (family), telework, working hour, time balance, mental involvement with personal & professional life adversities, satisfaction balance, job demands, supervisor's support, coworkers support, family support, flexitime, adequate maternity leaves, multiple roles, fairness or equality, satisfaction & fulfillment of role salience, a link among conflict & facilitation, perceived control of distinct positions. Some other work-life balance factors are likely age, gender, marital status, parental status, and experience. In this study, job satisfaction also has plenty of factors like social relationships, workers' reliability, working conditions, colleagues, challenging job, rewards, style of management, payments, working hours, schedule, benefits, stress, flexibility, rewards and recognition, salary, supervision, leadership, working condition, morality and ethics of people, autonomy, achievement, communication system, monetary benefits, training & technology usage, flexible office time, effort, hard work, commitment, ability, adaptability, tolerance, determination, enthusiasm, personal sacrifice, trust in superiors, support from coworkers & colleagues, skills, job security, esteem, salary, employee benefits, expenses, recognition, reputation, responsibilities, and sense of achievement, achievement, work itself, responsibility, opportunity, company policies (administration & procedures), interpersonal relationships, promotion & status, feedback, and evaluation.

In addition, work-life balance gives the organization and employees various benefits and policies. This study proved to be a milestone for the researchers and policymakers to properly comprehend the concepts of the web, employee retention, job satisfaction, and the relationship among them. In organizations, employers provide WLB support through web policies to their employees so they can be content with their work and stay in the organization for a more extended period. These policies also aid them in many ways, such as dependents' care and personal, parental & family leaves, flextime, job sharing, healthcare, sports centers, and entertainment programs, moving from full-time to a part-time job, flexible working hours, and working from home. Therefore, to keep employees stay loyal to the organizations, organizational policies have an imperative role. Organizations should observe their employees' moods, attitudes, behaviors, and work environment since these factors affect their performance (Silaban & Margaretha, 2021).

Further, other retention practices should also be considered by them to render satisfaction to employees at the professional level. There is a process-driven approach, i.e., the rethinking retention model, which aids in retaining the best employees in the workplace (Greene, 2013). After reviewing previous studies and theories of work-life balance and job satisfaction, considering other factors in the current research may aid future researchers. This investigation is solely based on previous literature reviews and theories. During the investigation, it is depicted that an empirical study has yet to be conducted to test the theoretical modal. Therefore, it is envisaged that future studies will deepen and expand on the present antecedents.

REFERENCES

Abdulla, J., Djebarni, R., & Mellahi, K. (2011). Determinants of job satisfaction in the UAE: A case study of the Dubai police. *Personnel Review*, *40*(1), 126–146. doi:10.1108/00483481111095555

Abuhashesh, M., Al-Dmour, R., & Masa'deh, R. E. (2019). Factors that affect employees job satisfaction and performance to increase customers' satisfactions. *Journal of Human Resources Management Research*, *2019*, 1–23.

Agus, A., & Selvaraj, R. (2020). The mediating role of employee commitment in the relationship between quality of work life and the intention to stay. *Employee Relations*, *42*(6), 1231–1248. doi:10.1108/ER-07-2019-0292

Al-Zoubi, M. T. (2012). The shape of the relationship between salary and job satisfaction: A field study. *Far East Journal of Psychology and Business*, *7*(1), 1–12.

Alkandari, Q., & Hammad, A. A. (2009). *Employees retention in private sector an exploratory study in the State Of Kuwait. Project paper submitted in partial fulfillment of the requirement for the degree of Master of Business Administration*. Open University Malaysia.

Aman-Ullah, A., Aziz, A., Ibrahim, H., Mehmood, W., & Aman-Ullah, A. (2023). The role of compensation in shaping employee's behaviour: A mediation study through job satisfaction during the Covid-19 pandemic. *Revista de Gestão*, *30*(2), 221–236. doi:10.1108/REGE-04-2021-0068

Andrade, M. S., & Westover, J. H. (2022). Job satisfaction–an international comparison of public and private sector employees. *International Journal of Public Administration*, 1–15. doi:10.1080/01900692.2022.2076696

Arif, B., & Farooqi, Y. A. (2014). Impact of work life balance on job satisfaction and organizational commitment among university teachers: A case study of University of Gujrat, Pakistan. *International Journal of Multidisciplinary Sciences and Engineering*, *5*(9), 24-29.

Arunashantha, A. (2019). The impact of work-life balance on job satisfaction: With special reference to ABC private limited in Sri Lanka. *American Journal of Humanities and Social Sciences Research*, *3*, 97–108.

Balmforth, K., & Gardner, D. (2006). Conflict and facilitation between work and family: Realizing the outcomes for organizations. *New Zealand Journal of Psychology*, *35*(2), 69–76.

Bartlett, C. A., & Ghoshal, S. (2002). Building competitive advantage through people. *MIT Sloan Management Review*, *43*(2), 16–25.

Bashir, B., & Gani, A. (2020). Testing the effects of job satisfaction on organizational commitment. *Journal of Management Development*, *39*(4), 525–542. doi:10.1108/JMD-07-2018-0210

Bell, A. S., Rajendran, D., & Theiler, S. (2012). Job stress, wellbeing, work-life balance and work-life conflict among Australian academics. *E-Journal of Applied Psychology : Clinical and Social Issues*, *8*(1), 25–37. doi:10.7790/ejap.v8i1.320

Beynon, M. J., Jones, P., Pickernell, D., & Packham, G. (2015). Investigating the impact of training influence on employee retention in small and medium enterprises: A regression-type classification and ranking believe simplex analysis on sparse data. *Expert Systems: International Journal of Knowledge Engineering and Neural Networks*, *32*(1), 141–154. doi:10.1111/exsy.12067

Bhalerao, S. K. (2013). Work life balance: The key driver of employee engagement. *ASM's International E-Journal of Ongoing Research in Management and IT*, *16*(1), 1–9.

Bharadwaj, S., Khan, N. A., & Yameen, M. (2022). Unbundling employer branding, job satisfaction, organizational identification and employee retention: A sequential mediation analysis. *Asia-Pacific Journal of Business Administration*, *14*(3), 309–334. doi:10.1108/APJBA-08-2020-0279

Biason, R. S. (2020). The effect of job satisfaction on employee retention. *International Journal of Economics. Commerce and Management*, *8*(3), 405–413.

Bocean, C. G., Popescu, L., Varzaru, A. A., Avram, C. D., & Iancu, A. (2023). Work–Life Balance and Employee Satisfaction during COVID-19 Pandemic. *Sustainability (Basel)*, *15*(15), 11631. doi:10.3390u151511631

Burgess, C. T., Adkins, J., & Suniga, N. (2021). Management and Leadership: A Practical Review of the Organizational Culture. Handbook of Research on Multidisciplinary Perspectives on Managerial and Leadership Psychology, 396-412.

Burke, R. J., & Greenglass, E. (1987). Work and family. In C. L. Cooper & I. T. Robertson (Eds.), *International review of industrial and organizational psychology*. Wiley.

Burke, R. J., & Greenglass, E. R. (1999). Work–family conflict, spouse support, and nursing staff well-being during organizational restructuring. *Journal of Occupational Health Psychology, 4*(4), 327-336.

Chaturvedi, A., & Sangwan, K. (2016). Impact of Job Satisfaction on Retention of Employees in BPO Industry. *International Multidisciplinary Research Journal*, *5*(11), 1–6.

Cheng, Z., Nielsen, I., & Cutler, H. (2019). Perceived job quality, work-life interference and intention to stay: Evidence from the aged care workforce in Australia. *International Journal of Manpower*, *40*(1), 17–35. doi:10.1108/IJM-08-2017-0208

Chi, H., Vu, T. V., Nguyen, H. V., & Truong, T. H. (2023). How financial and non–financial rewards moderate the relationships between transformational leadership, job satisfaction, and job performance. *Cogent Business & Management, 10*(1), 1–18. doi:10.1080/23311975.2023.2173850

Chu, H. C., & Kuo, T. (2015). Testing Herzberg's two-factor theory in educational settings in Taiwan. *The Journal of Human Resource and Adult Learning, 11*(1), 54–65.

Davis, K. (1989). *Human behavior at work: Organizational behavior*. McGraw-Hill Inc.

Deshpande, B., Arekar, K., Sharma, R., & Somaiya, S. (2012). Effect of employee satisfaction on organization performance: An empirical study in hotel industry. *Ninth AIMS International Conference on Management,* 1-4.

Dhir, S., Dutta, T., & Ghosh, P. (2020). Linking employee loyalty with job satisfaction using PLS–SEM modelling. *Personnel Review, 49*(8), 1695–1711. doi:10.1108/PR-03-2019-0107

Dutta, A. B., & Banerjee, S. (2014). Study of employee retention. *International Journal of Business Management & Research, 4*(1), 83–88.

Edwards, J. R., & Rothbard, N. P. (2000). Mechanisms linking work and family: Clarifying the relationship between work and family constructs. *Academy of Management Review, 25*(1), 178–199. doi:10.2307/259269

Furso, B. (2020). *The relationship between performance appraisal and teachers and teachers job satisfaction in secondary schools of west Arsi Zone, Oromia Regional State* [Doctoral dissertation]. Hawassa University.

Garg, P., & Yajurvedi, N. (2016). Impact of work-life balance practices on employees retention and organizational performance–A study on IT industry. *Indian Journal of Applied Research, 6*(8), 105–108.

Gayathiri, R., Ramakrishnan, L., Babatunde, S. A., Banerjee, A., & Islam, M. Z. (2013). Quality of work life–Linkage with job satisfaction and performance. *International Journal of Business and Management Invention, 2*(1), 1–8.

Girisha, M. C. (2019) Job satisfaction: a study at AT & S India Pvt Ltd Nanjangud (mysore District, Karnataka). *Proceedings of the SIMS Annual Research Conference, SIMSARC12.*

Glass, J. L., & Estes, S. B. (1997). The family responsive workplace. *Annual Review of Sociology, 23*(1), 289–313. doi:10.1146/annurev.soc.23.1.289

Gragnano, A., Simbula, S., & Miglioretti, M. (2020). Work–life balance: Weighing the importance of work–family and work–health balance. *International Journal of Environmental Research and Public Health, 17*(3), 907. doi:10.3390/ijerph17030907 PMID:32024155

Greene, K. L. (2013). *Better equipping reserve military intelligence analyst to meet the needs of the commander by championing a process-driven training model* [Doctoral dissertation]. Fort Leavenworth, KS: US Army Command and General Staff College.

Greenhaus, J. H., Collins, K. M., & Shaw, J. D. (2003). The relation between work–family balance and quality of life. *Journal of Vocational Behavior, 63*(3), 510–531. doi:10.1016/S0001-8791(02)00042-8

Guest, D. (2002). Human resource management, corporate performance and employee wellbeing: Building the worker into HRM. *The Journal of Industrial Relations, 44*(3), 335–358. doi:10.1111/1472-9296.00053

Hackman, J. R., & Oldham, G. R. (1976). Motivation through the design of work: Test of a theory. *Organizational Behavior and Human Performance, 16*(2), 250–279. doi:10.1016/0030-5073(76)90016-7

Hafeez, U., & Akbar, W. (2015). Impact of work-life balance on job satisfaction among school teachers of 21st century. *Australian Journal of Business and Management Research, 4*(11), 25–37. doi:10.52283/NSWRCA.AJBMR.20150411A03

Hashim, A., Azman, N. S., Ghani, M. A., & Sabri, M. F. M. (2016). The relationship between work-life balance and employee retention. *Proceeding of the 3rd International Conference on Management & Muamalah, 1*, 316-322.

Hemmasi, M., Graf, L. A., & Lust, J. A. (1992). Correlates of pay and benefit satisfaction: The unique case of public university faculty. *Public Personnel Management, 21*(4), 429–443. doi:10.1177/009102609202100402

Hermanto, B., Muhyi, H. A., & Purnomo, M. (2018). The effect of job characteristics on job satisfaction and its impact on employee performance. *Advances in Social Sciences Research Journal, 5*(9), 95–101.

Herzberg, F. (2008). *One more time: How do you motivate employees?* Harvard Business Review Press.

Herzberg, F., Mausner, B., & Snyderman, B. B. (1959). The Motivation to Work (2nd ed.). John Wiley & Sons.

Herzberg, F., Mausnes, B., Peterson, R. O., & Capwell, D. F. (1957). *Job attitudes: Review of research and opinion*. Psychological Service of Pittsburgh.

Hwang, J., & Jang, W. (2020). The effects of job characteristics on perceived organizational identification and job satisfaction of the Organizing Committee for the Olympic Games employees. *Managing Sport and Leisure, 25*(4), 290–306. doi:10.1080/23750472.2020.1723435

Jaharuddin, N. S., & Zainol, L. N. (2019). *The impact of work-life balance on job engagement and turnover intention*. The South East Asian Journal of Management. doi:10.21002eam.v13i1.10912

Johnson, J. T., Griffeth, R. W., & Griffin, M. (2000). Factors discriminating functional and dysfunctional salesforce turnover. *Journal of Business and Industrial Marketing, 15*(6), 399–415. doi:10.1108/08858620010349493

Jun, M., Cai, S., & Shin, H. (2006). TQM practice in maquiladora: Antecedents of employee satisfaction and loyalty. *Journal of Operations Management, 24*(6), 791–812. doi:10.1016/j.jom.2005.09.006

Kaliski, B. S. (2007). Encyclopedia of Business and Finance. Thompson Gale.

Kapur, R. (2018). *Factors influencing performance and job satisfaction of teachers in secondary schools in India.* Research Gate.

Karodia, A. M., Motlou, R. G., & Singh, S. (2016). An Evaluation of the Impact of Job Satisfaction on Employee Retention at Lonmin Rowland Shaft North West Province. *Kuwait Chapter of Arabian Journal of Business and Management Review, 5*(10), 15–49. doi:10.12816/0027366

Kaye, B., & Jordan-Evans, S. (2000). Retention: Tag, you're it! *Training and Development-Alexandria-American Society for Training and Development, 54*(4), 29-39.

Kelliher, C., & Anderson, D. (2010). Doing more with less? Flexible working practices and the intensification of work. *Human Relations, 63*(1), 83–106. doi:10.1177/0018726709349199

Kelly, R. F., & Voydanoff, P. (1985). Work/family role strain among employed parents. *Family Relations, 34*(3), 367–374. doi:10.2307/583575

Kim, J. H., Jung, S. H., Seok, B. I., & Choi, H. J. (2022). The Relationship among Four Lifestyles of Workers amid the COVID-19 Pandemic (Work–Life Balance, YOLO, Minimal Life, and Staycation) and Organizational Effectiveness: With a Focus on Four Countries. *Sustainability (Basel), 14*(21), 14059. doi:10.3390u142114059

Kreiner, G. E., Hollensbe, E. C., & Sheep, M. L. (2009). Balancing borders and bridges: Negotiating the work-home interface via boundary work tactics. *Academy of Management Journal, 52*(4), 704–730. doi:10.5465/amj.2009.43669916

Kulkarni, S. R., & Kulkarni, S. S. (2016). Work life balance and stress management: Two sides of a single coin. *Indian Journal of Commerce & Management Studies*, 16–19.

Kumar, G. V., & Janakiram, B. (2017). Theories of work life balance–a conceptual review. *International Research Journal of Management and Commerce, 4*(9), 184–192.

Kumar, K., & Chaturvedi, R. (2017). Work-life balance and job satisfaction from the perspective of multiple job holding women: Comparative analysis of generational cohorts. *International Journal of Advances in Management and Economics, 6*(5), 32–40.

Kundnani, N., & Mehta, P. (2016). Identifying the factors affecting work-life balance of employees in banking sector. *PARIPEX-Indian Journal of Research, 4*(6), 328–331.

Lamane-Harim, J., Cegarra-Leiva, D., & Sanchez-Vidal, M. E. (2023). Work–life balance supportive culture: A way to retain employees in Spanish SMEs. *International Journal of Human Resource Management, 34*(10), 2074–2106. doi:10.1080/09585192.2021.1878255

Lamprinou, V. D. I., Tasoulis, K., & Kravariti, F. (2021). The impact of servant leadership and perceived organisational and supervisor support on job burnout and work–life balance in the era of teleworking and COVID-19. *Leadership and Organization Development Journal, 42*(7), 1071–1088. doi:10.1108/LODJ-12-2020-0526

Lazar, I., Osoian, C., & Ratiu, P. I. (2010). The role of work-life balance practices in order to improve organizational performance. *European Research Studies, XIII*(1), 1–14. doi:10.35808/ersj/267

Locke, E. A. (1976). The nature and causes of job satisfaction. Handbook of industrial and organizational psychology. 1297-1349.

Majumder, J., Giri, A., & Gangopadhyay, S. (2019). Factors Affecting Work Life Balance of Employees in Indian Manufacturing Companies: An Empirical Analysis Using Structural Equation Modeling (SEM). *International Journal of Innovative Technology and Exploring Engineering, 8*(7), 1551–1555.

Mandhanya, Y. (2015). A study of impact of working environment of retention of employees (With special reference to Automobile sector)'. *Global Management Review*, *9*(4), 116–128.

Mesmer-Magnus, J. R., & Viswesvaran, C. (2006). How family-friendly work environments affect work/family conflict: A meta-analytic examination. *Journal of Labor Research*, *27*(4), 555–574. doi:10.100712122-006-1020-1

Moore, F. (2007). Work-life balance: Contrasting managers and workers in an MNC. *Employee Relations*, *29*(4), 385–399. doi:10.1108/01425450710759217

Morris, M. L., & Madsen, S. R. (2007). Advancing work—Life integration in individuals, organizations, and communities. *Advances in Developing Human Resources*, *9*(4), 439–454. doi:10.1177/1523422307305486

Mullins, L. (2005). *Management & Organisational Behaviour* (7th ed.). Prentice Hall / Pearson Education.

Munir, R., & Tobi, S. N. M. (2020). Understanding and measuring the determinants of employees' turnover intention. *Environ. Behav. Proc. J*, *5*(13), 395–403. doi:10.21834/e-bpj.v5i13.2092

Mustapha, N., & Zakaria, Z. C. (2013). The effect of promotion opportunity in influencing job satisfaction among academics in higher public institutions in Malaysia. *International Journal of Academic Research in Business & Social Sciences*, *3*(3), 20.

Naithani, D. P. (2009). Overview of work-life balance discourse and its relevance in current economic scenario. *Asian Social Science*, *6*(6), 148–155.

Naseem, A., Sheikh, S. E., & Malik, K. P. (2011). Impact of employee satisfaction on success of organization: Relation between customer experience and employee satisfaction. *International Journal of Multidisciplinary Sciences and Engineering, 2*(5), 41-46.

Newaz, M. T., & Zaman, F. K. (2012). Work-life balance: Is it still a new concept in private commercial banking sector of Bangladesh. *International Journal of Research Studies in Management*, *1*(2), 57–66. doi:10.5861/ijrsm.2012.95

Osakwe, R. N. (2003). *A Comparative study of job satisfaction between public and private secondary school in Delta State* [Unpublished Ph. D. thesis]. Faculty of Education, Delta State University.

Piotrkowski, C. S. (1979). *Work and the family system: A naturalistic study of working-class and lower-middle-class families.* Collier Macmillan.

Poelmans, S. A., Kalliath, T., & Brough, P. (2008). Achieving work–life balance: Current theoretical and practice issues. *Journal of Management & Organization*, *14*(3), 227–238. doi:10.5172/jmo.837.14.3.227

Poulose, S., & Dhal, M. (2020). Role of perceived work–life balance between work overload and career commitment. *Journal of Managerial Psychology*, *35*(3), 169–183. doi:10.1108/JMP-03-2018-0117

Rani, S. V., & Mariappan, S. (2011). Work/life balance reflections on employee satisfaction. *Serbian Journal of Management*, *6*(1), 85–96. doi:10.5937jm1101085R

Rathore, M. (2022). *Opinion on employee satisfaction in India as of December 2021*. Statista.

Rayos, J. L. T., & Borbon, N. M. D. (2022). Seafarers' attitude and cultural adaptation influencing the work motivation during the time of pandemic. *International Journal of Research*, *10*(1), 107–124.

Rice, B., Martin, N., Fieger, P., & Hussain, T. (2022). Older healthcare workers' satisfaction: Managing the interaction of age, job security expectations and autonomy. *Employee Relations*, *44*(2), 319–334. doi:10.1108/ER-07-2020-0346

Riley, D. (2006). *Turnover intentions: The mediation effects of job satisfaction, affective commitment and continuance commitment* [Doctoral dissertation]. The University of Waikato.

Rincy, V. M., & Panchanatham, N. (2014). Work life balance: A short review of the theoretical and contemporary concepts. *Continental Journal of Social Sciences*, *7*(1), 1–24.

Rojikinnor, R., Gani, A. J. A., Saleh, C., & Amin, F. (2022). The role of compensation as a determinant of performance and employee work satisfaction: a study at the PT Bank Rakyat Indonesia (Persero) Tbk. *Journal of Economic and Administrative Sciences*.

Saeed, K., & Farooqi, Y. A. (2014). Examining the relationship between work life balance, job stress and job satisfaction among university teachers (A case of University of Gujrat). *International Journal of Multidisciplinary Sciences and Engineering*, *5*(6), 9-15.

Sahito, Z., & Vaisanen, P. (2017). Factors affecting job satisfaction of teacher educators: Empirical evidence from the Universities of Sindh Province of Pakistan. *Journal of Teacher Education and Educators*, *6*(1), 5–30.

Sahito, Z., & Vaisanen, P. (2020). A literature review on teachers' job satisfaction in developing countries: Recommendations and solutions for the enhancement of the job. *Review of Education*, *8*(1), 3–34. doi:10.1002/rev3.3159

Saif, M. I., Malik, M. I., & Awan, M. Z. (2011). Employee work satisfaction and work-life balance: A Pakistani perspective. *Interdisciplinary Journal of Contemporary Research in Business*, *3*(5), 606–617.

Saikia, J. N. (2011). *A study of the work life balance among the academics of higher education institutions: A case study of Golaghat District*. Academic Press.

Samroodh, M., Anwar, I., Ahmad, A., Akhtar, S., Bino, E., & Ali, M. A. (2022). The indirect effect of job resources on employees' intention to stay: A serial mediation model with psychological capital and work–life balance as the mediators. *Sustainability (Basel)*, *15*(1), 551. doi:10.3390u15010551

Saufi, A. B. R., Permarupan, Y. P., Aidara, S., Zainol, B. R. N., & Khan, U. N. (2023). Sustaining Work Life Balance Toward Job Satisfaction and Employee Intention to Leave Universities. Academy of Management, (1). doi:10.5465/AMPROC.2023.14613abstract

Saufi, A. R., Aidara, S., Nawi, C. N. B., Permarupan, P. Y., Zainol, N. R. B., & Kakar, A. S. (2023). Turnover intention and its antecedents: The mediating role of work–life balance and the moderating role of job opportunity. *Frontiers in Psychology*, *14*, 1–15. doi:10.3390u15043397 PMID:37077854

Shobitha, P., & Sudarsan, N. (2014). Work life balance: A conceptual review. *International Journal of Advances in Management and Economics*, *3*(2), 1–17.

Silaban, H., & Margaretha, M. (2021). The impact work-life balance toward job satisfaction and employee retention: Study of millennial employees in Bandung city, Indonesia. *International Journal of Innovation and Economic Development*, *7*(3), 18–26. doi:10.18775/ijied.1849-7551-7020.2015.73.2002

Sirgy, M. J., Efraty, D., Siegel, P., & Lee, D. J. (2001). A new measure of quality of work life (QWL) based on need satisfaction and spillover theories. *Social Indicators Research*, *55*(3), 241–302. doi:10.1023/A:1010986923468

Staines, G. L. (1980). Spillover versus compensation: A review of the literature on the relationship between work and nonwork. *Human Relations*, *33*(2), 111–129. doi:10.1177/001872678003300203

Subha, T. (2013). A Study on Work Life Balance Among Women Faculties Working in Arts and Science Colleges with Special Reference to Coimbatore City. *Paripex-Indian Journal of Research*, *2*(12), 160–163.

Syptak, J. M., Marsland, D. W., & Ulmer, D. (1999). Job satisfaction: Putting theory into practice. *Family Practice Management*, *6*(9), 26–30.

Syuzairi, M., Pratiwi, M. A., & Apriansyah, F. (2023). Work-Life Balance and its Impact on Turnover Intention among Educators: Job Satisfaction as a Mediation. In *ICEMBA 2022: Proceedings of the International Conference on Economic, Management, Business and Accounting, ICEMBA 2022, 17 December 2022, Tanjungpinang, Riau Islands, Indonesia* (p. 258). European Alliance for Innovation. 10.4108/eai.17-12-2022.2333271

Thangaswamy, A., & Thiyagaraj, D. (2017). Theoretical concept of job satisfaction-a study. *International Journal of Research*, *5*(6), 464–470.

Too, R., & Kwasira, D. J. (2017). Evaluation of employee retention practices to generation y at cooperative bank in the county of Nairobi, Kenya. *European Journal of Business and Management, 9*(11), 43-53.

Toropova, A., Myrberg, E., & Johansson, S. (2021). Teacher job satisfaction: The importance of school working conditions and teacher characteristics. *Educational Review*, *73*(1), 71–97. doi:10.1080/00131911.2019.1705247

Tymon, W. G. Jr, Stumpf, S. A., & Smith, R. R. (2011). Manager support predicts turnover of professionals in India. *Career Development International*, *16*(3), 293–312. doi:10.1108/13620431111140174

Uddin, M., Ali, K. B., & Khan, M. A. (2020). Perceived social support (PSS) and work-life balance (WLB) in a developing country: The moderating impact of work-life policy. *Iranian Journal of Management Studies.*

Ueda, Y. (2012). The relationship between work-life balance programs and employee satisfaction: Gender differences in the moderating effect of annual income. *Journal of Business Administration Research*, *1*(1), 65. doi:10.5430/jbar.v1n1p65

Uka, A., & Prendi, A. (2021). Motivation as an indicator of performance and productivity from the perspective of employees. *Management & Marketing. Challenges for the Knowledge Society*, *16*(3), 268–285. doi:10.2478/mmcks-2021-0016

Vyas, A., & Shrivastava, D. (2017). Factors affecting work life balance-a review. *Pacific Business Review International*, *9*(7), 194–200.

Walster, E., Berscheid, E., & Walster, G. W. (1976). New directions in equity research. In Advances in experimental social psychology (Vol. 9, pp. 1–42). Academic Press.

Wangai, M. M. (2012). *Factors affecting job satisfaction among secondary school teachers of Mwatate district, Kenya. Project paper presented as a partial fulfillment for the requirement of the award of certificate in senior management course.* GTI EMBU.

White, C., & Maniam, B. (2020). Flexible working arrangements, worklife balance, and working women. *Journal of Business and Accounting, 13*(1), 59–73.

Wijayathunga, L. A. D. H. R., & Rathish, D. (2023). Non-financial Rewards and Job Satisfaction among State University Academics of Sri Lanka. *Sri Lankan Journal of Human Resource Management, 13*(1), 27–39. doi:10.4038ljhrm.v13i1.5676

Wise, S., Bond, S., & Meikle, A. (2003). *Work-life balance-literature and research review.* DTI and Fair Play.

Wood, J., Oh, J., Park, J., & Kim, W. (2020). The relationship between work engagement and work–life balance in organizations: A review of the empirical research. *Human Resource Development Review, 19*(3), 240–262. doi:10.1177/1534484320917560

Xu, L. (2009). View on work-family linkage and work-family conflict model. *International Journal of Business and Management, 4*(12), 229–233. doi:10.5539/ijbm.v4n12p229

Yadav, R. K., & Dabhade, N. (2013). A case study with overview of job satisfaction. *Indian Journal of Economics and Development, 1*(8), 136–146.

Zedeck, S. (1992). Introduction: Exploring the domain of work and family concerns. In Frontiers of Industrial and Organizational Psychology, Vol. 5; The Jossey-Bass Management Series and The Jossey-Bass Social and Behavioral Science Series. Work, families, and organizations (pp. 1–32). Jossey-Bass.

Zeng, C., & Chen, H. (2020). An exploration of the relationships between organizational dissent, employee burnout, and work-family balance: A cross-cultural comparison between China and Finland. *Communication Studies, 71*(4), 633–648. doi:10.1080/10510974.2020.1749864

Zheng, C., Kashi, K., Fan, D., Molineux, J., & Ee, M. S. (2016). Impact of individual coping strategies and organisational work–life balance programmes on Australian employee well-being. *International Journal of Human Resource Management, 27*(5), 501–526. doi:10.1080/09585192.2015.1020447

ADDITIONAL READING

Abdulaziz, A., Bashir, M., & Alfalih, A. A. (2022). The impact of work-life balance and work overload on teacher's organizational commitment: Do Job Engagement and Perceived Organizational support matter. *Education and Information Technologies, 27*(7), 9641–9663. doi:10.100710639-022-11013-8

Bagheri Hosseinabadi, M., Khanjani, N., Etemadinezhad, S., Samaei, S. E., Raadabadi, M., & Mostafaee, M. (2019). The associations of workload, individual and organisational factors on nurses' occupational injuries. *Journal of Clinical Nursing*, *28*(5-6), 902–911. doi:10.1111/jocn.14699 PMID:30357967

Duan, S. X., Deng, H., & Wibowo, S. (2023). Exploring the impact of digital work on work–life balance and job performance: A technology affordance perspective. *Information Technology & People*, *36*(5), 2009–2029. doi:10.1108/ITP-01-2021-0013

Kelliher, C., Richardson, J., & Boiarintseva, G. (2019). All of work? All of life? Reconceptualising work-life balance for the 21st century. *Human Resource Management Journal*, *29*(2), 97–112. doi:10.1111/1748-8583.12215

Oosthuizen, R. M., Coetzee, M., & Munro, Z. (2019). Work–life balance, job satisfaction and turnover intention amongst information technology employees. *Southern African Business Review*, *20*(1), 446–467. doi:10.25159/1998-8125/6059

Rojikinnor, R., Gani, A. J. A., Saleh, C., & Amin, F. (2022). The role of compensation as a determinant of performance and employee work satisfaction: a study at the PT Bank Rakyat Indonesia (Persero) Tbk. *Journal of Economic and Administrative Sciences*.

Shirmohammadi, M., Au, W. C., & Beigi, M. (2022). Remote work and work-life balance: Lessons learned from the covid-19 pandemic and suggestions for HRD practitioners. *Human Resource Development International*, *25*(2), 163–181. doi:10.1080/13678868.2022.2047380

KEY TERMS AND DEFINITIONS

Employee Retention: It is an organization's ability to prevent employees to quit/leave their jobs.

Job Satisfaction: It is a feeling of fulfillment or enjoyment that a person derives from his/her job.

Work-Life Balance: Work life balance is defined as the level of prioritization between personal and professional activities of a person's life and the level to which activities related to their job are present at home.

Compilation of References

1st Five Year Plan (1961-1966). (n.d.). Gross National Happiness Commission, Royal Government of Bhutan.

Abbass, K., Qasim, M. Z., Song, H., Murshed, M., Mahmood, H., & Younis, I. (2022). A review of the global climate change impacts, adaptation, and sustainable mitigation measures. *Environmental Science and Pollution Research International, 29*(28), 42539–42559. doi:10.100711356-022-19718-6 PMID:35378646

Abdel-Ghani, N. T.-S.-H. (2018). Utilization of by-pass cement kiln dust and air-cooled blast-furnace steel slag in the production of some "green" cement products. *HBRC Journal, 14*(3), 408-414.

Abdulla, J., Djebarni, R., & Mellahi, K. (2011). Determinants of job satisfaction in the UAE: A case study of the Dubai police. *Personnel Review, 40*(1), 126–146. doi:10.1108/00483481111095555

Abimbola, T., Lim, M., Hillestad, T., Xie, C., & Haugland, S. A. (2010). Innovative corporate social responsibility: The founder's role in creating a trustworthy corporate brand through "green innovation.". *Journal of Product and Brand Management, 19*(6), 440–451. doi:10.1108/10610421011085758

Abuhashesh, M., Al-Dmour, R., & Masa'deh, R. E. (2019). Factors that affect employees job satisfaction and performance to increase customers' satisfactions. *Journal of Human Resources Management Research, 2019*, 1–23.

Abzari, M., Shad, F. S., Sharbiyani, A. A., & Morad, A. P. (2013). Studying the effect of green marketing mix on market share increase. *European Online Journal of Natural and Social Sciences, 2*(3), 641–653.

Agus, A., & Selvaraj, R. (2020). The mediating role of employee commitment in the relationship between quality of work life and the intention to stay. *Employee Relations, 42*(6), 1231–1248. doi:10.1108/ER-07-2019-0292

Ahari, R. S. (2015). Thixotropy and organizational breakdown belongings of Self-consolidating concrete comprising several supplementary cementitious materials. *Cement and Concrete Composites, 59*, 26–37. doi:10.1016/j.cemconcomp.2015.03.009

Ahmed, R. R., Streimikiene, D., Qadir, H., & Streimikis, J. (2023). Effect of green marketing mix, green customer value, and attitude on green purchase intention: Evidence from the USA. *Environmental Science and Pollution Research International, 30*(5), 11473–11495. doi:10.100711356-022-22944-7 PMID:36094715

Ahmed, Z., Nathaniel, S. P., & Shahbaz, M. (2021). The criticality of information and communication technology and human capital in environmental sustainability: Evidence from Latin American and Caribbean countries. *Journal of Cleaner Production, 286*, 125529. doi:10.1016/j.jclepro.2020.125529

Aïtcin, P.-C. (2000). Cements of yesterday and today: Concrete of tomorrow. *Cement and Concrete Research, 30*(9), 1349–1359. doi:10.1016/S0008-8846(00)00365-3

Aithal, P., & Shubhrajyotsna, A. (2016). Nanotechnology Innovations and Commercialization – Opportunities, Challenges & Reasons for Delay. *Proceedings of National Conference on Changing Perspectives of Management, IT, and Social Sciences in Contemporary Environment, 14*, 1-12. 10.5815/ijem.2016.06.02

Aithal, P., & Shubhrajyotsna, A. (2018). Nanotechnology based Innovations and Human Life Comfortability –Are we Marching towards Immortality? *International Journal of Applied Engineering and Management Letters*, *2*(1), 1–8. doi:10.5281/zenodo.1485048

Alam, M. S., Apergis, N., Paramati, S. R., & Fang, J. (2021). The impacts of R&D investment and stock markets on clean-energy consumption and CO2 emissions in OECD economies. *International Journal of Finance & Economics*, *26*(4), 4979–4992. doi:10.1002/ijfe.2049

Alawattage, C., & Fernando, S. (2017). Postcoloniality in corporate social and environmental accountability. *Accounting, Organizations and Society*, *60*, 1–20. doi:10.1016/j.aos.2017.07.002

Aldrich, H. (2014). *Stand up and be counted: Why social science should stop using the qualitative/quantitative dichotomy.* https://blogs.lse.ac.uk/impactofsocialsciences/2014/11/28/stand-up-and-becounted-social-science-qualitative-quantitative-dichotom

Al-Jawaldeh, A., Nabhani, M., Taktouk, M., & Nasreddine, L. (2022). Climate Change and Nutrition: Implications for the Eastern Mediterranean Region. *International Journal of Environmental Research and Public Health*, *19*(24), 17086. doi:10.3390/ijerph192417086 PMID:36554966

Alkandari, Q., & Hammad, A. A. (2009). *Employees retention in private sector an exploratory study in the State Of Kuwait. Project paper submitted in partial fulfillment of the requirement for the degree of Master of Business Administration.* Open University Malaysia.

Al-Mulali, U., Ozturk, I., & Lean, H. H. (2015). The influence of economic growth, urbanization, trade openness, financial development, and renewable energy on pollution in Europe. *Natural Hazards*, *79*(1), 621–644. doi:10.100711069-015-1865-9

Alqahtany, A. (2022). Green Roofs as an Approach to Enhance Urban Sustainability: A Study of Public Perception in Riyadh, Saudi Arabia. *Buildings*, *2022*(12), 2202. doi:10.3390/buildings12122202

Alsmadi, S. (2007). Green Marketing and the Concern over the Environment: Measuring Environmental Consciousness of Jordanian Consumers. *Journal of Promotion Management*, *13*(3-4), 339–361. doi:10.1080/10496490802306905

Alvarez, R. (2016). *Computational social science*. Cambridge University Press. doi:10.1017/CBO9781316257340

Al-Zoubi, M. T. (2012). The shape of the relationship between salary and job satisfaction: A field study. *Far East Journal of Psychology and Business*, *7*(1), 1–12.

Alzubi, E., & Akkerman, R. (2022). Sustainable supply chain management practices in developing countries: An empirical study of Jordanian manufacturing companies. *Cleaner Production Letters*, *2*, 100005. doi:10.1016/j.clpl.2022.100005

Aman-Ullah, A., Aziz, A., Ibrahim, H., Mehmood, W., & Aman-Ullah, A. (2023). The role of compensation in shaping employee's behaviour: A mediation study through job satisfaction during the Covid-19 pandemic. *Revista de Gestão*, *30*(2), 221–236. doi:10.1108/REGE-04-2021-0068

Anderson, R., Amodeo, M., & Harzfeld, J. (2010). *Changing business cultures from within.* 2010 State of the World Report. The World Watch Institute. doi:10.1016/j.jenvman.2006.12.032

Andrade, M. S., & Westover, J. H. (2022). Job satisfaction–an international comparison of public and private sector employees. *International Journal of Public Administration*, 1–15. doi:10.1080/01900692.2022.2076696

Andrews, T. (2012). What is Social Constructionism? *The Grounded Theory Review, 11*(1), 39–46.

Ansar, N. (2013). Impact of Green Marketing on Consumer Purchase Intention. *Mediterranean Journal of Social Sciences, 4*(11), 650–655. doi:10.5901/mjss.2013.v4n11p650

Ansu-Mensah, P. (2021). Green product awareness effect on green purchase intentions of university students': An emerging market's perspective. *Futur Bus J., 7*(1), 48. doi:10.118643093-021-00094-5

Anton, S. G., & Nucu, A. E. A. (2020). The effect of financial development on renewable energy consumption. A panel data approach. *Renewable Energy, 147*, 330–338. doi:10.1016/j.renene.2019.09.005

Anwar, M., Miyagawa, T., & Gaweesh, M. (2000). Using rice husk ash as a cement replacement material in concrete. *Waste Management (New York, N.Y.), 1*, 671–684.

Aparecida, M., Schettino, S., Nilson, J., & Holanda, F. (2015). Characterization of sugarcane bagasse ash waste for Its Use in Ceramic Floor Tile. *Procedia Materials Science, 8*, 190–196. doi:10.1016/j.mspro.2015.04.063

Aribi, Z. A., & Arun, G. T. (2015). Corporate Social Responsibility and Islamic Financial Institutions (IFIs): Management Perceptions from IFIs in Bahrain. *Journal of Business Ethics, 129*(4), 785–794. doi:10.100710551-014-2132-9

Arif, B., & Farooqi, Y. A. (2014). Impact of work life balance on job satisfaction and organizational commitment among university teachers: A case study of University of Gujrat, Pakistan. *International Journal of Multidisciplinary Sciences and Engineering, 5*(9), 24-29.

Arnould, E. J., & Cayla, J. (2015). Consumer fetish: Commercial ethnography and the sovereign consumer. *Organization Studies, 36*(October), 1361–1386. doi:10.1177/0170840615580012

Arora, A. (2019). *Essential insights into the compressive and flexural response of binder- and aggregate-optimized ultra-high-performance concrete (UHPC)* (Vol. 98). Elsevier Ltd.

Arora, A., Aguayo, M., Hansen, H., Castro, C., Federspiel, E., Mobasher, B., & Neithalath, N. (2018). Microstructural padding- and rheology-based binder collection and depiction for Ultra-high-Performance Concrete (UHPC). *Cement and Concrete Research, 103*(February), 179–190. doi:10.1016/j.cemconres.2017.10.013

Arrow, K. J., Chenery, B. H., Minhas, S. B., & Solow, M. R. (1961). Capital-Labor Substitution and Economic Efficiency. *The Review of Economics and Statistics, 43*(3), 225–250. doi:10.2307/1927286

Arunashantha, A. (2019). The impact of work-life balance on job satisfaction: With special reference to ABC private limited in Sri Lanka. *American Journal of Humanities and Social Sciences Research, 3*, 97–108.

Asibey, M., Mintah, F., Adutwum, I., Wireko-Gyebi, R., Tagnan, J., Yevugah, L., Agyeman, K., & Abdul-Salam, J. (2022). Beyond rhetoric: Urban planning climate change resilience conundrum in Accra, Ghana. *Cities (London, England), 131*, 1–12. doi:10.1016/j.cities.2022.103950

Atanasova, A. (2021). Re-examining utopia in contemporary consumption: Conceptualization and implications for marketing. *AMS Review, 11*(1-2), 23–29. doi:10.100713162-021-00193-0

Audretsch, D. B. (1998). Agglomeration and the Location of Innovative Activity. *Oxford Review of Economic Policy, 14*(2), 18-28.

Awan, U. (2011). Green Marketing: Marketing Strategies for the Swedish Energy Companies. *International Journal of Industrial Marketing, 1*(2), 1–19. doi:10.5296/ijim.v1i2.1008

Ayub, M. (2007). *Understanding Islamic Finance* (1st ed.). John Wiley & Son's Ltd.

Azevedo, S. G., Carvalho, H., & Machado, V. C. (2011). The influence of green practices on supply chain performance: A case study approach. *Transportation Research Part E, Logistics and Transportation Review*, *47*(6), 850–871. doi:10.1016/j.tre.2011.05.017

Baer, H., & Singer, M. (2014). *The anthropology of climate change: An integrated critical perspective*. Routledge. doi:10.4324/9781315818702

Bajrami, A. (2022). From nature to man: Environmental anthropology in the Anthropocene. *Journal of Biological Research (Thessaloniki)*, *95*(1), 10377. doi:10.4081/jbr.2022.10377

BakonK. A.EliasN. F.AbusamhadanaG. A. O. (n.d.). Culture and Digital Divide Influence on E-Government Success of Developing Countries: A Literature Review. *Journal of Theoretical and Applied Information Technology, 98*(9), 1362-1378. https://ssrn.com/abstract=3762293

Baldassarri, D., & Abascal, M. (2017). Field experiments across the social sciences. *Annual Review of Sociology*, *43*(1), 41–73. doi:10.1146/annurev-soc-073014-112445

Balmforth, K., & Gardner, D. (2006). Conflict and facilitation between work and family: Realizing the outcomes for organizations. *New Zealand Journal of Psychology*, *35*(2), 69–76.

Baltagi, B. H., Feng, Q., & Kao, C. (2012). A Lagrange Multiplier test for cross-sectional dependence in a fixed effects panel data model. *Journal of Econometrics*, *170*(1), 164–177. doi:10.1016/j.jeconom.2012.04.004

Bandhavya, G. B., Sandeep, K., & Bindhushree, G. B. (2017). An experimental study on partial replacement of cement with egg shell powder in concrete. Academic Press.

Banerjee. (2014). Advantages of green technology. *Recent Res. Sci. Technol., 6*, 97–100.

Bano, S., Liu, L., & Khan, A. (2022). Dynamic influence of aging, industrial innovations, and ICT on tourism development and renewable energy consumption in BRICS economies. *Renewable Energy*, *192*, 431–442. doi:10.1016/j.renene.2022.04.134

Bansal, P. (2005). Evolving sustainably: A longitudinal study of corporate sustainable development. *Strategic Management Journal*, *26*(3), 197–218. doi:10.1002mj.441

Barakji, Y. A., Rønnstad, A. T. M., Christensen, M. O., Zachariae, C., Wienholtz, N. K. F., Halling, A. S., Maul, J. T., Thomsen, S. F., Egeberg, A., & Thyssen, J. P. (2022). Assessment of Frequency of Rosacea Subtypes in Patients With Rosacea: A Systematic Review and Meta-analysis. *JAMA Dermatology*, *158*(6), 617–625. doi:10.1001/jamadermatol.2022.0526 PMID:35385049

Barne, C., Alexis, N. E., Bernstein, J. A., Cohn, J. R., Demain, J. G., Horner, E., Levetin, E., Nei, A., & Phipatanakul, W. (2013). Climate change and our environment: The effect on respiratory and allergic disease. *The Journal of Allergy and Clinical Immunology. In Practice*, *1*(2), 137–141. doi:10.1016/j.jaip.2012.07.002 PMID:23687635

Barnes, J., Dove, M., Lahsen, M., Mathews, A., McElwee, P., McIntosh, R., Moore, F., O'Reilly, J., Orlove, B., Puri, R., Weiss, H., & Yager, K. (2013). Contribution of anthropology to the study of climate change. *Nature Climate Change*, *3*(6), 541–544. doi:10.1038/nclimate1775

Barnet, M. L. (2007). Stakeholder influence capacity and the variability of financial returns to corporate social responsibility. *Academy of Management Review*, *33*(3), 794–816. doi:10.5465/amr.2007.25275520

Barney, J. (1991). Firm Resources and Sustained Competitive Advantage. *Journal of Management*, *17*(1), 99–120. doi:10.1177/014920639101700108

Barrett, B., Charles, J. W., & Temte, J. L. (2015). Climate change, human health, and epidemiological transition. *Preventive Medicine, 70*, 69–75. doi:10.1016/j.ypmed.2014.11.013 PMID:25434735

Barrett, S. (2009). *Anthropology*. University of Toronto Press. doi:10.3138/9781442697362

Bartlett, C. A., & Ghoshal, S. (2002). Building competitive advantage through people. *MIT Sloan Management Review, 43*(2), 16–25.

Bartlett, M. S. (1951). The effect of standardization on a Chi-square approximation in factor analysis. *Biometrika, 38*(3/4), 337–344. doi:10.2307/2332580

Baschera, W., & Schneider, I. (2022). *Can sustainable urban planning determine people's happiness and well-being?* Brazilian Journal of Management / Revista de Administração da UFSM. doi:10.5902/1983465969433

Bashir, B., & Gani, A. (2020). Testing the effects of job satisfaction on organizational commitment. *Journal of Management Development, 39*(4), 525–542. doi:10.1108/JMD-07-2018-0210

Baykara, H. C., Cornejo, M. H., Murillo, R., Gavilanes, A., Paredes, C., & Elsen, J. (2017). Preparation, characterization and reaction kinetics of green cement: Ecuadorian natural mordenite-based geopolymers. *Materials and Structures, 50*(3), 1–12. doi:10.161711527-017-1057-z

Bell, A. S., Rajendran, D., & Theiler, S. (2012). Job stress, wellbeing, work-life balance and work-life conflict among Australian academics. *E-Journal of Applied Psychology : Clinical and Social Issues, 8*(1), 25–37. doi:10.7790/ejap.v8i1.320

Benhelal, E. Z., Zahedi, G., & Hashim, H. (2012). A novel design for green and economical cement manufacturing. *Journal of Cleaner Production, 22*(1), 60–66. doi:10.1016/j.jclepro.2011.09.019

Berardi, U., GhaffarianHoseini, A. H., & GhaffarianHoseini, A. (2014). State-of-the-art analysis of the environmental benefits of green roofs. *Applied Energy, 115*, 411–428. doi:10.1016/j.apenergy.2013.10.047

Berlin, G. (2020). *Capital of Green Trends: How Berlin leads the Way in Urban Sustainability*. Available online: https://about.visitberlin.de/en/greenberlin

Berlin, N. (2020). *Stadt der grünen Trends-Wie Berlin den Weg der urbanen Nachhaltigkeit geht*. Available online: https://about.visitberlin.de/nachhaltiges-berlin

Bernard, H. (2011). *Research methods in anthropology: Qualitative and quantitative approaches*. AltaMira Press.

Bernardini, C. & Irvine, K. N. (2007). The 'nature' of urban sustainability: Private or public green spaces. *Sustainable Development and Planning III, 2*, 661-674. doi:10.2495/SDP070642

Beynon, M. J., Jones, P., Pickernell, D., & Packham, G. (2015). Investigating the impact of training influence on employee retention in small and medium enterprises: A regression-type classification and ranking believe simplex analysis on sparse data. *Expert Systems: International Journal of Knowledge Engineering and Neural Networks, 32*(1), 141–154. doi:10.1111/exsy.12067

Bezencon, V. & Blili, S. (2010). Ethical products and consumer involvement: what's new? *European Journal of Marketing, 44*(9/10), 1305-1321.

Bezirtzoglou, C., Dekas, K., & Charvalos, E. (2011). Climate changes, environment and infection: Facts, scenarios and growing awareness from the public health community within Europe. *Anaerobe, 17*(6), 337–340. doi:10.1016/j.anaerobe.2011.05.016 PMID:21664978

Bhalerao, S. K. (2013). Work life balance: The key driver of employee engagement. *ASM's International E-Journal of Ongoing Research in Management and IT, 16*(1), 1–9.

Bharadwaj, S., Khan, N. A., & Yameen, M. (2022). Unbundling employer branding, job satisfaction, organizational identification and employee retention: A sequential mediation analysis. *Asia-Pacific Journal of Business Administration, 14*(3), 309–334. doi:10.1108/APJBA-08-2020-0279

Bhardwaj. (2015). The Advantages and Disadvantages of Green Technology. *J. Basic Appl. Eng. Res., 2*, 1957–1960.

Bhutan Tourism Monitor. (2019). Publication of the Tourism Council of Bhutan.

Biason, R. S. (2020). The effect of job satisfaction on employee retention. *International Journal of Economics. Commerce and Management, 8*(3), 405–413.

Bilharz, J. A. (1998). *The Allegany Senecas and Kinzua Dam: Forced Relocation through Two Generations.* University of Nebraska Press.

Birch, L. L., Gunder, L., Grimm-Thomas, K., & Laing, D. G. (1998). Infants' consumption of a new food enhances acceptance of similar foods. *Appetite, 30*(3), 283–295. doi:10.1006/appe.1997.0146 PMID:9632459

Bissing-Olson, M. J., Iyer, A., Fielding, K. S., & Zacher, H. (2013). Relationships between daily affect and pro-environmental behavior at work: The moderating role of pro-environmental attitude. *Journal of Organizational Behavior, 175*(2), 156–175. doi:10.1002/job.1788

Blomquist, J., & Westerlund, J. (2013). Testing slope homogeneity in large panels with serial correlation. *Economics Letters, 121*(3), 374–378. doi:10.1016/j.econlet.2013.09.012

Bocean, C. G., Popescu, L., Varzaru, A. A., Avram, C. D., & Iancu, A. (2023). Work–Life Balance and Employee Satisfaction during COVID-19 Pandemic. *Sustainability (Basel), 15*(15), 11631. doi:10.3390u151511631

Bock, G. W., Zmud, R. W., Kim, Y. G., & Lee, J. N. (2005). Behavioral intention formation in knowledge sharing: Examining the roles of extrinsic motivators, social-psychological forces, and organizational climate. *Management Information Systems Quarterly, 29*(1), 87–112. doi:10.2307/25148669

Boesch, M., & Hellweg, S. (2010). Identifying improvement potentials in cement production with life cycle assessment. *Environmental Science & Technology, 44*(23), 9143–9149. doi:10.1021/es100771k PMID:21047057

Boiral, O., & Paillé, P. (2012). Organizational citizenship behaviour for the environment: Measurement and validation. *Journal of Business Ethics, 109*(4), 431–445. doi:10.100710551-011-1138-9

Boliko, M. C. (2019). FAO and the Situation of Food Security and Nutrition in the World. *Journal of Nutritional Science and Vitaminology, 65*(Supplement), S4–S8. doi:10.3177/jnsv.65.S4 PMID:31619643

Bollen, K. A. (2011). Evaluating Effect, Composite, and Causal Indicators in Structural Equation Models. *Management Information Systems Quarterly, 35*(2), 359–372. doi:10.2307/23044047

Bollen, K. A., Harden, J. J., Ray, S., & Zavisca, J. (2014). BIC and Alternative Bayesian Information Criteria in the Selection of Structural Equation Models. *Structural Equation Modeling, 21*(1), 1–19. doi:10.1080/10705511.2014.856691 PMID:31360054

Booth, M. (2018). Climate Change and the Neglected Tropical Diseases. *Advances in Parasitology, 100*, 39–126. doi:10.1016/bs.apar.2018.02.001 PMID:29753342

Bose, S., & Khan, H. Z. (2022). Sustainable development goals (SDGs) reporting and the role of country-level institutional factors: An international evidence. *Journal of Cleaner Production, 335*, 130290. doi:10.1016/j.jclepro.2021.130290

Brady, H. (2019). The challenge of big data and data science. *Annual Review of Political Science, 22*(1), 297–323. doi:10.1146/annurev-polisci-090216-023229

Brands, R. A., & Fernandez-Mateo, I. (2017). Leaning out: How negative recruitment experiences shape women's decisions to compete for executive roles. *Administrative Science Quarterly*, *62*(3), 405–442. doi:10.1177/0001839216682728

Bratanova, B., Loughnan, S., & Gatersleben, B. (2012). The moral circle as a common motivational cause of cross-situational pro-environmentalism. *European Journal of Social Psychology*, *42*(5), 539–545. doi:10.1002/ejsp.1871

Breusch, T. S., & Pagan, A. R. (1980). The lagrange multiplier test and its applications to model specification in econometrics. *The Review of Economic Studies*, *47*(1), 239–253. doi:10.2307/2297111

Brooks, C. (2008). *Introductory Econometrics for Finance* (2nd ed.). Cambridge University Press. doi:10.1017/CBO9780511841644

Brudermann, T., & Sangkakool, T. (2017). Green roofs in temperate climate cities in Europe - An analysis of key decision factors. *Urban Forestry & Urban Greening*, *21*, 224–234. doi:10.1016/j.ufug.2016.12.008

Brundtland, G. (1987). *Our Common Future: The World Commission on Environment and Development.* Oxford University Press.

Bryant, F. B., & Yarnold, P. R. (1995). Principal-components analysis and confirmatory factor analysis. In L. G. Grimm & P. R. Yarnold (Eds.), *Reading and understanding multivariate statistics* (pp. 99–136). American Psychological Association.

Bu, J., Ding, R., Zhou, L., Chen, X., & Shen, E. (2022). Epidemiology of Psoriasis and Comorbid Diseases: A Narrative Review. *Frontiers in Immunology*, *13*, 880201. doi:10.3389/fimmu.2022.880201 PMID:35757712

Bunge, M. (1999). *Buscar la filosofía en las ciencias sociales*. Siglo XXI.

Burakov, D., & Freidin, M. (2017). Financial development, economic growth and renewable energy consumption in Russia: A vector error correction approach. *International Journal of Energy Economics and Policy*, *7*(6), 39.

Burgess, C. T., Adkins, J., & Suniga, N. (2021). Management and Leadership: A Practical Review of the Organizational Culture. Handbook of Research on Multidisciplinary Perspectives on Managerial and Leadership Psychology, 396-412.

Burke, R. J., & Greenglass, E. R. (1999). Work–family conflict, spouse support, and nursing staff well-being during organizational restructuring. *Journal of Occupational Health Psychology, 4*(4), 327-336.

Burke, R. J., & Greenglass, E. (1987). Work and family. In C. L. Cooper & I. T. Robertson (Eds.), *International review of industrial and organizational psychology*. Wiley.

Burszta-Adamiak, E., & Fiałkiewicza, W. (2019). Review of green roof incentives as motivators for the expansion of green infrastructure in European cities. *Scientific Review – Engineering and Environmental Sciences*, *28*(4), 641–652. DOI doi:10.22630/PNIKS.2019.28.4.5

Burszta-Adamiak, E., Stańczyk, J., & Łomotowski, J. (2019). Hydrological performance of green roofs in the context of the meteorological factors during the 5-year monitoring period. *Water and Environment Journal : the Journal / the Chartered Institution of Water and Environmental Management*, *33*(1), 144–154. doi:10.1111/wej.12385

Butt, A. (2016). *What is Green Technology and Its Benefits?* https://usgreentechnology.com/green-technology/

Cai, Y., Sam, C. Y., & Chang, T. (2018). Nexus between clean energy consumption, economic growth and CO2 emissions. *Journal of Cleaner Production*, *182*, 1001–1011. doi:10.1016/j.jclepro.2018.02.035

Caminade, C., McIntyre, K. M., & Jones, A. E. (2019). Impact of recent and future climate change on vector-borne diseases. *Annals of the New York Academy of Sciences*, *1436*(1), 157–173. doi:10.1111/nyas.13950 PMID:30120891

Camodeca, R., & Almici, A. (2021). Digital Transformation and Convergence toward the 2030 Agenda's Sustainability Development Goals: Evidence from Italian Listed Firms. *Sustainability, 13*(21), 11831. doi:10.3390/su132111831

Carp, L. (2012). Can stock market development boost economic growth? Empirical evidence from emerging markets in Central and Eastern Europe. *Procedia Economics and Finance*, *3*, 438–444. doi:10.1016/S2212-5671(12)00177-3

Casey, K., Lichrou, M., & O'Malley, L. (2020). Prefiguring sustainable living: An ecovillage story. *Journal of Marketing Management*, *36*(17–18), 1658–1679. doi:10.1080/0267257X.2020.1801800

Cashdan, E. (1998). Adaptiveness of food learning and food aversions children. *Social Sciences Information. Information Sur les Sciences Sociales*, *37*(4), 613–632. doi:10.1177/053901898037004003

Castro, A. J. L., & Orellana, F. (2019). Green roofs as a strategy for urban green innovation spaces. *Sustainable Cities and Society*, *44*, 697–705. doi:10.1016/j.scs.2018.10.034

Centobelli, P., Cerchione, R., Chiaroni, D., del Vecchio, P., & Urbinati, A. (2020). Designing business models in circular economy: A systematic literature review and research agenda. *Business Strategy and the Environment*, *29*(4), 1734–1749. doi:10.1002/bse.2466

Çeştepe, H., Çetin, M., Avcı, P., & Bahtiyar, B. (2022). The link between technological innovation and financial development: Evidence from selected OECD countries. *International Journal of Finance & Economics*, 1–17.

Chandler, D., Rothe, D., & Muller, F. (2022). Relaciones Internacionales en el Antropoceno. *Relaciones Internacionales*, *50*, 107–126. doi:10.15366/relacionesinternacionales2022.50.005

Chang, C. (2011). Feeling ambivalent about going green – Implication For Green. *Journal of Advertising*, *40*(4), 19–31. doi:10.2753/JOA0091-3367400402

Chang, C. L., Ilomäki, J., Laurila, H., & McAleer, M. (2020). Causality between CO2 emissions and stock markets. *Energies*, *13*(11), 2893. doi:10.3390/en13112893

Chang, N. J., & Fong, C. M. (2010). Green product quality, green corporate image, green customer satisfaction, and green customer loyalty. *African Journal of Business Management*, *4*(13), 2336–2344.

Chang, T. W., Chen, Y. S., Yeh, Y. L., & Li, H. X. (2020). Sustainable consumption models for customers: Investigating the significant antecedents of green purchase behavior from the perspective of information asymmetry. *Journal of Environmental Planning and Management*, *64*(9), 1668–1688. doi:10.1080/09640568.2020.1837087

Chapra, M. U. (2000). *The Future of Economics: An Islamic Perspective*. The Islamic Foundation.

Chaturvedi, A., & Sangwan, K. (2016). Impact of Job Satisfaction on Retention of Employees in BPO Industry. *International Multidisciplinary Research Journal*, *5*(11), 1–6.

Chen, T. B., & Chai, L. T. (2010). Attitude towards the environment and green products: Consumers perspective. *Management Science and Engineering, 4*(2), 27-39.

Chen, A., Lin, W.-T., & Huang, R. (2011). Article. *Materials & Design*, *32*, 635–642.

Chen, C. (2001). Design for the environment: A quality-based model for green product development. *Management Science*, *47*(2), 250–263. doi:10.1287/mnsc.47.2.250.9841

Chen, C., Habert, G., Bouzidi, Y., & Jullien, A. (2010a). Environmental impact of cement production: Detail of the different processes and cement plant variability evaluation. *Journal of Cleaner Production*, *18*(5), 478–485. doi:10.1016/j.jclepro.2009.12.014

Cheng, Z., Nielsen, I., & Cutler, H. (2019). Perceived job quality, work-life interference and intention to stay: Evidence from the aged care workforce in Australia. *International Journal of Manpower*, *40*(1), 17–35. doi:10.1108/IJM-08-2017-0208

Chen, H. Q., Qin, R., Chow, C. L., & Lau, D. (2023). Recycling thermoset plastic waste for manufacturing green cement mortar. *Cement and Concrete Composites*, *137*, 104922. doi:10.1016/j.cemconcomp.2022.104922

Chen, Y. S. (2008). The driver of green innovation and green image–green core competence. *Journal of Business Ethics*, *81*(3), 531–543. doi:10.100710551-007-9522-1

Chen, Y. S., & Chang, C. H. (2013). Greenwash and green trust: The mediation effects of green consumer confusion and green perceived risk. *Journal of Business Ethics*, *114*(3), 489–500. doi:10.100710551-012-1360-0

Chen, Y. S., & Chang, C. H. (2013). The determinants of green product development performance: Green dynamic capabilities, green transformational leadership, and green creativity. *Journal of Business Ethics*, *116*(1), 107–119. doi:10.100710551-012-1452-x

Chen, Y. S., Lai, S. B., & Wen, C. T. (2006). The influence of green innovation performance on corporate advantage in Taiwan. *Journal of Business Ethics*, *67*(4), 331–339. doi:10.100710551-006-9025-5

Chen, Y. S., Lin, S. H., Lin, C. Y., Hung, S. T., Chang, C. W., & Huang, C. W. (2020). Improving green product development performance from green vision and organizational culture perspectives. *Corporate Social Responsibility and Environmental Management*, *27*(1), 222–231. doi:10.1002/csr.1794

Chi, H., Vu, T. V., Nguyen, H. V., & Truong, T. H. (2023). How financial and non–financial rewards moderate the relationships between transformational leadership, job satisfaction, and job performance. *Cogent Business & Management*, *10*(1), 1–18. doi:10.1080/23311975.2023.2173850

Chitra, K. (2007). In search of the Green Consumers: A perceptual Study. *Journal of Services Research*, *7*(1), 173–191.

Chu, H. C., & Kuo, T. (2015). Testing Herzberg's two-factor theory in educational settings in Taiwan. *The Journal of Human Resource and Adult Learning*, *11*(1), 54–65.

Connolly, C., Keil, R., & Ali, S. H. (2021). Extended urbanisation and the spatialities of infectious disease: Demographic change, infrastructure and governance. *Urban Studies (Edinburgh, Scotland)*, *58*(2), 245–263. doi:10.1177/0042098020910873

Consulting, D. (2017). *12 mal Berliner Leben, 12 mal Berlin Erleben. Konzept für einen stadtverträglichen Berlin-Tourismus 2018+*. VisitBerlin. https://about.visitberlin.de/sites/default/files/2018 02/Tourismuskonzept_Berlin_Studie_2017.pdf

Coolen, H., & Meesters, J. (2012). Private and public green spaces: Meaningful but different settings. *Journal of Housing and the Built Environment*, *27*(1), 49–67. doi:10.100710901-011-9246-5

Cronbach, L. (1951). Coefficient alpha and the internal structure of tests. *Psychometrika*, *16*(3), 297–334. doi:10.1007/BF02310555

D'Amato, G., Chong-Neto, H. J., Monge Ortega, O. P., Vitale, C., Ansotegui, I., Rosario, N., Haahtela, T., Galan, C., Pawankar, R., Murrieta-Aguttes, M., Cecchi, L., Bergmann, C., Ridolo, E., Ramon, G., Gonzalez Diaz, S., D'Amato, M., & Annesi-Maesano, I. (2020). The effects of climate change on respiratory allergy and asthma induced by pollen and mold allergens. *Allergy*, *75*(9), 2219–2228. doi:10.1111/all.14476 PMID:32589303

D'Souza, C., Lamb, P., & Taghian, M. (2006). An empirical study on the influence of environmental labels on consumers. *Corporate Communications*, *11*(2), 162–173. doi:10.1108/13563280610661697

Dahl, U., & Persson, S. (2008). *Purchasing of environmental friendly computers*. Jonkoping: Bachelor Thesis within Business Administration

Damtoft, J., Lukasik, J., Herfort, D., Sorrentino, D., & Gartner, E. (2008). Sustainable development and climate change initiatives. *Cement and Concrete Research*, *38*(2), 115–127. doi:10.1016/j.cemconres.2007.09.008

Davis, K. (1989). *Human behavior at work: Organizational behavior*. McGraw-Hill Inc.

de Freitas Netto, S. V., Sobral, M. F. F., Ribeiro, A. R. B., & Soares, G. R. L. (2020). Concepts and forms of greenwashing: A systematic review. *Environmental Sciences Europe*, *32*(1), 19. doi:10.118612302-020-0300-3

De Melo, M. A. C. (1997). *Processo de Planejamento e as Inovações Tecnológicas e Sociais: uma perspectiva sócio-ecológica*. Anais do 5o. Seminário de Modernização Tecnológica.

De Melo, A. (2001). *The Innovation Systems of Latin America and the Caribbean*. Inter-American Development Bank. Agosto.

De Melo, M. A. C. (2002). *Enriquecendo a Atuação de Incubadora de Emrpesas. Tecnologia e Inovação: experiências de gestão nas micro e pequenas empresas*. PGT/USP.

Deci, E. L., & Ryan, R. M. (2000). The "what" and "why" of goal pursuits: Human needs and the self-determination of behavior. *Psychological Inquiry*, *11*(4), 227–268. doi:10.1207/S15327965PLI1104_01

Deloitte. (2017). *Rewriting the rules for the digital age 2017 Deloitte Global Human Capital Trends*. www.deloitte.com

Denyer, S. (2007). *Bhutan Wonders If TV Really Brings Happiness*. Reuters. www.reuters.com/article/us-bhutan-television-idUSSP29261020070513

Deshpande, B., Arekar, K., Sharma, R., & Somaiya, S. (2012). Effect of employee satisfaction on organization performance: An empirical study in hotel industry. *Ninth AIMS International Conference on Management*, 1-4.

DeVellis, R. F. (2016). *Scale development: Theory and applications* (Vol. 26). Sage Publications.

DeWalt, K., DeWalt, B., & Wayland, C. (1998). Participant observation. In H. R. Bernard (Ed.), *Handbook of methods in cultural anthropology* (pp. 259–299). Altamira.

Dhir, S., Dutta, T., & Ghosh, P. (2020). Linking employee loyalty with job satisfaction using PLS–SEM modelling. *Personnel Review*, *49*(8), 1695–1711. doi:10.1108/PR-03-2019-0107

Diamond, J., & Robinson, J. (2010). *Natural experiments of history*. The Belknap Press of Harvard University Press.

Díaz, D., Kristoff, J., Castro, M., Miceli, J., Castro, D., Quinteros, R., & Guerrero, S. (2007). *Exploraciones en antropología y complejidad*. Editorial Sb.

Ditlevsen, K., Sandøe, P., & Lassen, J. (2019). Healthy food is nutritious, but organic food is healthy because it is pure: The negotiation of healthy food choices by Danish consumers of organic food. *Food Quality and Preference*, *71*(January), 46–53. doi:10.1016/j.foodqual.2018.06.001

Doytch, N., & Narayan, S. (2016). Does FDI influence renewable energy consumption? An analysis of sectoral FDI impact on renewable and non-renewable industrial energy consumption. *Energy Economics*, *54*, 291–301. doi:10.1016/j.eneco.2015.12.010

Drechsler, W. (2019). The Reality and Diversity of Buddhist Economics. *American Journal of Economics and Sociology*, *78*(2), 523–560. doi:10.1111/ajes.12271

Dubey. (1996). Buddhism in Bhutan and Sri Lanka: A Comparative Study. In *Bhutan: Society and Polity*. New Delhi: Indus Publishing Company.

Dunne, J. (2015). First Thoughts Toward a Buddhist Economics. In M. Ricard & T. Singer (Eds.), *Caring Economics: Conversations on Altruism and Compassion Between Scientists, Economists and the Dalai Lama*. Picador.

Dunning, T. (2012). *Natural experiments in the social sciences*. Cambridge University Press. doi:10.1017/CBO9781139084444

Durrenberger, P. (2007) The Anthropology of Organized Labor in the United States. *Ann Review of Anthropology, 36*, 73-88. https://www.annualreviews.org/doi/pdf/10.1146/annurev.anthro.36.081406.094433

Dutta, A. B., & Banerjee, S. (2014). Study of employee retention. *International Journal of Business Management & Research*, *4*(1), 83–88.

Ebi, K. L., Capon, A., Berry, P., Broderick, C., de Dear, R., Havenith, G., Honda, Y., Kovats, R. S., Ma, W., Malik, A., Morris, N. B., Nybo, L., Seneviratne, S. I., Vanos, J., & Jay, O. (2021). Hot weather and heat extremes: Health risks. *Lancet*, *398*(10301), 698–708. doi:10.1016/S0140-6736(21)01208-3 PMID:34419205

Edwards, J. R., & Rothbard, N. P. (2000). Mechanisms linking work and family: Clarifying the relationship between work and family constructs. *Academy of Management Review*, *25*(1), 178–199. doi:10.2307/259269

Egerton, F. N. (2001). A history of the ecological sciences: Early Greek origins. *Bulletin of the Ecological Society of America*, *82*(1), 93–97.

Elinwa, A., & Mahmood, Y. (2002). Ash from timber waste as cement replacement material. *Cement and Concrete Composites*, *24*(2), 219–222. doi:10.1016/S0958-9465(01)00039-7

Elisabetta, R., & Claudio, V. (2018). Investments in big data analytics and firm performance: An empirical investigation of direct and mediating effects. *International Journal of Production Research*, *56*(15), 5206–5221. doi:10.1080/00207543.2018.1427900

Elkington, H., & Makower. (1988). *The green consumers*. New York: Penguin Books.

Elkington, J. (1994). Towards the sustainable corporation: Win-win-win business strategies for sustainable development. *Cal. Management Review, 36*(3), 90-100.

Ellwanger, J. H., Kulmann-Leal, B., Kaminski, V. L., Valverde-Villegas, J. M., Veiga, A. B. G. D., Spilki, F. R., Fearnside, P. M., Caesar, L., Giatti, L. L., Wallau, G. L., Almeida, S. E. M., Borba, M. R., Hora, V. P. D., & Chies, J. A. B. (2020). Beyond diversity loss and climate change: Impacts of Amazon deforestation on infectious diseases and public health. *Anais da Academia Brasileira de Ciências*, *92*(1), e20191375. doi:10.1590/0001-3765202020191375 PMID:32321030

El-Sayed, A., & Kamel, M. (2020). Climatic changes and their role in emergence and re-emergence of diseases. *Environmental Science and Pollution Research International*, *27*(18), 22336–22352. doi:10.100711356-020-08896-w PMID:32347486

Engle, R. F., & Sheppard, K. (2001). *Theoretical and Empirical Properties of Dynamic Conditional Correlation MVGARCH*. Economic Working Paper Series 8554, University of California.

Epskamp, S. (2015). semPlot: Unified visualizations of structural equation models. *Structural Equation Modeling*, *22*(3), 474–483. doi:10.1080/10705511.2014.937847

Eriksen, T. H. (2016). *Overheating: An Anthropology of Accelerated Change*. Pluto. doi:10.2307/j.ctt1cc2mxj

Escalera, J., & Ruiz, E. (2011). Socioecological Resilience: Contributions and challenges from Anthropology. *Revista de Antropología Social, 20*, 109–135. doi:10.5209/rev_RASO.2011.v20.36264

Escobar, A. (2020). *Pluriversal Politics: The Real and the Possible*. Duke University Press.

ESCWA. (2021). *International Dialogue on the Role of Islamic Social Financing in Achieving the SDGs.* Available online: https: //www.unescwa.org/news/international-dialogue-role-islamic-social-financing-achieving-sdgs

Esperland, D., de Weerd, L., & Mercer, J. B. (2022, December). Health effects of voluntary exposure to cold water - a continuing subject of debate. *International Journal of Circumpolar Health, 81*(1), 2111789. doi:10.1080/22423982.2022.2111789 PMID:36137565

Eyraud, L., Clements, B., & Wane, A. (2013). Green investment: Trends and determinants. *Energy Policy, 60*, 852–865. doi:10.1016/j.enpol.2013.04.039

Fabrigar, L. R., Wegener, D. T., MacCallum, R. C., & Strahan, E. J. (1999). Evaluating the use of exploratory factor analysis in psychological research. *Psychological Methods, 4*(3), 272–299. doi:10.1037/1082-989X.4.3.272

Fan, H., & Zeng, L. (2011). *Implementation of Green Marketing Strategy in China: A Study of the Green Food Industry* [MSc thesis]. Dept. Bus. Administration Econ. Study, Univ. Gavle.

Fernandes, A. A. R., & Solimun. (2017). The mediating effect of strategic orientation and innovations on the effect of environmental uncertainties on the performance of business in the Indonesian aviation industry. *International Journal of Law and Management, 59*(6), 1269–1278. doi:10.1108/IJLMA-10-2016-0087

Fialkiewicz, W., Burszta-Adamiak, E., Kolonko-Wiercik, A., Manzardo, A., Loss, A., Mikovits, C., & Scipioni, A. (2018). Simplified direct water footprint model to support urban water management. *Water (Basel), 10*(5), 630. doi:10.3390/w10050630

Figge & Thorpe. (2023). Circular economy, operational eco-efficiency, and sufficiency. An integrated view. *Ecological Economics, 204*(B), 107692.

Figueiredo, B., & Scaraboto, D. (2016). The systemic creation of value through circulation in collaborative consumer networks. *The Journal of Consumer Research, 43*(4), 509–533. Advance online publication. doi:10.1093/jcr/ucw038

Fiske, S. J., Crate, S. A., Crumley, C., Galvin, K., Lazrus, H., Lucero, L., Oliver-Smith, A., Orlove, B., Strauss, S., & Wilk, R. R. (2014). *Changing the atmosphere. Anthropology and climate change. Final report of the AAA global climate change task force*. American Anthropological Association.

Fong, C. M., & Chang, N. J. (2012). The impact of green learning orientation on proactive environmental innovation capability and firm performance. *African Journal of Business Management, 6*(32), 727–735.

Fuess, S., & Millea, M. (2002). Do Employers Pay Efficiency Wage? Evidence from Japan. *Journal of Labor Research, 23*(2), 279–292. doi:10.100712122-002-1007-5

Fung, K. L. (2018). *Expanding the green network on rooftops: A study of integrating green roofs as a part of urban green innovation space planning* (Order No. 10932095). Available from ProQuest One Academic. (2124999351). Retrieved from http://wdg.biblio.udg.mx:2048/login?url=https://www.proquest.com/dissertations-theses/expanding-green-network-on-rooftops-study/docview/2124999351/se-2?accountid=28915

Furso, B. (2020). *The relationship between performance appraisal and teachers and teachers job satisfaction in secondary schools of west Arsi Zone, Oromia Regional State* [Doctoral dissertation]. Hawassa University.

Galdeano-Gómez, E., Céspedes-Lorente, J., & Martínez-del-Río, J. (2008). Environmental performance and spillover effects on productivity: Evidence from horticultural firms. *Journal of Environmental Management*, *88*(4), 1552–1561. doi:10.1016/j.jenvman.2007.07.028 PMID:17825476

Garg, P., & Yajurvedi, N. (2016). Impact of work-life balance practices on employees retention and organizational performance–A study on IT industry. *Indian Journal of Applied Research*, *6*(8), 105–108.

Gault, R. (1907). A history of the questionnaire method of research in psychology. *Research in Psychology*, *14*(3), 366–383. doi:10.1080/08919402.1907.10532551

Gayathiri, R., Ramakrishnan, L., Babatunde, S. A., Banerjee, A., & Islam, M. Z. (2013). Quality of work life–Linkage with job satisfaction and performance. *International Journal of Business and Management Invention*, *2*(1), 1–8.

Geng, X., Yu, Z., Zhang, D., Li, C., Yuan, Y., & Wang, X. (2022). The influence of local background climate on the dominant factors and threshold-size of the cooling effect of urban parks. *The Science of the Total Environment*, *823*, 153806. doi:10.1016/j.scitotenv.2022.153806 PMID:35150695

Geraie, M. S., & Rad, F. M. (2015). Mediator role of the organizational identity green in relationship between total quality management and perceived innovation with sustainable competitive advantage. *International Journal of Biology, Pharmacy and Allied Sciences*, *4*, 266–276.

Geweke, J. (1982). Measurement of Linear Dependence and Feedback between Multiple Time Series. *Journal of the American Statistical Association*, *77*(378), 304–313. doi:10.1080/01621459.1982.10477803

Giannetti, B. F., Agostinho, F., Almeida, C. M. V. B., & Huisingh, D. (2015, January). A Review of Limitations of GDP and Alternative Indices to Monitor Human Wellbeing and to Manage Eco-System Functionality. *Journal of Cleaner Production*, *87*(1), 11–25. doi:10.1016/j.jclepro.2014.10.051

Giner, C. (2004). J. Antropología urbana. *Ariel.*

Girisha, M. C. (2019) Job satisfaction: a study at AT & S India Pvt Ltd Nanjangud (mysore District, Karnataka). *Proceedings of the SIMS Annual Research Conference, SIMSARC12.*

Githeko, A. K., Lindsay, S. W., Confalonieri, U. E., & Patz, J. A. (2000). Climate change and vector-borne diseases: A regional analysis. *Bulletin of the World Health Organization*, *78*(9), 1136–1147. PMID:11019462

Givel, M., & Figueroa, L. (2015). *Early Happiness Policy as a Government Mission of Bhutan-A Survey of the Bhutanese Unwritten Constitution From 1619 to 1729.* Academic Press.

Glass, J. L., & Estes, S. B. (1997). The family responsive workplace. *Annual Review of Sociology*, *23*(1), 289–313. doi:10.1146/annurev.soc.23.1.289

Gluckman, M. (1968). The Utility of the Equilibrium Model in the Study of Social Change. *American Anthropologist*, *70*(2), 219–237. doi:10.1525/aa.1968.70.2.02a00010

Goda, Z., Foda, M. & Elsayyad, N. (2023). Using Green Roofs for Social Housing to Improve Energy Consumption in New Cities. (An Applied Study of Social Housing in Egypt's New Cairo City). *Future Cities and Environment*, *9*(1), 2, 1–11. doi:10.5334/fce.165

Gohari, A., Gohari, A., & Bin, A. (2023). Importance of green roof criteria for residential and governmental buildings: A multi-criteria decision analysis. *Environmental Science and Pollution Research International*, *30*(2), 3707–3725. doi:10.100711356-022-22472-4 PMID:35953748

Gong, Y., Li, X., & Li, Z. (2020). How to design green roofs to enhance urban green innovation space? Evidence from Beijing. *Habitat International*, *107*, 102–113. doi:10.1016/j.habitatint.2020.102113

González Álvarez, M. D. (1997). Processos de Planejamento nos Pólos Tecnológicos: um enfoque adaptativo. Tese de Doutorado, Departamento de Engenharia Industrial, PUC-Rio, Rio de Janeiro.

Gozgor, G., Mahalik, M. K., Demir, E., & Padhan, H. (2020). The impact of economic globalization on renewable energy in the OECD countries. *Energy Policy*, *139*, 111365. doi:10.1016/j.enpol.2020.111365

Grace, J., & Bollen, K. (2007). Representing general theoretical concepts in structural equation models: The role of composite variables. *Environmental and Ecological Statistics*, *15*(2), 191–213. doi:10.100710651-007-0047-7

Gragnano, A., Simbula, S., & Miglioretti, M. (2020). Work–life balance: Weighing the importance of work–family and work–health balance. *International Journal of Environmental Research and Public Health*, *17*(3), 907. doi:10.3390/ijerph17030907 PMID:32024155

Gravano, A. (2007b). Claves para la facilitación organizacional del proceso participativo en la planificación urbano-ambiental metropolitana. *Revista Regional de Trabajo Social*, *21*(40), 9-22.

Gravano, A. (2008a) *¿Vecinos o ciudadanos? el fenómeno NIMBY (Not In My Back Yard) o SPAN (Sí, Pero No Aquí) del imaginario urbano en un proceso de participación social y su tratamiento desde la facilitación organizacional antropológica*. Trabajo presentado en II Congreso de la Asociación Latinoamericana de Antropología (ALA), Universidad de Costa Rica, Simposio "Imaginarios urbanos y participación social".

Gravano, A. (1992). Antropología Práctica, muestra y posibilidades de la Antropología Organizacional. *Publicar en Antropología y Ciencias Sociales, Buenos Aires*, (1), 95–126.

Gravano, A. (2005). *El barrio en la teoría social*. Espacio Editorial.

Gravano, A. (2006). Imaginarios regionales y circularidad en la planificación: El caso del TOAR. *Intersecciones*, (7), 305–323.

Gray, K. (2022). Climate change, human health, and health informatics: A new view of connected and sustainable digital health. *Frontiers in Digital Health*, *4*, 869721. doi:10.3389/fdgth.2022.869721 PMID:35373178

Greene, K. L. (2013). *Better equipping reserve military intelligence analyst to meet the needs of the commander by championing a process-driven training model* [Doctoral dissertation]. Fort Leavenworth, KS: US Army Command and General Staff College.

Greene, J., Caracelli, V., & Graham, W. (1989). Toward a conceptual framework for mixed-method evaluation designs. *Educational Evaluation and Policy Analysis*, *11*(3), 255–274. doi:10.3102/01623737011003255

Greenhaus, J. H., Collins, K. M., & Shaw, J. D. (2003). The relation between work–family balance and quality of life. *Journal of Vocational Behavior*, *63*(3), 510–531. doi:10.1016/S0001-8791(02)00042-8

Grimmer, J., Roberts, M., & Stewart, B. (2021). Machine learning for social science: An agnostic approach. *Annual Review of Political Science*, *24*(1), 395–419. doi:10.1146/annurev-polisci-053119-015921

Guerreiro & Pacheco. (2021). *How Green Trust, Consumer Brand Engagement and Green Word-of-Mouth Mediate Purchasing Intentions*. doi:10.3390/su13147877

Guest, D. (2002). Human resource management, corporate performance and employee wellbeing: Building the worker into HRM. *The Journal of Industrial Relations*, *44*(3), 335–358. doi:10.1111/1472-9296.00053

Gunn, A., & Mintrom, M. (2016). Higher Education Policy Change in Europe: Academic Research Funding and the Impact Agenda. *European Education, 48*(4), 241–257. doi:10.1080/10564934.2016.1237703

Gunn, P., Burak, H., Van Geel, P., & Baldwin, C. (2022). Development of resistor-capacitor and finite difference models to evaluate green roof thermal performance. *Building and Environment, 226*, 1–15. doi:10.1016/j.buildenv.2022.109700

Gupta, J., & Vegelin, C. (2016). Sustainable development goals and inclusive development. *International Environmental Agreement: Politics, Law and Economics, 16*(3), 433–448. doi:10.100710784-016-9323-z

Habert, G. (2014). Assessing the environmental impact of conventional and 'green' cement production. Woodhead Publishing.

Habib, A. (2013). Financial Inclusion and Islamic Finance: Organizational Formats, Products, Outreach and Sustainability. In Economic Development and Islamic Finance. World Bank Publications.

Habib, A., Mohieldin, M., Verbeek, J., & Aloulmagd, F. (2015). *On Sustainable Development Goals and the Role of Islamic Finance.* Policy Research Working Paper 7266, World Bank Group.

Habiba, U., Xinbang, C., & Ahmad, R. I. (2021). The influence of stock market and financial institution development on carbon emissions with the importance of renewable energy consumption and foreign direct investment in G20 countries. *Environmental Science and Pollution Research International, 28*(47), 67677–67688. doi:10.100711356-021-15321-3 PMID:34263400

Hackman, J. R., & Oldham, G. R. (1976). Motivation through the design of work: Test of a theory. *Organizational Behavior and Human Performance, 16*(2), 250–279. doi:10.1016/0030-5073(76)90016-7

Hadavi, S. (2017). Direct and indirect effects of the physical aspects of the environment on mental well-being. *Environment and Behavior, 49*(10), 1071–1104. doi:10.1177/0013916516679876

Hafeez, U., & Akbar, W. (2015). Impact of work-life balance on job satisfaction among school teachers of 21st century. *Australian Journal of Business and Management Research, 4*(11), 25–37. doi:10.52283/NSWRCA.AJBMR.20150411A03

Haines, S. (2019b). Managing expectations: Articulating expertise in climate services for agriculture in Belize. *Climatic Change, 157*(1), 43–59. doi:10.100710584-018-2357-1

Hair Jr, J., Sarstedt, M., Hopkins, L., & Kuppelwieser, V. (2014). Partial least squares structural equation modeling (PLS-SEM): An emerging tool in business research. *European Business Review, 26*(2), 106–121. doi:10.1108/EBR-10-2013-0128

Haller, T. (2019). The Different Meanings of Land in the Age of Neoliberalism: Theoretical Reflections on Commons and Resilience Grabbing from a Social Anthropological Perspective. *Land (Basel), 2019*(8), 104. doi:10.3390/land8070104

Haniffa, R., & Hudaib, M. (2007). Exploring the Ethical Identity of Islamic Banks via Communication in Annual Reports. *Journal of Business Ethics, 76*(1), 97–116. doi:10.100710551-006-9272-5

Hanners, U. (1986). *Exploración de la ciudad.* Fondo de Cultura Económica.

Hannerz, U. (2004). *Foreign New: Exploring the World of Foreign Correspondents.* University of Chicago Press.

Hansel, N. N., McCormack, M. C., & Kim, V. (2016). The Effects of Air Pollution and Temperature on COPD. *COPD, 13*(3), 372–379. doi:10.3109/15412555.2015.1089846 PMID:26683097

Hartmut Resonanz, R. (2016). *Eine Soziologie der Weltbeziehung.* Suhrkamp.

Hashem, T., & Al-Rifai, N. (2011). The Influence of Applying Green Marketing Mix by Chemical Industries Companies in Three Arab States in West Asia on Consumer's Mental Image. *The International Journal of Social Sciences (Islamabad)*, *2*(3), 92–101.

Hashim, A., Azman, N. S., Ghani, M. A., & Sabri, M. F. M. (2016). The relationship between work-life balance and employee retention. *Proceeding of the 3rd International Conference on Management & Muamalah, 1*, 316-322.

Hassine, M. B., & Harrathi, N. (2017). The causal links between economic growth, renewable energy, financial development and foreign trade in gulf cooperation council countries. *International Journal of Energy Economics and Policy*, *7*(2), 76–85.

Hausfather, Z. (2019). *Factcheck: How electric vehicles help to tackle climate change.* https://www.carbonbrief.org/factcheck-how-electric-vehicles-help-to-tackle-climate-change

Hautala-Kankaanpää, T. (2022). The impact of digitalization on firm performance: Examining the role of digital culture and the effect of supply chain capability. *Business Process Management Journal*, *28*(8), 90–109. doi:10.1108/BPMJ-03-2022-0122

Helliwell, J. F., Layard, R., Sachs, J. D., De Neve, J.-E., Aknin, L. B., & Wang, S. (Eds.). (2022). *World Happiness Report 2022*. Sustainable Development Solutions Network.

Hemmasi, M., Graf, L. A., & Lust, J. A. (1992). Correlates of pay and benefit satisfaction: The unique case of public university faculty. *Public Personnel Management*, *21*(4), 429–443. doi:10.1177/009102609202100402

Hemsley-Brown, T., & Laing, A. (2019). Green roofs and sustainable urban development: An exploratory review of potential barriers to adoption. *Sustainable Cities and Society*, *50*, 101648. doi:10.1016/j.scs.2019.101648

Hermanto, B., Muhyi, H. A., & Purnomo, M. (2018). The effect of job characteristics on job satisfaction and its impact on employee performance. *Advances in Social Sciences Research Journal*, *5*(9), 95–101.

Herzberg, F., Mausner, B., & Snyderman, B. B. (1959). The Motivation to Work (2nd ed.). John Wiley & Sons.

Herzberg, F. (2008). *One more time: How do you motivate employees?* Harvard Business Review Press.

Herzberg, F., Mausnes, B., Peterson, R. O., & Capwell, D. F. (1957). *Job attitudes: Review of research and opinion.* Psychological Service of Pittsburgh.

Hirst, K. (2019). *Is anthropology a science?* ThoughtCo. https://www.thoughtco.com/isanthropology-a-science-3971060

History of GNH. (n.d.). GNH Centre Bhutan. www.gnhcentrebhutan.org/history-of-gnh/

Hodge, T. (1997). Toward a conceptual framework for assessing progress toward sustainability. *Social Indicators Research*, *40*(1/2), 5–98. doi:10.1023/A:1006847209030

Hofman, J., Sharma, A., & Watts, D. (2017). Prediction and explanation in social systems. *Science*, *355*(6324), 486–488. doi:10.1126cience.aal3856 PMID:28154051

Hofman, J., Watts, D., Athey, S., Garip, F., Griffiths, T., Kleinberg, J., Margetts, H., Mullainathan, S., Salganik, M., Vazire, S., Vespignani, A., & Yarkoni, T. (2021). Integrating explanation and prediction in computational social science. *Nature*, *595*(7866), 181–188. doi:10.103841586-021-03659-0 PMID:34194044

Holling, C. S. (1973). *Resilience and Stability of Ecological Systems. Annual Review of Ecology and Systematics*, 4. https://www.jstor.org/stable/2096802

Holmes, D. R., & Marcus, G. E. (2006). Fast capitalism: para-ethnography and the rise of the symbolic analyst. In M. Fisher & G. Downey (Eds.), *Frontiers of Capital: Ethnographic Reflections on the New Economy*. Duke University Press.

Holmes, D. R., & Marcus, G. E. (2021). How do we collaborate? An updated manifesto. In D. Boyer & G. Marcus (Eds.), *Collaborative Anthropology Today: A Collection of Exceptions*. Cornell University Press. doi:10.7591/cornell/9781501753343.003.0002

Holzinger, K. J., & Swineford, F. (1937). The Bi-factor method. *Psychometrika*, *2*(1), 41–54. doi:10.1007/BF02287965

Horn, J. L. (1965). A rationale and test for the number of factors in factor analysis. *Psychometrika*, *30*(2), 179–185. doi:10.1007/BF02289447 PMID:14306381

Hossnofsky, V., & Junge, S. (2019). Does the market reward digitalization efforts? Evidence from securities analysts' investment recommendations. *Journal of Business Economics*, *89*(8/9), 965-994. doi:10.1007/s11573-019-00949-y

Hotez, P. J. (2018). Human Parasitology and Parasitic Diseases: Heading Towards 2050. *Advances in Parasitology*, *100*, 29–38. doi:10.1016/bs.apar.2018.03.002 PMID:29753341

Huang, H., Long, R., Chen, H., Sun, K., Sun, Q., & Li, Q. (2023). Why don't more people engage in green practices in China? A policy-oriented approach to promoting green transformation in five consumption areas. *Environmental Impact Assessment Review*, *101*, 107099. doi:10.1016/j.eiar.2023.107099

Huang, J., Kong, F., Yin, H., Middel, A., Liu, H., & Meadows, M. E. (2023). Green roof effects on urban building surface processes and energy budgets. *Energy Conversion and Management*, *287*, 1–13. doi:10.1016/j.enconman.2023.117100

Huang, Y., Haseeb, M., Usman, M., & Ozturk, I. (2022). Dynamic association between ICT, renewable energy, economic complexity and ecological footprint: Is there any difference between E-7 (developing) and G-7 (developed) countries? *Technology in Society*, *68*, 101853. doi:10.1016/j.techsoc.2021.101853

Huang, Y., Li, B., Biswas, A., & Li, Z. (2022). Factors dominating the horizontal and vertical variability of soil water vary with climate and plant type in loess deposits. *The Science of the Total Environment*, *811*, 152172. doi:10.1016/j.scitotenv.2021.152172 PMID:34883182

Huang, Y., Song, H., Wang, Z., Cheng, Y., Liu, Y., Hao, S., Li, N., Wang, Y., Wang, Y., Zhang, X., Sun, B., Li, Y., & Yao, X. (2022, October 22). Heat and outpatient visits of skin diseases - A multisite analysis in China, 2014-2018. *Heliyon*, *8*(10), e11203. doi:10.1016/j.heliyon.2022.e11203 PMID:36339999

Hu, L., & Bentler, P. M. (1998). Fit indices in covariance structure modeling: Sensitivity to under parameterized model misspecification. *Psychological Methods*, *3*(4), 424–453. doi:10.1037/1082-989X.3.4.424

Hu, L., & Bentler, P. M. (1999). Cut off criteria for fit indexes in covariance structure analysis: Conventional criteria versus new alternatives. *Structural Equation Modeling*, *6*(1), 1–55. doi:10.1080/10705519909540118

Hwang, J., & Jang, W. (2020). The effects of job characteristics on perceived organizational identification and job satisfaction of the Organizing Committee for the Olympic Games employees. *Managing Sport and Leisure*, *25*(4), 290–306. doi:10.1080/23750472.2020.1723435

Hyatt, S. (2021). *Teaching Urban Anthropology in a time of COVID*. National Association for the Practice of Anthropology. https://scholarworks.iupui.edu/bitstream/handle/1805/29883/Hyatt2021Teaching-pp.pdf?sequence=1

Hylland, T. (2022). The Sustainability of an Anthropology of the Anthropocene. *Sustainability*, *2022*(14), 3674. doi:10.3390u14063

Ihnatenko, M., & Novak, N. (2018). Development of regional programs for the development of agrarian enterprises with organic production based on the European and international experience. *Baltic Journal of Economic Studies*, *4*(4), 126–133. doi:10.30525/2256-0742/2018-4-4-126-133

Imbabi, M. S., Carrigan, C., & McKenna, S. (2012). Trends and developments in green cement and concrete technology. *International Journal of Sustainable Built Environment*, *1*(2), 194–216. doi:10.1016/j.ijsbe.2013.05.001

Interface. (2019). *2019 Annual Report*. Author.

International Monetary Fund. (1995). Bhutan: Background Paper. doi:10.5089/9781451806212.002

Ionaşcu, I., Ionaşcu, M., Nechita, E., Săcărin, M., & Minu, M. (2022). Digital Transformation, Financial Performance and Sustainability: Evidence for European Union Listed Companies. *Amfiteatru Economic*, *24*(59), 94–109. doi:10.24818/EA/2022/59/94

Jaharuddin, N. S., & Zainol, L. N. (2019). *The impact of work-life balance on job engagement and turnover intention.* The South East Asian Journal of Management. doi:10.21002eam.v13i1.10912

Jain, S. K., & Kaur, G. (2004). Green marketing: An Indian perspective. *Decision (Washington, D.C.)*, *31*(2), 161–209.

Jamei, E., Chau, H., Seyedmahmoudian, M., Mekhilef, S., & Alzahraa, F. (2023). Green roof and energy – role of climate and design elements in hot and temperate climates. *Heliyon*, *9*(5), 1–16. doi:10.1016/j.heliyon.2023.e15917 PMID:37215798

Janson, C., Henderson, R., Löfdahl, M., Hedberg, M., Sharma, R., & Wilkinson, A. J. K. (2020). Carbon footprint impact of the choice of inhalers for asthma and COPD. *Thorax*, *75*(1), 82–84. doi:10.1136/thoraxjnl-2019-213744 PMID:31699805

Jansson, H. B., Bernin, D., & Ramser, K. (2015). Silicate species of water glass and insights for alkali-activated green cement. *AIP Advances*, *5*(6), 067167. doi:10.1063/1.4923371

Jeevan, P., & Bhargav, V. (2016). A Study on Green Packaging - A Case Study Approach with Reference to Dell Inc. *International Educational Scientific Research Journal, 2*(7).

Jha, M. K. (2023, March 29). *The Financial Express*. https://www.financialexpress.com/business/defence-doklam-crisis-bhutan-china-talks-raise-concerns-for-india-3026076/

Jiang, C., & Chang, Y. (2022). Clean energy projects in Asian economies: Does FDI and stock market matter for sustainable development? *Ekonomska Istrazivanja*, *35*(1), 5843–5857. doi:10.1080/1331677X.2022.2038650

Ji, Q., & Zhang, D. (2019). How much does financial development contribute to renewable energy growth and upgrading of energy structure in China? *Energy Policy*, *128*, 114–124. doi:10.1016/j.enpol.2018.12.047

Johnson, S. K. (2019). *Leaking Talent How People of Color are Pushed Out of Environmental Organizations How People of Color are Pushed Out of Environmental Organizations*. https://diversegreen.org/research/leaking-talent/

Johnson, J. T., Griffeth, R. W., & Griffin, M. (2000). Factors discriminating functional and dysfunctional salesforce turnover. *Journal of Business and Industrial Marketing*, *15*(6), 399–415. doi:10.1108/08858620010349493

Johnson, S. K. (2017, August 17th). What 11 CEOs have learned about championing diversity. *Harvard Business Review*.

Johnson, T. F., Isaac, N. J. B., Paviolo, A., & González-Suárez, M. (2023). Socioeconomic factors predict population changes of large carnivores better than climate change or habitat loss. *Nature Communications*, *14*(1), 74. doi:10.103841467-022-35665-9 PMID:36693827

Jorgensen, D. (2015). Participant observation. In R. Scott & S. Kosslyn (Eds.), *Emerging trends in the social and behavioral sciences* (pp. 1–15). Wiley. doi:10.1002/9781118900772.etrds0247

Joshi, M., Goraya, H., Joshi, A., & Bartter, T. (2020). Climate change and respiratory diseases: A 2020 perspective. *Current Opinion in Pulmonary Medicine*, *26*(2), 119–127. doi:10.1097/MCP.0000000000000656 PMID:31851023

Joshi, Y., & Rahman, Z. (2015). Factors Affecting Green Purchase Behaviour and Future Research Directions. *International Strategic Management Review*, *3*(1-2), 128–143. doi:10.1016/j.ism.2015.04.001

Jun, M., Cai, S., & Shin, H. (2006). TQM practice in maquiladora: Antecedents of employee satisfaction and loyalty. *Journal of Operations Management*, *24*(6), 791–812. doi:10.1016/j.jom.2005.09.006

Kaiser, H. F. (1974). An index of factorial simplicity. *Psychometrika*, *39*(1), 31–36. doi:10.1007/BF02291575

Kaiser, H. F. (1992). On Cliff's Formula, the Kaiser-Guttman Rule, and the Number of Factors. *Perceptual and Motor Skills*, *74*(2), 595–598. doi:10.2466/pms.1992.74.2.595 PMID:1408626

Kalandides, A., & Grésillon, B. (2021). The Ambiguities of "Sustainable" Berlin. *Sustainability (Basel)*, *2021*(13), 1666. doi:10.3390u13041666

Kaliski, B. S. (2007). Encyclopedia of Business and Finance. Thompson Gale.

Kamla, R., & Rammal, G. H. (2013). Social Reporting by Islamic banks: Does Social Justice Matter? *Accounting, Auditing & Accountability Journal*, *26*(6), 911–945. doi:10.1108/AAAJ-03-2013-1268

Kam-Sing Wong, S. (2012). The influence of green product competitiveness on the success of green product innovation: Empirical evidence from the Chinese electrical and electronics industry. *European Journal of Innovation Management*, *15*(4), 468–490. doi:10.1108/14601061211272385

Kao, C. (1999). Spurious regression and residual-based tests for co-integration in panel data. *Journal of Econometrics*, *90*(1), 1–44. doi:10.1016/S0304-4076(98)00023-2

Kapur, R. (2018). *Factors influencing performance and job satisfaction of teachers in secondary schools in India.* Research Gate.

Karodia, A. M., Motlou, R. G., & Singh, S. (2016). An Evaluation of the Impact of Job Satisfaction on Employee Retention at Lonmin Rowland Shaft North West Province. *Kuwait Chapter of Arabian Journal of Business and Management Review*, *5*(10), 15–49. doi:10.12816/0027366

Kautish, P., Paul, J., & Sharma, R. (2019). The moderating influence of environmental consciousness and recycling intentions on green purchase behavior. *Journal of Cleaner Production*, *228*, 1425–1436. doi:10.1016/j.jclepro.2019.04.389

Kawulich, B. (2005). La observación participante como método de recolección de datos. *Forum Qualitative Social Research*, *68*(2), 43.

Kaye, B., & Jordan-Evans, S. (2000). Retention: Tag, you're it! *Training and Development-Alexandria-American Society for Training and Development*, *54*(4), 29-39.

Kelliher, C., & Anderson, D. (2010). Doing more with less? Flexible working practices and the intensification of work. *Human Relations*, *63*(1), 83–106. doi:10.1177/0018726709349199

Kelly, R. F., & Voydanoff, P. (1985). Work/family role strain among employed parents. *Family Relations*, *34*(3), 367–374. doi:10.2307/583575

Kenton. (2020). *Green tech.* investopedia.com

Kershaw, R. (2001). *Monarchy in South-East Asia: The Faces of Tradition in Transition.* Routledge.

Khaira, N., Triyonggo, Y., & Sukmawati, A. (2023). Examining the Mediating Role of Digital Culture on Digital Talent in the Telecommunication Sector. Jurnal Aplikasi Manajemen, 21(2), 439–454. doi:10.21776/ub.jam.2023.021.02.13

Khan, S., Magd, H., Ansari, M., Sharma, P., Tirwa, I., & Bhuyan, U. (2022). *Analysis of Regional Tourists in Bhutan before Covid-19*. .2022.v4i1.39-58 doi:10.56411/anusandhan

Khan, A., Sicen, L., Khan, B., & Salman, N. (2021). On the influence of demographic structure and industrial growth on environmental quality. *Journal of Environmental Management*, *288*, 112453. doi:10.1016/j.jenvman.2021.112453 PMID:33827026

Khandelwal, U., & Yadav, S. K. (2014). Green marketing and sustainable development: Marketing professionals' attitude towards green marketing. *International Journal of Marketing and Business Communication*, *3*(1).

Khan, H., Weili, L., & Khan, I. (2022). Examining the effect of information and communication technology, innovations, and renewable energy consumption on CO2 emission: Evidence from BRICS countries. *Environmental Science and Pollution Research International*, *29*(31), 47696–47712. doi:10.100711356-022-19283-y PMID:35184242

Khan, M. D., Thi Vu, H. H., Lai, Q. T., & Ahn, J. W. (2019). Aggravation of Human Diseases and Climate Change Nexus. *International Journal of Environmental Research and Public Health*, *16*(15), 2799. doi:10.3390/ijerph16152799 PMID:31390751

Kim, J. H., Jung, S. H., Seok, B. I., & Choi, H. J. (2022). The Relationship among Four Lifestyles of Workers amid the COVID-19 Pandemic (Work–Life Balance, YOLO, Minimal Life, and Staycation) and Organizational Effectiveness: With a Focus on Four Countries. *Sustainability (Basel)*, *14*(21), 14059. doi:10.3390u142114059

Kim, J.-O., & Mueller, C. W. (1978). *Introduction to factor analysis. What it is and how to do it*. Sage.

Kim, J.-O., & Mueller, C. W. (1987). *Factor analysis. Statistical methods and practical issues*. Sage.

Kim, J., & Park, K. (2016). Financial development and deployment of renewable energy technologies. *Energy Economics*, *59*, 238–250. doi:10.1016/j.eneco.2016.08.012

Kim, Y. (2011). Understanding green purchase: The influence of collectivism, personal values and environmental attitudes, and the moderating effect of perceived consumer effectiveness. *Seoul Journal of Business*, *17*(1), 65–92. doi:10.35152nusjb.2011.17.1.003

Kırgız, M. S. (2018). Green cement composite concept reinforced by graphite nano-engineered particle suspension for infrastructure renewal material. *Composites. Part B, Engineering*, *154*, 423–429. doi:10.1016/j.compositesb.2018.09.012

Klemm, W., Zinngrebe, Y., & Heidenreich, S. (2019). Green roofs as part of urban green infrastructure for innovative cities. *Environmental Innovation and Societal Transitions*, *31*, 50–56. doi:10.1016/j.eist.2018.11.003

Kline, R. B. (1998). *Principles and practice of structural equation modeling*. Guilford Press.

Köhler, M., & Poll, P. H. (2010). Long-term performance of selected old Berlin greenroofs in comparison to younger extensive green roofs in Berlin. *Ecological Engineering*, *36*(5), 722–729. doi:10.1016/j.ecoleng.2009.12.019

Kolb, Ch. (1998). Review of Bilharz, Joy A., *The Allegany Senecas and Kinzua Dam: Forced Relocation through Two Generations. H-AmIndian, H-Net Reviews*. https://www.h-net.org/reviews/showrev.php?id=2366

Kolomiets, T. V., & Tomashuk, I. V. (2021). Entrepreneurship and development of rural areas in Ukraine. *Colloquium-journal*, *9*(96), 29–42.

Korkmaz, S., Goksuluk, D., & Zararsiz, G. (2014). MVN: An R Package for Assessing Multivariate Normality. *The R Journal*, *6*(2), 151–162. doi:10.32614/RJ-2014-031

Kose & Ohnsorge. (Eds.). (2020). A Decade after the Global Recession: Lessons and Challenges for Emerging and Developing Economies. World Bank. doi:10.1596/978-1-4648-1527-0

Kosova, R., Qendraj, D.H., Xhafaj, E. (2022). Meta-Analysis ELECTRE III and AHP in Evaluating and Ranking the Urban Resilience. *Journal of Environmental Management and Tourism, 3*(59), 756 - 768. doi:10.14505/jemt.v13.3(59).15

Kramar, R., & Syed, J. (2012). *Human Resource Management in a Global Context: A Critical Approach Making critical sense of human resource management in a globalized world.* https://www.researchgate.net/publication/295867671

Kramer, J. M. (2021). From predator to parasite: On private property and our ecological disaster. *Journal of Economic Issues, 55*(June), 416–422. doi:10.1080/00213624.2021.1908804

Krauß, W. (2020). Narratives of change and the co-development of climate services for action. *Climate Risk Management, 28*, 100217. doi:10.1016/j.crm.2020.100217

Kreiner, G. E., Hollensbe, E. C., & Sheep, M. L. (2009). Balancing borders and bridges: Negotiating the work-home interface via boundary work tactics. *Academy of Management Journal, 52*(4), 704–730. doi:10.5465/amj.2009.43669916

Kucher, A. (2019). *Sustainable soil management in the formation of competitiveness of agricultural enterprises.* Academic Publishing House «Talent». doi:10.13140/RG.2.2.19554.07366

Kucher, L., Heldak, M., & Orlenko, A. (2018). Project management in organic agricultural production. *Agricultural and Resource Economics, 4*(3), 104–128. doi:10.22004/ag.econ.281753

Kuensel, Bhutan's Daily. (2022). *Phuentsholing Hoteliers Say SDF Could Affect Their Businesses.* Kuensel Online. kuenselonline.com/phuentsholing-hoteliers-say-sdf-could-affect-their-businesses/

Kuensel, Bhutan's Daily. (2023a). *Govt. Waives SDF for Border Towns.* Kuensel Online. kuenselonline.com/govt-waives-sdf-for-border-towns/

Kuensel, Bhutan's Daily. (2023b). *Inflation Drops to 4.33 Percent in January.* Kuensel Online. kuenselonline.com/inflation-drops-to-4-33-percent-in-january/

Kuhn, T. S. (1962/1996). *The Structure of Scientific Revolutions* (3rd ed.). University of Chicago Press.

Kulkarni, S. R., & Kulkarni, S. S. (2016). Work life balance and stress management: Two sides of a single coin. *Indian Journal of Commerce & Management Studies*, 16–19.

Kumar, G. V., & Janakiram, B. (2017). Theories of work life balance–a conceptual review. *International Research Journal of Management and Commerce, 4*(9), 184–192.

Kumar, K., & Chaturvedi, R. (2017). Work-life balance and job satisfaction from the perspective of multiple job holding women: Comparative analysis of generational cohorts. *International Journal of Advances in Management and Economics, 6*(5), 32–40.

Kumar, V., & Christodoulopoulou, A. (2014). Sustainability and branding: An integrated perspective. *Industrial Marketing Management, 43*(1), 6–15. doi:10.1016/j.indmarman.2013.06.008

Kuna, M., & Dreslerová, D. (2007). Landscape archaeology and "community areas" in the archaeology of central Europe. In D. Hicks, L. McAtackney, & G.J. Fairclough (Eds.), Envisioning landscape: Situations and standpoints in archaeology and heritage. Left Coast Press.

Kundnani, N., & Mehta, P. (2016). Identifying the factors affecting work-life balance of employees in banking sector. *PARIPEX-Indian Journal of Research, 4*(6), 328–331.

Labour Force Survey Report. (2022). National Statistics Bureau, Bhutan.

Ladyman, J. (1998). What is structural realism? *Studies in History and Philosophy of Science*, *29*(3), 409–424. doi:10.1016/S0039-3681(98)80129-5

Lafortune, G., Fuller, G., Moreno, J., Schmidt-Traub, G., & Kroll, V. (2018). *SDG Index and dashboards. Detailed Methodological Paper*. Bertelsmann Stiftung and Sustainable Development Solutions Network Sustainable Development Solutions Network.

Lake, I. R., & Barker, G. C. (2018). Climate Change, Foodborne Pathogens and Illness in Higher-Income Countries. *Current Environmental Health Reports*, *5*(1), 187–196. doi:10.100740572-018-0189-9 PMID:29446033

Lamane-Harim, J., Cegarra-Leiva, D., & Sanchez-Vidal, M. E. (2023). Work–life balance supportive culture: A way to retain employees in Spanish SMEs. *International Journal of Human Resource Management*, *34*(10), 2074–2106. doi:10.1080/09585192.2021.1878255

Lambert, A. (2019). Psychotic, acritical and precarious? A Lacanian exploration of the neoliberal consumer subject. *Marketing Theory*, *19*(3), 329–346. doi:10.1177/1470593118796704

Lamprinou, V. D. I., Tasoulis, K., & Kravariti, F. (2021). The impact of servant leadership and perceived organisational and supervisor support on job burnout and work–life balance in the era of teleworking and COVID-19. *Leadership and Organization Development Journal*, *42*(7), 1071–1088. doi:10.1108/LODJ-12-2020-0526

Lanzara, F. (2021). Islamic Finance and Sustainable Development Goals. A Bibliometric Analysis from 2000 to 2021. *European Journal of Islamic Finance*, *18*, 1–19.

Lapassade, G. (1986). *Grupos, organizaciones e instituciones: la transformación de la burocracia*. Gedisa.

Latour, B. (2013). *An inquiry into modes of existence: An anthropology of the moderns* (C. Porter, Trans.). Harvard University Press.

Latour, B. (2017). Anthropology at the Time of the Anthropocene: A Personal View of What Is to Be Studied. In M. Brightman & J. Lewis (Eds.), *The Anthropology of Sustainability*. Palgrave Studies in Anthropology of Sustainability. doi:10.1057/978-1-137-56636-2_2

Law, K., Wong, C., & Mobley, W. (1998). Toward a Taxonomy of Multidimensional Constructs. *Academy of Management Review*, *23*(4), 741–755. doi:10.2307/259060

Lawley, D., & Maxwell, A. (1973). Regression and Factor Analysis. *Biometrika*, *60*(2), 331–338.

Lazar, I., Osoian, C., & Ratiu, P. I. (2010). The role of work-life balance practices in order to improve organizational performance. *European Research Studies*, *XIII*(1), 1–14. doi:10.35808/ersj/267

Lee, C. C., Chen, M. P., & Yuan, Z. (2023). Is information and communication technology a driver for renewable energy? *Energy Economics*, *124*, 106786. doi:10.1016/j.eneco.2023.106786

Leonidou, C. N., Katsikeas, C. S., & Morgan, N. A. (2013). "Greening" the marketing mix: Do firms do it and does it pay off? *Journal of the Academy of Marketing Science*, *41*(2), 151–170. doi:10.100711747-012-0317-2

Leonidou, L. C., Leonidou, C. N., Palihawadana, D., & Hultman, M. (2011). Evaluating the green advertising practices of international firms: A trend analysis. *International Marketing Review*, *28*(1), 6–33. doi:10.1108/02651331111107080

Lester, R. K. (2005). *Universities, Innovation, and the Competitiveness of Local Economies A Summary Report from the Local Innovation Systems Project – Phase I*. Industrial Performance Center Massachusetts Institute of Technology 13 December 2005 MIT Industrial Performance Center Working Paper 05-010.

Levine, D. S., & Strube, M. J. (2012). Environmental Attitudes, Knowledge, Intentions and Behaviors among College Students'. *The Journal of Social Psychology*, *152*(3), 308–326. doi:10.1080/00224545.2011.604363 PMID:22558826

Levy, K., Smith, S. M., & Carlton, E. J. (2018). Climate Change Impacts on Waterborne Diseases: Moving Toward Designing Interventions. *Current Environmental Health Reports*, *5*(2), 272–282. doi:10.100740572-018-0199-7 PMID:29721700

Lhendup, K., & Panda, B. (2020). The Fable of Regional Tourists' Loyalty to Bhutan. *Journal of Tourism and Hospitality Management*, *8*(2). Advance online publication. doi:10.15640/jthm.v8n2a6

Liao, S. H., Chang, J. C., Cheng, S. C., & Kuo, C. M. (2004). Employee relationship and knowledge sharing: A case study of a Taiwanese finance and securities firm. *Knowledge Management Research and Practice*, *2*(1), 24–34. doi:10.1057/palgrave.kmrp.8500016

Liao, Y.-K., Wu, W.-Y., & Pham, T.-T. (2020). Examining the moderating effects of green marketing and green psychological benefits on customers' green attitude, value and purchase intention. *Sustainability (Basel)*, *12*(18), 7461. doi:10.3390u12187461

Li, J., Yin, Z., Li, Y., Sun, S., Li, M., & Li, Y. (2021). The application of green roofs in urban green innovation spaces: A case study in Jinan City, China. *Environmental Science and Pollution Research International*, *28*(3), 3525–3537. doi:10.100711356-020-11544-2 PMID:32892283

Lindblom, C. E. (1959). The Science of Muddling through. *Public Administration Review*, *19*(2), 79–88. doi:10.2307/973677

Lin, Y. H., & Chen, Y. S. (2017). Determinants of green competitive advantage: The roles of green knowledge sharing, green dynamic capabilities, and green service innovation. *Quality & Quantity*, *51*(4), 1663–1685. doi:10.100711135-016-0358-6

Liu, J. (2023). Sustainability and natural resources management in developed countries: The role of financial inclusion and human development. *Resources Policy, 80*, 103143.

Liu, D., Jiang, K., Shalley, C. E., Keem, S., & Zhou, J. (2016). Motivational mechanisms of employee creativity: A meta-analytic examination and theoretical extension of the creativity literature. *Organizational Behavior and Human Decision Processes*, *137*, 236–263. doi:10.1016/j.obhdp.2016.08.001

Liu, J., Chen, J., & Tao, Y. (2015). Innovation Performance in New Product Development Teams in China's Technology Ventures: The Role of Behavioral Integration Dimensions and Collective Efficacy. *Journal of Product Innovation Management*, *32*(1), 29–44. doi:10.1111/jpim.12177

Liu, Y. Z., Zhang, Z., Hou, G., & Yan, P. (2021). Preparation of sustainable and green cement-based composite binders with high-volume steel slag powder and ultrafine blast furnace slag powder. *Journal of Cleaner Production, 289*, 125–133. doi:10.1016/j.jclepro.2020.125133

Liu, Y., Sohail, M. T., Khan, A., & Majeed, M. T. (2022). Environmental benefit of clean energy consumption: Can BRICS economies achieve environmental sustainability through human capital? *Environmental Science and Pollution Research International*, *29*(5), 6766–6776. doi:10.100711356-021-16167-5 PMID:34460087

Li, X., & Norford, L. (2016). Evaluation of cool roof and vegetations in mitigating urban heat island in a tropical city, Singapore. *Urban Climate*, *16*, 59–74. doi:10.1016/j.uclim.2015.12.002

Llanos, M. (2009). *Epistemología de las ciencias sociales*. UNMSM.

Llonch, P., Haskell, M. J., Dewhurst, R. J., & Turner, S. P. (2017). Current available strategies to mitigate greenhouse gas emissions in livestock systems: An animal welfare perspective. *Animal*, *11*(2), 274–284. doi:10.1017/S1751731116001440 PMID:27406001

Locke, E. A. (1976). The nature and causes of job satisfaction. Handbook of industrial and organizational psychology. 1297-1349.

Loebbecke, C., & Picot, A. (2015). Reflections on societal and business model transformation arising from digitization and big data analytics: A research agenda. *The Journal of Strategic Information Systems, 24*(3), 149-157. . doi:10.1016/j.jsis.2015.08.002

Lopez-Gamero, M., Molina-Azorın, J., & Claver-Cortes, E. (2009). The whole relationship between environmental variables and firm performance: competitive advantage and firm resources as media to variables. *Journal of Environmental Management, 90*(10), 3110–3121.

López-Gamero, M. D., Claver-Cortés, E., & Molina-Azorín, J. F. (2008). Complementary resources and capabilities for an ethical and environmental management: A qual/quan study. *Journal of Business Ethics, 82*(3), 701–732. doi:10.100710551-007-9587-x

Lourau, R. (1979). *El análisis institucional.* Amorrortu.

Lüdecke, D., Ben-Shachar, M., Patil, I., & Makowski, D. (2020). parameters: Extracting, Computing and Exploring the Parameters of Statistical Models using R. *Journal of Open Source Software, 5*(53), 2445. doi:10.21105/joss.02445

Luechauer. (n.d.). *Gross National Happiness of Bhutan and its False Promises.* https://www.gsdmagazine.org/the-false-promises-of-bhutans-gross-national-happiness/

Lu, L. C., Chiu, S. Y., Chiu, Y. H., & Chang, T. H. (2022). Sustainability efficiency of climate change and global disasters based on greenhouse gas emissions from the parallel production sectors - A modified dynamic parallel three-stage network DEA model. *Journal of Environmental Management, 317*, 115401. doi:10.1016/j.jenvman.2022.115401 PMID:35660833

Lundvall, B.-Å. (Ed.). (1992). National Systems of Innovation: towards a theory of innovation and interactive learning. Printer Publisher.

Maali, B., Casson, P., & Napier, C. (2006). Social Reporting by Islamic Banks. *Abacus, 42*(2), 266–289. doi:10.1111/j.1467-6281.2006.00200.x

Maas, C. J., & Hox, J. J. (2005). Sufficient sample sizes for multilevel modeling. *Methodology: European Journal of Research Methods for the Behavioral and Social Sciences, 1*(3), 86–92. doi:10.1027/1614-2241.1.3.86

MacCallum, R. C., Widaman, K. F., Zhang, S., & Hong, S. (1999). Sample size in factor analysis. *Psychological Methods, 4*(1), 84–99. doi:10.1037/1082-989X.4.1.84

MacCallum, R., Widaman, K., Preacher, K., & Hong, S. (2001). Sample Size in Factor Analysis: The Role of Model Error. *Multivariate Behavioral Research, 36*(4), 36. doi:10.1207/S15327906MBR3604_06 PMID:26822184

Maheshwari, A., & Malhotra, D. G. (2011). Green marketing: A study on Indian youth. *International Journal of Management and Strategy, 2*(3).

Mahmoud, T. O., Ibrahim, S. B., Ali, A. H., & Bleady, A. (2017). The Influence of Green Marketing Mix on Purchase Intention: The Mediation Role of Environmental Knowledge. *International Journal of Scientific and Engineering Research, 8*(9), 1040–1048. doi:10.14299/ijser.2017.09.006

Ma, J., Guo, Y., Gao, J., Tang, H., Xu, K., Liu, Q., & Xu, L. (2022). Climate Change Drives the Transmission and Spread of Vector-Borne Diseases: An Ecological Perspective. *Biology (Basel), 11*(11), 1628. doi:10.3390/biology11111628 PMID:36358329

Majumder, J., Giri, A., & Gangopadhyay, S. (2019). Factors Affecting Work Life Balance of Employees in Indian Manufacturing Companies: An Empirical Analysis Using Structural Equation Modeling (SEM). *International Journal of Innovative Technology and Exploring Engineering*, *8*(7), 1551–1555.

Malhi, Y., Franklin, J., Seddon, N., Solan, M., Turner, M. G., Field, C. B., & Knowlton, N. (2020). Climate change and ecosystems: threats, opportunities and solutions. *Philosophical transactions of the Royal Society of London. Series B, Biological sciences, 375*(1794), 20190104. https://doi.org/ doi:10.1098/rstb.2019.0104

Malinowski, B. (1961). *Argonauts of the Western Pacific*. Routledge.

Mandhanya, Y. (2015). A study of impact of working environment of retention of employees (With special reference to Automobile sector)'. *Global Management Review*, *9*(4), 116–128.

Mangun, D., & Thurston, D. L. (2002). Incorporating component reuse, remanufacture, and recycle into product portfolio design. *Engineering Management, IEEE Transactions on, 49*(4), 479-490.

Maran, T., & Kull, K. (2014). Ecosemiotics: Main principles and current developments. *Geografiska Annaler. Series B, Human Geography*, *96*(1), 41–50. doi:10.1111/geob.12035

Marchi, V. D. (2012). Environmental innovation and R&D cooperation: Empirical evidence from Spanish manufacturing firms. *Research Policy*, *41*(3), 614–623. doi:10.1016/j.respol.2011.10.002

Maruani, T., & Amit-Cohen, I. (2007). Open Space planning models: A review of approaches and methods. *Landscape and Urban Planning*, *81*(1–2), 1–13. doi:10.1016/j.landurbplan.2007.01.003

Masaki, K., & Tshering, J. (2021). Exploring the origins of Bhutan's gross national happiness. *Journal of South Asian Development*, *16*(2), 273–292. doi:10.1177/09731741211039049

Massawe, E., & Geiser, K. (2012). The dilemma of promoting green products: What we know and don't know about biobased metalworking fluids. *Journal of Environmental Health*, *74*(8), 8–16. PMID:22533119

Mayo, E. (1945). *The social problems of an Industrial Civilization*. Harvard University Press.

Mazar, N., & Zhong, C. B. (2010). Do green products make us better people? *Psychological Science*, *21*(4), 494–498. doi:10.1177/0956797610363538 PMID:20424089

Mazur, K. V., & Tomashuk, I. V. (2019). Governance, and regulation as an indispensable condition for developing the potential of rural areas. *Baltic Journal of Economic Studies*, *5*(5), 67–78. doi:10.30525/2256-0742/2019-5-5-67-78

McDonald. (2012). *2012 Annual Report*. Author.

McDonald, S., & Oates, C. J. (2006). Sustainability: Consumer Perceptions and Marketing Strategies. *Business Strategy and the Environment*, *15*(3), 157–170. doi:10.1002/bse.524

McGrath, R. E. (2005). Conceptual complexity and construct validity. *Journal of Personality Assessment*, *85*(2), 112–124. doi:10.120715327752jpa8502_02 PMID:16171410

McGraw, K. O., & Wong, S. P. (1996). Forming inferences about some intraclass correlation coefficients. *Psychological Methods*, *1*(1), 30–46. doi:10.1037/1082-989X.1.1.30

McIntosh, R. J., Tainter, J. A., & McIntosh, S. K. (Eds.). (2000). *The Way the Wind Blows: Climate Change, History, and Human Activity*. Columbia University Press.

McMichael, A. J., Campbell-Lendrum, D. H., Corvalán, C. F., Ebi, K. L., Githeko, A., Scheraga, J. D., & Woodward, A. (2003). *Climate change and human health: risks and responses*. World Health Organization.

Meerow, S., & Newell, J. P. (2017). Detroit, Spatial planning for multifunctional green infrastructure: Growing resilience in. *Landscape and Urban Planning*, *72*, 62–75. doi:10.1016/j.landurbplan.2016.10.005

Meeus, J. H. A. (1995). Pan-European landscapes. *Landscape and Urban Planning*, *31*(1-3), 57–79. doi:10.1016/0169-2046(94)01036-8

Mell, I. C. (2010). *Green infrastructure: concepts, perceptions, and its use in spatial planning* [Doctoral Thesis]. School of Architecture, Planning and Landscape, Newcastle University.

Mentens, J., Raes, D., & Hermy, M. (2006). Green roofs as a tool for solving the rainwater runoff problem in the urbanized 21st century? *Landscape and Urban Planning*, *77*(3), 217–226. doi:10.1016/j.landurbplan.2005.02.010

Mesimaki, M., Hauru, K., Kotze, D. J., & Lehvavirta, S. (2017). Neo-spaces for urban livibility? Urbanities' versatile mental image of green roofs in the Helsinki metropolitan area, Finland. *Land Use Policy*, *61*, 587–600. doi:10.1016/j.landusepol.2016.11.021

Mesmer-Magnus, J. R., & Viswesvaran, C. (2006). How family-friendly work environments affect work/family conflict: A meta-analytic examination. *Journal of Labor Research*, *27*(4), 555–574. doi:10.100712122-006-1020-1

Mills, C. W. (1959). *The sociological imagination*. Oxford University Press.

Miroshnychenko, I., Barontini, R., & Testa, F. (2017). Green practices and financial performance: A global outlook. *Journal of Cleaner Production*, *147*, 340–351. doi:10.1016/j.jclepro.2017.01.058

Mir, R., & Mir, A. (2009). From the Colony to the Corporation: Studying Knowledge Transfer Across International Boundaries. *Group & Organization Management*, *34*(1), 90–113. doi:10.1177/1059601108329714

Mishra, A. P. (2013). Green cement for sustainable concrete using marble dust. *International Journal of Chemtech Research*, *5*(2), 616–622.

Mohd Suki, N., & Mohd Suki, N. (2019). Examination of peer influence as a moderator and predictor in explaining green purchase behaviour in a developing country. *Journal of Cleaner Production*, *228*, 833–844. Advance online publication. doi:10.1016/j.jclepro.2019.04.218

Mohd, H. A. & Norhidayah, S. (2016), Sustainable food production: insights of Malaysian halal small and medium sized enterprises. *International Journal of Production Economics, 181*(part B), 303–314. . doi:10.1016/j.ijpe.2016.06.003

Molyneux, D., Bush, S., Bannerman, R., Downs, P., Shu'aibu, J., Boko-Collins, P., Radvan, I., Wohlgemuth, L., & Boyton, C. (2021). Neglected tropical diseases activities in Africa in the COVID-19 era: The need for a "hybrid" approach in COVID-endemic times. *Infectious Diseases of Poverty*, *10*(1), 1. doi:10.118640249-020-00791-3 PMID:33397494

Moore, F. (2007). Work-life balance: Contrasting managers and workers in an MNC. *Employee Relations*, *29*(4), 385–399. doi:10.1108/01425450710759217

Moore, H. L. (2017). What Can Sustainability Do for Anthropology? In M. Brightman & J. Lewis (Eds.), *The Anthropology of Sustainability*. Palgrave Studies in Anthropology of Sustainability. doi:10.1057/978-1-137-56636-2_4

Morris, M. L., & Madsen, S. R. (2007). Advancing work—Life integration in individuals, organizations, and communities. *Advances in Developing Human Resources*, *9*(4), 439–454. doi:10.1177/1523422307305486

Morrison, D. F. (1990). *Multivariate Statistical Methods*. McGraw-Hill.

Mouchet, J., & Carnevale, P. (1997). Impact des transformations de l'environnement sur les maladies à transmission vectorielle [Impact of changes in the environment on vector-transmitted diseases]. *Sante (Montrouge, France)*, *7*(4), 263–269. PMID:9410453

Mubarak, M. F., Shaikh, F. A., Mubarik, M., Samo, K. A., & Mastoi, S. (2019). Impact of digital transformation on business performance: A study of Pakistani SMEs. *Engineering, Technology & Applied Scientific Research*, *9*(6), 5056–5061.

Mukonza, C. (2021). *Green Marketing in Emerging Markets Strategic and Operational Perspectives*. Springer Nature Switzerland AG. doi:10.1007/978-3-030-74065-8

Mullins, L. (2005). *Management & Organisational Behaviour* (7th ed.). Prentice Hall / Pearson Education.

Munir, R., & Tobi, S. N. M. (2020). Understanding and measuring the determinants of employees' turnover intention. *Environ. Behav. Proc. J*, *5*(13), 395–403. doi:10.21834/e-bpj.v5i13.2092

Murray, N. (2017). Urban disaster risk governance. A systemic review. PPI-Centre. UCL Institute of Education.

Murshed, M. (2020). An empirical analysis of the non-linear impacts of ICT-trade openness on renewable energy transition, energy efficiency, clean cooking fuel access and environmental sustainability in South Asia. *Environmental Science and Pollution Research International*, *27*(29), 36254–36281. doi:10.100711356-020-09497-3 PMID:32556989

Mustapha, N., & Zakaria, Z. C. (2013). The effect of promotion opportunity in influencing job satisfaction among academics in higher public institutions in Malaysia. *International Journal of Academic Research in Business & Social Sciences*, *3*(3), 20.

Mu, X., Fang, C., Yang, Z., & Guo, X. (2022). Impact of the COVID-19 Epidemic on Population Mobility Networks in the Beijing–Tianjin–Hebei Urban Agglomeration from a Resilience Perspective. *Land (Basel)*, *2022*(11), 675. doi:10.3390/land11050675

Nahapiet, J. E., & Ghoshal, S. (1998). Social Capital, Intellectual Capital, and the Organizational Advantage. *Academy of Management Review*, *33*(2), 242–266. doi:10.2307/259373

Naithani, D. P. (2009). Overview of work-life balance discourse and its relevance in current economic scenario. *Asian Social Science*, *6*(6), 148–155.

Namkung, Y., & Jang, S. (2013). Effects of restaurant green practices on brand equity formation: Do green practices really matter? *International Journal of Hospitality Management*, *33*(2), 85–95. doi:10.1016/j.ijhm.2012.06.006

Nantichas. (2017). *Philosophy of Sufficiency Economy*. www.chaipat.or.th/eng/concepts-theories/sufficiency-economy-new-theory.html#:~:text=Sufficiency%20Economy%20is%20a%20philosophy

Naqi, A. J., & Jang, J. (2019). Recent progress in green cement technology utilizing low-carbon emission fuels and raw materials: A review. *Sustainability (Basel)*, *11*(2), 537. doi:10.3390u11020537

Naseem, A., Sheikh, S. E., & Malik, K. P. (2011). Impact of employee satisfaction on success of organization: Relation between customer experience and employee satisfaction. *International Journal of Multidisciplinary Sciences and Engineering, 2*(5), 41-46.

National Statistical Bureau. (2004). *Bhutan National Accounts Statistics*. National Statistical Bureau.

Newaz, M. T., & Zaman, F. K. (2012). Work-life balance: Is it still a new concept in private commercial banking sector of Bangladesh. *International Journal of Research Studies in Management*, *1*(2), 57–66. doi:10.5861/ijrsm.2012.95

Newspaper, Bhutan's Daily. (2022). *Tourism Development: Looking beyond SDF*. Kuensel Online. kuenselonline.com/tourism-development-looking-beyond-sdf/

Newspaper, Bhutan's Daily. (2023). *Bhutan to Join WTO*. Kuensel Online, kuenselonline.com/bhutan-to-join-wto/

Nguyen-Viet, B. (2023). The impact of green marketing mix elements on green customer based brand equity in an emerging market. *Asia-Pacific Journal of Business Administration, 15*(1), 96–116. doi:10.1108/APJBA-08-2021-0398

Nießen, D., Danner, D., Spengler, M., & Lechner, C. (2020). Big Five personality traits predict successful transitions from school to vocational education and training: A large-scale study. *Frontiers in Psychology, 11*, 1827. doi:10.3389/fpsyg.2020.01827 PMID:32903700

Nikendei, C., Bugaj, T. J., Nikendei, F., Kühl, S. J., & Kühl, M. (2020). Klimawandel: Ursachen, Folgen, Lösungsansätze und Implikationen für das Gesundheitswesen [Climate change: Causes, consequences, solutions and public health care implications]. *Zeitschrift für Evidenz, Fortbildung und Qualität im Gesundheitswesen, 156-157*, 59–67. doi:10.1016/j.zefq.2020.07.008 PMID:32859556

Nogué, S., & Sala, M. (2018). The use of green roofs in urban green innovation spaces: A case study in Barcelona. *Urban Forestry & Urban Greening, 33*, 19–29. doi:10.1016/j.ufug.2018.02.008

Nonato, S. (2021). Digital Culture And Digital Teaching Resources: An Overview Of Teaching During The Covid-19 Pandemics. *Revista Práxis Educacional, 17*(45), 1–25.

Norton, T. A., Parker, S. L., Zacher, H., & Ashkanasy, N. M. (2015). Employee green behavior: A theoretical framework, multilevel review, and future research agenda. *Organization & Environment, 28*(1), 103–125. doi:10.1177/1086026615575773

Nost, E. (2019). Climate services for whom? The political economics of contextualizing climate data in Louisiana's coastal master plan. *Climatic Change, 157*(1), 27–42. doi:10.100710584-019-02383-z

Nunnally, J., & Bernstein, I. (1994). *Psychometric Theory New York* (3rd ed.). McGraw Hill.

Nwankpa, J. K., & Roumani, Y. (2016). IT Capability and Digital Transformation: A Firm Performance Perspective. In *Association for Information Systems (AIS), 37 International Conference on Information Systems. 11-14 December 2016, Dublin, Ireland.* Association for Information Systems (AIS).

Odum, E. P. (1957). *Fundamentals of ecology*. W. B. Saunders Company.

OECD. (1997). Government Venture Capital for Technology-Based Firms. Ducument OCDE/GD (97) 201. Paris: OECD.

Ogden, N. H., & Gachon, P. (2019). Climate change and infectious diseases: What can we expect? *Canada communicable disease report = Releve des maladies transmissibles au Canada, 45*(4), 76–80. doi:10.14745/ccdr.v45i04a01

Olimid, A. P., Georgescu, C. M., & Gherghe, C. L. (2022). Influences of Covid-19 Crisis on Resilience Theories: An analysis of Community, Societal and Governance Resilience. *Revue des Sciences Politiques, 73*, 38–51.

OmariK. (2019). The Effects of E-Commerce and Digital Transformation in Today's Business/Lebanon Case Studies. *Proceedings of the Ninth International Conference on Engaged Management Scholarship.* doi:10.2139/ssrn.3454124

Ones, D. S., & Dilchert, S. (2012). Employee green behaviors. In D. S. S. E. Jackson (Ed.), *Managing human resource for environmental sustainability*. Jossey-Bass.

Orsato, R. J. (2006). Competitive environmental strategies: When does it pay to be green? *California Management Review, 48*(2), 127–143. doi:10.2307/41166341

Ortner, S. (2010). Access: Reflections on studying up in Hollywood. *Ethnography, 11*(2), 211–233. doi:10.1177/1466138110362006

Ortner, S. (2013). *Not Hollywood: Independent Film at the Twilight of the American Dream.* Duke University Press.

Osakwe, R. N. (2003). *A Comparative study of job satisfaction between public and private secondary school in Delta State* [Unpublished Ph. D. thesis]. Faculty of Education, Delta State University.

Ottman, J., Stafford, E. R., & Hartman, C. L. (2006). Avoiding Green Marketing Myopia: Ways to Improve Consumer Appeal for Environmentally Preferable Products. *Environment, 48*(5), 22–36. doi:10.3200/ENVT.48.5.22-36

Öztürk, L. (2002). Dijital uçurumun küresel boyutları. *Ege Academic Review, 2*(1), 1–10.

Pacheco, S. E., Guidos-Fogelbach, G., Annesi-Maesano, I., Pawankar, R., D' Amato, G., Latour-Staffeld, P., Urrutia-Pereira, M., Kesic, M. J., & Hernandez, M. L. (2021). Climate change and global issues in allergy and immunology. *The Journal of Allergy and Clinical Immunology, 148*(6), 1366–1377. doi:10.1016/j.jaci.2021.10.011 PMID:34688774

Pagiaslis, A., & Krontalis, A. K. (2014). Green Consumption Behavior Antecedents: Environmental Concern, Knowledge, and Beliefs. *Psychology and Marketing, 31*(5), 335–348. doi:10.1002/mar.20698

Palomino, J. C., & Durán, P. (2020). The influence of green roofs on the urban heat island effect: A review of the current state of knowledge. *Urban Climate, 34*, 100696. doi:10.1016/j.uclim.2020.100696

Pao, H. T., & Fu, H. C. (2013). Renewable energy, non-renewable energy and economic growth in Brazil. *Renewable & Sustainable Energy Reviews, 25*, 381–392. doi:10.1016/j.rser.2013.05.004

Paramati, S. R., Alam, M. S., & Apergis, N. (2018). The role of stock markets on environmental degradation: A comparative study of developed and emerging market economies across the globe. *Emerging Markets Review, 35*, 19–30. doi:10.1016/j.ememar.2017.12.004

Paramati, S. R., Apergis, N., & Ummalla, M. (2017a). Financing clean energy projects through domestic and foreign capital: The role of political cooperation among the EU, the G20 and OECD countries. *Energy Economics, 61*, 62–71. doi:10.1016/j.eneco.2016.11.001

Paramati, S. R., Mo, D., & Gupta, R. (2017b). The effects of stock market growth and renewable energy use on CO2 emissions: Evidence from G20 countries. *Energy Economics, 66*, 360–371. doi:10.1016/j.eneco.2017.06.025

Paramati, S. R., Ummalla, M., & Apergis, N. (2016). The effect of foreign direct investment and stock market growth on clean energy use across a panel of emerging market economies. *Energy Economics, 56*, 29–41. doi:10.1016/j.eneco.2016.02.008

Páramo, D., & Ramírez, E. (2017). *Etnomarketing, la dimensión cultural del marketing*. Klasse Editorial.

Parker, B., Segev, S., & Pinto, J. (2009). *What it means to go green: Consumer perception of green brands and dimensions of greenness*. Florida International University.

Patz, J. A., Epstein, P. R., Burke, T. A., & Balbus, J. M. (1996). Global climate change and emerging infectious diseases. *Journal of the American Medical Association, 275*(3), 217–223. doi:10.1001/jama.1996.03530270057032 PMID:8604175

Pedro Pereira Luzio, J., & Lemke, F. (2013). Exploring green consumers' product demands and consumption processes. *European Business Review, 25*(3), 281–300. doi:10.1108/09555341311314825

Pedroni, P. (2004). Panel cointegration: Asymptotic and finite sample properties of pooled time series tests with an application to the ppp hypothesis. *Econometric Theory, 20*(3), 597–625. doi:10.1017/S0266466604203073

Pega, F., Shaw, C., Rasanathan, K., Yablonski, J., Kawachi, I., & Hales, S. (2015). Climate change, cash transfers and health. *Bulletin of the World Health Organization, 93*(8), 559–565. doi:10.2471/BLT.14.150037 PMID:26478613

Pereira, Á., & Vence, X. (2012). Key business factors for eco-innovation: An overview of recent firm-level empirical studies. *Cuadernos de Gestión, 12*, 73–103. doi:10.5295/cdg.110308ap

Pesaran, M. H. (2004). *General diagnostic tests for cross section dependence in panels*. University of Cambridge, Faculty of Economics, Cambridge Working Papers in Economics, No.0435.

Pesaran, M. H. (2007). A simple panel unit root test in the presence of cross-section dependence. *Journal of Applied Econometrics*, *22*(2), 265–312. doi:10.1002/jae.951

Pesaran, M. H., Ullah, A., & Yamagata, T. (2008). A bias-adjusted LM test of error cross section independence. *The Econometrics Journal*, *11*(1), 105–127. doi:10.1111/j.1368-423X.2007.00227.x

Pesaran, M. H., & Yamagata, T. (2008). Testing slope homogeneity in large panels. *Journal of Econometrics*, *142*(1), 50–93. doi:10.1016/j.jeconom.2007.05.010

Peterson, N., & Broad, K. (2009). Climate and weather discourse in anthropology: From determinism to uncertain futures. In *Anthropology and climate change: From encounters to actions* (pp. 70–86). Left Coast Press.

Pfeifer, J. (2006). Laws of nature. In S. Sarkar & J. Pfeifer (Eds.), The philosophy of science: An encyclopedia (pp. 439–444). Academic Press.

Phair, J. W. (2006). Green chemistry for sustainable cement production and use. *Green Chemistry*, *8*(9), 763–780. doi:10.1039/b603997a

Pimienta-Bueno, J.A. (1999). *Parque de Inovação Tecnológica e Cultural da Gávea: A Visão da PUC-Rio*. Documento Interno.

Piotrkowski, C. S. (1979). *Work and the family system: A naturalistic study of working-class and lower-middle-class families*. Collier Macmillan.

Pizarro, C. (2014). La entrevista etnográfica como práctica discursiva: Análisis de caso sobre las pistas meta-discursivas y la emergencia de categorías nativas. *Revista de Antropologia*, *57*(1), 461–496. doi:10.11606/2179-0892.ra.2014.87770

Poelmans, S. A., Kalliath, T., & Brough, P. (2008). Achieving work–life balance: Current theoretical and practice issues. *Journal of Management & Organization*, *14*(3), 227–238. doi:10.5172/jmo.837.14.3.227

Polonsky, M. J. (2011). Transformative green marketing: Impediments and opportunities. *The Journal of Business*, *64*(12), 1311–1319.

Popp, D. (2010). *The Role of Green Technology Transfer in Climate Policy*, https://www.resourcesmag.org/common-resources/the-role-of-green-technology-transfer-in-climate-policy/

Porter, M. (2001). Clusters of Innovation: regional foundations of U.S competitiveness. Council on Competitiveness. Monitor Group.

Poulose, S., & Dhal, M. (2020). Role of perceived work–life balance between work overload and career commitment. *Journal of Managerial Psychology*, *35*(3), 169–183. doi:10.1108/JMP-03-2018-0117

Pradhan, R. P., Arvin, M., Nair, M., Bennett, S., & Bahmani, S. (2017). ICT-finance-growth nexus: Empirical evidence from the Next-11 countries. *Cuadernos de Economía (Santiago, Chile)*, *40*(113), 115–134. doi:10.1016/j.cesjef.2016.02.003

Prüss-Üstün, A., & Corvalán, C. (2006). *Preventing disease through healthy environments: towards an estimate of the environmental burden of disease*. World Health Organization.

Purzycki, B., Pisor, A., Apicella, C., Atkinson, Q., Cohen, E., Henrich, J., McElreath, R., McNamara, R., Norenzayan, A., Willard, A., & Xygalatas, D. (2018). The cognitive and cultural foundations of moral behavior. *Evolution and Human Behavior*, *39*(5), 490–501. doi:10.1016/j.evolhumbehav.2018.04.004

Qamruzzaman, M., & Wei, J. (2018). Financial innovation, stock market development, and economic growth: An application of ARDL model. *International Journal of Financial Studies*, *6*(3), 69. doi:10.3390/ijfs6030069

Qiu, L., Chu, L., Zhou, R., Xu, H., & Yuan, S. (2023). How do carbon, stock, and renewable energy markets interact: Evidence from Europe. *Journal of Cleaner Production*, *407*, 137106. doi:10.1016/j.jclepro.2023.137106

Radcliffe-Brown, A. R. (1940). On Social Structure. *J. R. Anthropol. Inst. Great Br. Irel.*, *1940*(70), 1–12.

Rahbar, E., & Wahid, N. A. (2011). Investigation of green marketing tools' effect on consumers' purchase behavior. *Business Strategy Series*, *12*(2), 73–83. doi:10.1108/17515631111114877

Rahman, S. U., Chen, S., Saud, S., Bano, S., & Haseeb, A. (2019). The nexus between financial development, globalization, and environmental degradation: Fresh evidence from Central and Eastern European Countries. *Environmental Science and Pollution Research International*, *26*(24), 24733–24747. doi:10.100711356-019-05714-w PMID:31240660

Rai, M. (2023). 15,354 Tourists Arrive in 2023 so Far. *The Bhutanese*. thebhutanese.bt/15354-tourists-arrive-in-2023-so-far/#:~:text=There%20has%20been%20a%20total

Rani, S. V., & Mariappan, S. (2011). Work/life balance reflections on employee satisfaction. *Serbian Journal of Management*, *6*(1), 85–96. doi:10.5937jm1101085R

Ranjan, R. K., & Kushwaha, R. (2017). Impact of Green Marketing Strategies on Consumer Purchase Behaviour. *Review of Management*, *7*(3), 9–22.

Rathore, M. (2022). *Opinion on employee satisfaction in India as of December 2021*. Statista.

Raudeliuniene, J., Davidaviˇciene, V., & Jakubaviˇcius, A. (2018). Knowledge management process model. *Entrepreneurship and Sustainable Issues*, *5*, 542–554.

Ray, C., & Ming, X. (2020). Climate Change and Human Health: A Review of Allergies, Autoimmunity and the Microbiome. *International Journal of Environmental Research and Public Health*, *17*(13), 4814. doi:10.3390/ijerph17134814 PMID:32635435

Rayos, J. L. T., & Borbon, N. M. D. (2022). Seafarers' attitude and cultural adaptation influencing the work motivation during the time of pandemic. *International Journal of Research*, *10*(1), 107–124.

Raza, S. A., Qamar, S., & Ahmed, M. (2023). Asymmetric role of non-renewable energy consumption, ICT, and financial development on ecological footprints: Evidence from QARDL approach. *Environmental Science and Pollution Research International*, *30*(8), 20746–20764. doi:10.100711356-022-23549-w PMID:36255586

Razmi, S. F., Bajgiran, B. R., Behname, M., Salari, T. E., & Razmi, S. M. J. (2020). The relationship of renewable energy consumption to stock market development and economic growth in Iran. *Renewable Energy*, *145*, 2019–2024. doi:10.1016/j.renene.2019.06.166

Redman, C. (2014). Should sustainability and resilience be combined or remain distinct pursuits? *Ecology and Society*, *19*(2), 190–202. doi:10.5751/ES-06390-190237

Reguera, A. (2009). *Empresarios de ayer, de hoy y de siempre. Un recorrido latinoamericano por sus formas espacio-temporales América Latina en la historia económica, num 32, jul-dic, México*. https://www.scielo.org.mx/scielo.php?script=sci_arttext&pid=S1405-22532009000200003&lng=es&nrm=iso

Rehman, A. A., & Pickup, F. (2018). Zakat for the SDGs. *Islamic Finance News*, *15*(34), 1–14.

Reisinger, A., & Clark, H. (2018). How much do direct livestock emissions contribute to global warming? *Global Change Biology*, *24*(4), 1749–1761. doi:10.1111/gcb.13975 PMID:29105912

Renwick, D. W. S., Redman, T., & Maguire, S. (2013). Green Human Resource Management: A Review and Research Agenda. *International Journal of Management Reviews*, *15*(1), 1–14. doi:10.1111/j.1468-2370.2011.00328.x

Revelle, W. (2020). psych: Procedures for Psychological, Psychometric, and Personality Research. Northwestern University, Evanston, Illinois. R package version 2.0.9

Revelle, W. (2021). *psych: Procedures for Psychological, Psychometric, and Personality Research.* Northwestern University, Evanston, Illinois. R package version 2.1.3. https://CRAN.R-project.org/package=psych

Revelle, W., & Rocklin, T. (1979). Very Simple Structure: An Alternative Procedure For Estimating The Optimal Number Of Interpretable Factors. *Multivariate Behavioral Research*, *14*(4), 403–414. doi:10.120715327906mbr1404_2 PMID:26804437

Reynoso, C. (1995). Hacia la perfección del consenso: Los lugares comunes de la antropología. *Intersecciones*, *1*, 51–72.

Ricci, F., Scafarto, V., Ferri, S., & Tron, A. (2020). Value relevance of digitalization: The moderating role of corporate sustainability. An empirical study of Italian listed companies. *Journal of Cleaner Production, 276*, 1-8. . doi:10.1016/j.jclepro.2020.123282

Rice, B., Martin, N., Fieger, P., & Hussain, T. (2022). Older healthcare workers' satisfaction: Managing the interaction of age, job security expectations and autonomy. *Employee Relations*, *44*(2), 319–334. doi:10.1108/ER-07-2020-0346

Richardson, R. (2018a). *Depleting Earth's Resources.* https://msutoday.msu.edu/news/2018/depleting-earths-resources/

Richardson, R. (2018b). *Yes, humans are depleting Earth's resources, but 'footprint' estimates don't tell the full story.* https://theconversation.com/yes-humans-are-depleting-earths-resources-but-footprint-estimates-dont-tell-the-full-story-100705

Riley, D. (2006). *Turnover intentions: The mediation effects of job satisfaction, affective commitment and continuance commitment* [Doctoral dissertation]. The University of Waikato.

Rincy, V. M., & Panchanatham, N. (2014). Work life balance: A short review of the theoretical and contemporary concepts. *Continental Journal of Social Sciences*, *7*(1), 1–24.

Ritter, A., Borchardt, M., Vaccaro, G., Pereira, G., & Almeida, F. (2015). Motivations for promoting the consumption of green products in an emerging country: exploring attitudes of Brazilian consumers. *Journal of Cleaner Production, 106*, 507–520.

Rivera-Camino. (2006). Re-evaluating green marketing strategy: A stakeholder perspective. *European Journal of Marketing*, *41*(11/12), 1328–1358.

Roca, A. (2008). *La comprensión de la tecnología en la antropología.* Actas VII ESOCITE UFRJ, Jornadas Latino-Americanas de Estudos Sociais das Ciências e das Tecnologías.

Roethlisberger, F., & Dickson, W. (1939). *Management and the Worker.* Harvard Univ. Press.

Rojikinnor, R., Gani, A. J. A., Saleh, C., & Amin, F. (2022). The role of compensation as a determinant of performance and employee work satisfaction: a study at the PT Bank Rakyat Indonesia (Persero) Tbk. *Journal of Economic and Administrative Sciences.*

Romanello, M., van Daalen, K., Anto, J. M., Dasandi, N., Drummond, P., Hamilton, I. G., Jankin, S., Kendrovski, V., Lowe, R., Rocklöv, J., Schmoll, O., Semenza, J. C., Tonne, C., & Nilsson, M. (2021). Tracking progress on health and climate change in Europe. *The Lancet. Public Health*, *6*(11), e858–e865. doi:10.1016/S2468-2667(21)00207-3 PMID:34562381

Roni, S., Merga, M., & Morris, J. (2020). *Conducting quantitative research in education.* Springer. doi:10.1007/978-981-13-9132-3

Rossati, A. (2017). Global Warming and Its Health Impact. *The International Journal of Occupational and Environmental Medicine*, *8*(1), 7–20. doi:10.15171/ijoem.2017.963 PMID:28051192

Rowe, D. B. (2011). Green Roofs as a means of pollution abatement. *Environmental Pollution*, *159*(8-9), 2100–2110. doi:10.1016/j.envpol.2010.10.029 PMID:21074914

Royal Government of Bhutan (RGoB). (2012). *Bhutan: In Pursuit of Sustainable Development—National Report for the United Nations Conference on Sustainable Development 2012*. Royal Government of Bhutan.

Royal Monetary Authority of Bhutan, Annual Report 2008/09

Royal Monetary Authority of Bhutan, Annual Report 2020

Royal Monetary Authority of Bhutan, Annual Report 2022

Rupa, R. A., & Saif, A. N. M. (2022). Impact of Green Supply Chain Management (GSCM) on Business Performance and Environmental Sustainability: Case of a Developing Country. *Business Perspectives and Research*, *10*(1), 140–163. doi:10.1177/2278533720983089

SachsJ. (n.d.). https://www.aljazeera.com/opinions/2011/9/10/the-economics-of-happiness

Sadorsky, P. (2010). The impact of financial development on energy consumption in emerging economies. *Energy Policy*, *38*(5), 2528–2535. doi:10.1016/j.enpol.2009.12.048

Sadorsky, P. (2011). Financial development and energy consumption in central and Eastern European frontier economies. *Energy Policy*, *39*(2), 999–1006. doi:10.1016/j.enpol.2010.11.034

Saeed, K., & Farooqi, Y. A. (2014). Examining the relationship between work life balance, job stress and job satisfaction among university teachers (A case of University of Gujrat). *International Journal of Multidisciplinary Sciences and Engineering*, *5*(6), 9-15.

Sahito, Z., & Vaisanen, P. (2017). Factors affecting job satisfaction of teacher educators: Empirical evidence from the Universities of Sindh Province of Pakistan. *Journal of Teacher Education and Educators*, *6*(1), 5–30.

Sahito, Z., & Vaisanen, P. (2020). A literature review on teachers' job satisfaction in developing countries: Recommendations and solutions for the enhancement of the job. *Review of Education*, *8*(1), 3–34. doi:10.1002/rev3.3159

Saif, M. I., Malik, M. I., & Awan, M. Z. (2011). Employee work satisfaction and work-life balance: A Pakistani perspective. *Interdisciplinary Journal of Contemporary Research in Business*, *3*(5), 606–617.

Saikia, J. N. (2011). *A study of the work life balance among the academics of higher education institutions: A case study of Golaghat District*. Academic Press.

Sairally, S. (2007). Evaluating the 'Social Responsibility' of Islamic Finance: Learning from the Experience of Socially Responsible Investment Funds. In M. Iqbal, S.S. Ali, & D. Muljawan (Eds.), Advances in Islamic Economics and Finance. Islamic Research & Training Institute, Islamic Development Bank Group.

Samroodh, M., Anwar, I., Ahmad, A., Akhtar, S., Bino, E., & Ali, M. A. (2022). The indirect effect of job resources on employees' intention to stay: A serial mediation model with psychological capital and work–life balance as the mediators. *Sustainability (Basel)*, *15*(1), 551. doi:10.3390u15010551

Sangay. (2016). *Gross National Happiness – Royal Bhutanese Embassy, New Delhi*. Royal Bhutanese Embassy. www.mfa.gov.bt/rbedelhi/bhutan-at-glance/gross-national-happiness/

Santini, M., Tran, T., & Beath, A. (2017). *Investment Climate Assessment of Bhutan: Removing Constraints to Private Sector Development to Enable the Creation of More and Better Jobs*. World Bank Group. http://documents.worldbank.org/curated/en/673381506689928071/pdf/120144-ICA-PUBLIC-P153221-Bhutan- Competitiveness-Policy-Note-Web.pdf

Sarracino, F., Greyling, T., Peroni, C., O'Connor, K., & Rossouw, S. (2021). *A year of pandemic: levels, changes and validity of well-being data from Twitter. Evidence from ten countries*. Academic Press.

Sarstedt, M., Ringle, C., & Hair, J. (2017). Partial Least Squares. *Structural Equation Modeling*, 1–40. Advance online publication. doi:10.1007/978-3-319-05542-8_15-1

Saud, S., Chen, S., Danish, H. A., & Haseeb, A. (2019). Impact of financial development and economic growth on environmental quality: An empirical analysis from belt and road initiative (BRI) countries. *Environmental Science and Pollution Research International*, *26*(3), 2253–2269. doi:10.100711356-018-3688-1 PMID:30456608

Saud, S., Chen, S., Haseeb, A., & Sumayya. (2020). The role of financial development and globalization in the environment: Accounting ecological footprint indicators for selected one-belt-one-road initiative countries. *Journal of Cleaner Production*, *250*, 119518. doi:10.1016/j.jclepro.2019.119518

Saufi, A. B. R., Permarupan, Y. P., Aidara, S., Zainol, B. R. N., & Khan, U. N. (2023). Sustaining Work Life Balance Toward Job Satisfaction and Employee Intention to Leave Universities. Academy of Management, (1). doi:10.5465/AMPROC.2023.14613abstract

Saufi, A. R., Aidara, S., Nawi, C. N. B., Permarupan, P. Y., Zainol, N. R. B., & Kakar, A. S. (2023). Turnover intention and its antecedents: The mediating role of work–life balance and the moderating role of job opportunity. *Frontiers in Psychology*, *14*, 1–15. doi:10.3390u15043397 PMID:37077854

Savickiene, J. & Miceikiene, A. (2018). Sustainable economic development assessment model for family farms. *AgriculturalEconomics – Czech*, *64*, 527–535. doi:10.17221/310/2017-AGRICECON

Saxenian, A. (1994). *Regional Advantage: culture and competition in Silicon Valley and Route 128*. Harvard Business Press.

Sbia, R., Shahbaz, M., & Hamdi, H. (2014). A contribution of foreign direct investment, clean energy, trade openness, carbon emissions and economic growth to energy demand in UAE. *Economic Modelling*, *36*, 191–197. doi:10.1016/j.econmod.2013.09.047

Schelter, B., Winterhalder, M., Eichler, M., Peifer, M., Hellwig, B., Guschlbauer, B., Lucking, C., Dahlhaus, R., & Timmer, J. (2006). Testing for Directed Influences among Neural Signals using Partial Directed Coherence. *Journal of Neuroscience Methods*, *152*(2), 210–219. doi:10.1016/j.jneumeth.2005.09.001 PMID:16269188

Schlapfer, A., & Marinova, D. (2001). Local Innovation Systems: nature, importance and role. *Conference Proceedings: International Summer Academy on Technological Studies: User Involvement in Technological Innovation*.

Schmid, J., & Leiman, J. M. (1957). The development of hierarchical factor solutions. *Psychometrika*, *22*(1), 53–61. doi:10.1007/BF02289209

Schmuck, D., Matthes, J., Naderer, B., & Beaufort, M. (2017). The Effects of Environmental Brand Attributes and Nature Imagery in Green Advertising. *Environmental Communication*, *12*(3), 414–429. doi:10.1080/17524032.2017.1308401

Schnegg, M., O'Brian, C. I., & Sievert, I. J. (2021). It's our fault: A global comparison of different ways of explaining climate change. *Human Ecology: an Interdisciplinary Journal*, *49*(3), 327–339. doi:10.100710745-021-00229-w

Schumacker, R. E., & Lomax, R. G. (2010). *A beginner's guide to structural equation modeling* (3rd ed.). Routledge Academic.

Scott, S. V., Van Reenen, J., & Zachariadis, M. (2017). The long-term effect of digital innovation on bank performance: An empirical study of SWIFT adoption in financial services. *Research Policy*, *46*(5), 984–1004. doi:10.1016/j.respol.2017.03.010

Secondi, L., & Pirani, E. (2011). Eco-Friendly Attitudes: What European Citizens Say and What they do. *International Journal of Environmental Research*, *5*(1), 67–84.

See, H. (2022). The Two Faces of Gross National Happiness: Can Bhutan's Nation-Building Strategy Also Be a Sustainable Alternative Development Paradigm? *Journal of Contemporary Asia*, *52*(3), 452–470. doi:10.1080/00472336.2021.1933139

Sehgal, S., & Garg, P. (2020). The Human Challenges of the Digital World: The Paradigm of Employee Expectations in the Banking Sector. *First Pan IIT International Management Conference – 2018*. doi:10.2139/ssrn.3743389

Seldon, P. (2017). Regional Tourists Turn into Litterbugs in Bhutan. *The Bhutanese*. thebhutanese.bt/regional-tourists-turn-into-litterbugs-in-bhutan/.

Semenza, J. C., Suk, J. E., Estevez, V., Ebi, K. L., & Lindgren, E. (2012). Mapping climate change vulnerabilities to infectious diseases in Europe. *Environmental Health Perspectives*, *120*(3), 385–392. doi:10.1289/ehp.1103805 PMID:22113877

Setoguchi, S., Leddin, D., Metz, G., & Omary, M. B. (2022). Climate change, health, and health care systems: A global perspective. *Gastroenterology*, *162*(6), 1549–1555. doi:10.1053/j.gastro.2022.02.037 PMID:35247463

Shafique, M., Kim, R., & Rafiq, M. (2018). Green roof benefits, opportunities, and challenges– A review. *Renewable and Sustainable Energy Reviews, 90*, 757-773.

Shahbaz, M., Destek, M. A., & Polemis, M. L. (2018). Do foreign capital and financial development affect clean energy consumption and carbon emissions? Evidence from BRICS and Next-11 countries. *SPOUDAI-Journal of Economics and Business*, *68*(4), 20–50.

Shah, S. P. (2004). Development of 'green'cement for sustainable concrete using cement kiln dust and fly ash. *Proceedings of the International Workshop on Sustainable Development and Concrete Technology*, 15-23.

Shaikh, S. A., & Ismail, A. G. (2017). Role of Zakat in Sustainable Development Goals. *International Journal of Zakat*, *17*(2), 1–9. doi:10.37706/ijaz.v2i2.21

Shamdasani, P., Chon-Lin, G. O., & Richmond, D. (1993). Exploring green consumers in an oriental culture: Role of personal and marketing mix factors. *Advances in Consumer Research. Association for Consumer Research (U. S.)*, *20*(1), 488–493.

Shapiro, S. S., & Wilk, M. B. (1965). An analysis of variance test for normality (complete samples). *Biometrika*, *52*(3–4), 591–611. doi:10.1093/biomet/52.3-4.591

Sharma, Y. (2011). Changing Consumer Behaviour with Respect to Green Marketing: A Case Study of Consumer Durables and Retailing. *International Journal of Multidisciplinary Research*, *1*(4), 152–162.

Shetty, S. K., & Bhat, K. S. (2022). Green supply chain management practices implementation and sustainability–A review. *Materials Today: Proceedings*, *52*(3), 735–740. doi:10.1016/j.matpr.2021.10.135

Shevchenko, A., & Petrenko, O. (2020). Current state of micro and small agribusiness in Ukraine. *Agricultural and Resource Economics*, *6*(1), 146–160. doi:10.51599/are.2020.06.01.10

Shevchenko, T., Saidani, M., Ranjbari, M., Kronenberg, J., Danko, Y., & Laitala, K. (2023). Consumer behavior in the circular economy: Developing a product-centric framework. *Journal of Cleaner Production*, *384*, 135568. doi:10.1016/j.jclepro.2022.135568

Shil, P. (2012). Evolution and Future of Environmental Marketing. *Asia Pacific J. Marketing Manage. Rev*, *1*(3), 74–81.

Shiu, E., Hair, J., Bush, R., & Ortinau, D. (2009). *Marketing Research* (European Edition). McGraw-Hill Higher Edition.

Shobitha, P., & Sudarsan, N. (2014). Work life balance: A conceptual review. *International Journal of Advances in Management and Economics*, *3*(2), 1–17.

Shoeb, A., & Nisar, T. (2015) Green Human Resource Management: Policies and practices, *Cogent Business & Management, 2*(1), 1030817-103. https://www.tandfonline.com/doi/full/10.1080/23311975.2015.1030817 doi:10.1080/23311975.2015.1030817

Shrivastava, S. R. (2017). A systematic literature review on green manufacturing concepts in cement industries. *International Journal of Quality & Reliability Management*, *34*(1), 68–90. doi:10.1108/IJQRM-02-2014-0028

Shrout, P. E., & Fleiss, J. L. (1979). Intraclass correlations: Uses in assessing rater reliability. *Psychological Bulletin*, *86*(2), 420–3428. doi:10.1037/0033-2909.86.2.420 PMID:18839484

Shwekat, K. W., & Wu, H.-C. (2018). Benefit-cost analysis model of using class fly ash-based green cement in masonry units. *Journal of Cleaner Production*, *198*, 443–451. doi:10.1016/j.jclepro.2018.06.229

Signorelli, A. (1999). *Antropología Urbana*. UAM. doi:10.1080/0267257X.2014.925958

Silaban, H., & Margaretha, M. (2021). The impact work-life balance toward job satisfaction and employee retention: Study of millennial employees in Bandung city, Indonesia. *International Journal of Innovation and Economic Development*, *7*(3), 18–26. doi:10.18775/ijied.1849-7551-7020.2015.73.2002

Singh, P., & Siddiqui, A. A. (2021). Innovation, ICT penetration, trade and economic growth in developing and developed countries: a VECM approach. Competitiveness Review. doi:10.1108/CR-05-2021-0074

Singh, G. (2013). Green: The New Colour of Marketing in India. *ASCI Journal of Management*, *42*(2), 57–72.

Singh, S. K., Del Giudice, M., Chierici, R., & Graziano, D. (2020). Green innovation and environmental performance: The role of green transformational leadership and green human resource management. *Technological Forecasting and Social Change*, *150*, 119762. doi:10.1016/j.techfore.2019.119762

Sirgy, M. J., Efraty, D., Siegel, P., & Lee, D. J. (2001). A new measure of quality of work life (QWL) based on need satisfaction and spillover theories. *Social Indicators Research*, *55*(3), 241–302. doi:10.1023/A:1010986923468

Skydan, O., Nykolyuk, O., Pyvovar, P., & Martynchuk, I. (2020). Methodological approach to the evaluation of agricultural business system flexibility. *Management Theory and Studies for Rural Business and Infrastructure Development*, *41*(4), 444–462. doi:10.15544/mts.2019.36

Snow, C. P. (1998). *The Two Cultures*. Cambridge University Press.

Sohail, M. T., Majeed, M. T., Shaikh, P. A., & Andlib, Z. (2022). Environmental costs of political instability in Pakistan: Policy options for clean energy consumption and environment. *Environmental Science and Pollution Research International*, *29*(17), 1–10. doi:10.100711356-021-17646-5 PMID:34837622

Soonthonsmai, V. (2007). *Environmental or Green Marketing As Global Global Competitive Edge: Concept, Synthesis And Implications*. EABR (Business) &ETLC (Teaching) *Conference Proceedings*, Venice, Italy.

Spearman, C. (1904). 'General intelligence,' objectively determined and measured. *The American Journal of Psychology*, *15*(2), 201–293. doi:10.2307/1412107

Spielmann, N. (2020). Green is the New White: How Virtue Motivates Green Product Purchase. *Journal of Business Ethics*. Advance online publication. doi:10.100710551-020-04493-6

Spiewak, R. (2023). Diseases from the Spectrum of Dermatitis and Eczema: Can "Omics" Sciences Help with Better Systematics and More Accurate Differential Diagnosis? *International Journal of Molecular Sciences*, *24*(13), 10468. doi:10.3390/ijms241310468 PMID:37445645

Staines, G. L. (1980). Spillover versus compensation: A review of the literature on the relationship between work and nonwork. *Human Relations*, *33*(2), 111–129. doi:10.1177/001872678003300203

Stark, K., Niedrig, M., Biederbick, W., Merkert, H., & Hacker, J. (2009). Die Auswirkungen des Klimawandels. Welche neuen Infektionskrankheiten und gesundheitlichen Probleme sind zu erwarten? [Climate changes and emerging diseases. What new infectious diseases and health problem can be expected?]. *Bundesgesundheitsblatt, Gesundheitsforschung, Gesundheitsschutz*, *52*(7), 699–714. doi:10.100700103-009-0874-9 PMID:19536444

Stoner, A. M. (2021). Things are getting worse on our way to catastrophe: Neoliberal environmentalism, repressive desublimation, and the autonomous ecoconsumer. *Critical Sociology*, *47*(May), 491–506. doi:10.1177/0896920520958099

Stovin, V., Vesuviano, G., & Kasmin, H. (2012). The hydrological performance of a green roof test bed under UK climatic conditions. *Journal of Hydrology (Amsterdam)*, *414*, 148–161. doi:10.1016/j.jhydrol.2011.10.022

Strang, V. (2017). The Gaia complex: Ethical Challenges to an Anthropocentric 'Common Future. In B. Marc & L. Jerome (Eds.), *The Anthropology of Sustainability* (pp. 255–283). Palgrave. doi:10.1057/978-1-137-56636-2_12

Stubbs, W., & Cocklin, C., (2008). Conceptualizing a sustainability business model. *Organizational Environment, 21*(2), 103-127.

Su, C., Pang, L., Tao, R., Shao, X. & Umar, M. (2022). Renewable energy and technological innovation: Which one is the winner in promoting net-zero emissions? *Technological Forecasting & Social Change,* 1 – 11. doi:10.1016/j.techfore.2022.121798

Subha, T. (2013). A Study on Work Life Balance Among Women Faculties Working in Arts and Science Colleges with Special Reference to Coimbatore City. *Paripex-Indian Journal of Research*, *2*(12), 160–163.

Su, C., Liu, F., Stefea, P., & Umar, M. (2023). Does technology innovation help to achieve carbon neutrality? *Economic Analysis and Policy*, *78*, 1–14. doi:10.1016/j.eap.2023.01.010

Suki, N. (2016). Consumer environmental concern and green product purchase in Malaysia: Structural effects of consumption values. *Journal of Cleaner Production*, *132*, 204–214. doi:10.1016/j.jclepro.2015.09.087

Sumeda. (2022). Explained: Bhutan's Strategy to Levy Sustainable Development Fee on Tourists. *The Hindu*. www.thehindu.com/news/international/bhutan-strategy-daily-tax-tourists-sustainable-development-fee-tax/article65943766.ece

Sundbo, J. (1998). *The Theory of Innovation: entrepreneurs, technology, and strategy*. Edward Elgar.

Sun, H., Teh, P.-L., & Linton, J. D. (2018). Impact of environmental knowledge and product quality on student attitude toward products with recycled/remanufactured content: Implications for environmental education and green manufacturing. *Business Strategy and the Environment*, *27*(7), 935–945. Advance online publication. doi:10.1002/bse.2043

Sun, Y., & Wang, S. (2019). Understanding consumers' intentions to purchase green products in the social media marketing context. *Asia Pacific Journal of Marketing and Logistics*, *32*(4), 860–878. doi:10.1108/APJML-03-2019-0178

Sustainable Development Goals | United Nations Development Programme. (n.d.). *UNDP*. Retrieved May 11, 2023, from https://www.undp.org/bhutan/sustainable-development-goals

Sustainable Development Goals | United Nations in Bhutan. (2023). bhutan.un.org/en/sdgs

Sutton, R. (2014). Aesthetics for green roofs and green walls. *The Journal of Living Architecture*, *1*(2), 1–20. doi:10.46534/jliv.2014.01.02.001

Svensson, G., & Wagner, B. (2012). Business sustainability and E-footprints on Earth's life and ecosystems: Generic models. *European Business Review*, *24*(6), 543–552. doi:10.1108/09555341211270555

Syptak, J. M., Marsland, D. W., & Ulmer, D. (1999). Job satisfaction: Putting theory into practice. *Family Practice Management*, *6*(9), 26–30.

Syuzairi, M., Pratiwi, M. A., & Apriansyah, F. (2023). Work-Life Balance and its Impact on Turnover Intention among Educators: Job Satisfaction as a Mediation. In *ICEMBA 2022: Proceedings of the International Conference on Economic, Management, Business and Accounting, ICEMBA 2022, 17 December 2022, Tanjungpinang, Riau Islands, Indonesia* (p. 258). European Alliance for Innovation. 10.4108/eai.17-12-2022.2333271

Tainter, J. A. (1988). *The Collapse of Complex Societies*. Cambridge University Press.

Tainter, J. A. (2008). Collapse and Sustainability: Rome, the Maya, and the Modern World. *Archeological Papers of the American Anthropological Association*, *24*(1), 201–214. doi:10.1111/apaa.12038

Talan, A., Rao, A., Sharma, G. D., Apostu, S. A., & Abbas, S. (2023). Transition towards clean energy consumption in G7: Can financial sector, ICT and democracy help? *Resources Policy*, *82*, 103447. Advance online publication. doi:10.1016/j.resourpol.2023.103447

Tan, P. Y., Wong, N. H., Tan, T. K., & Wong, K. W. (2019). Green roof innovation in Singapore: Policies, performance, and potential. *Journal of Environmental Management*, *233*, 128–135. doi:10.1016/j.jenvman.2018.12.060

Tarantino-Curseri, S. (2018). Antropología, etología, sociología, y psicología, 4 disciplinas útiles en el ámbito empresarial. *Management Review*, *3*(3). Advance online publication. doi:10.18583/umr.v3i3.126

Tarek, F., Hassou, N., Benchekroun, M. N., Boughribil, S., Hafid, J., & Ennaji, M. M. (2019). Impact of rotavirus and hepatitis A virus by worldwide climatic changes during the period between 2000 and 2013. *Bioinformation*, *15*(3), 194–200. doi:10.6026/97320630015194 PMID:31354195

Taşbaşı, D. (2022). *Araştırma: "Yenilenebilir enerji 2050 yılına kadar küresel enerji karışımının yüzde 50'sini oluşturacak"*. https://temizenerji.org/2022/10/17/arastirma-yenilenebilir-enerji-2050-yilina-kadar-kuresel-enerji-karisiminin-yuzde-50sini-olusturacak/

Thangaswamy, A., & Thiyagaraj, D. (2017). Theoretical concept of job satisfaction-a study. *International Journal of Research*, *5*(6), 464–470.

The Economist Newspaper. (n.d.). Why is Finland so happy? *The Economist*. https://www.economist.com/the-economist-explains/2018/03/26/why-is-finland-so-happy

Thite, M., & Russell, B. (2010). Work organization, human resource practices and employee retention in Indian call centers. *Asia Pacific Journal of Human Resources*, *48*(3), 356–374. doi:10.1177/1038411110381623

Thomas, M. B. (2020). Epidemics on the move: Climate change and infectious disease. *PLoS Biology*, *18*(11), e3001013. doi:10.1371/journal.pbio.3001013 PMID:33232329

Thurstone, L. L. (1947). *Multiple factor analysis*. University of Chicago Press.

Tiwari, A. K., Nasreen, S., & Anwar, M. A. (2022). Impact of equity market development on renewable energy consumption: Do the role of FDI, trade openness and economic growth matter in Asian economies? *Journal of Cleaner Production, 334*, 130244. Advance online publication. doi:10.1016/j.jclepro.2021.130244

Tomashuk, I. V., & Baldynyuk, V. M. (2021). Identification of problems and prospects of rural infrastructure development of Ukraine. *Colloquium Journal,* 13(100), 58-70. doi:10.24412/2520-6990-2021-13100-58-70

Too, R., & Kwasira, D. J. (2017). Evaluation of employee retention practices to generation y at cooperative bank in the county of Nairobi, Kenya. *European Journal of Business and Management, 9*(11), 43-53.

Toropova, A., Myrberg, E., & Johansson, S. (2021). Teacher job satisfaction: The importance of school working conditions and teacher characteristics. *Educational Review, 73*(1), 71–97. doi:10.1080/00131911.2019.1705247

Trawick, P., & Alf, H. (2015). Revisiting the Image of Limited Good: On Sustainability, Thermodynamics, and the Illusion of Creating Wealth. *Current Anthropology, 56*(1), 1–27. doi:10.1086/679593

Trist, E. L. (1976). A Concept of Organizational Ecology: an invited address to the three Melbourne universities. Academic Press.

Trujillo, M. L. N. (2021). Notes for a critical and ecological view of patriarchal capitalism in the web of life. *Capital and Class, 45*(March), 21–32. doi:10.1177/0309816820929115

Tsing, A. L. (2015). *The mushroom at the end of the world.* Princeton University Press.

Tymon, W. G. Jr, Stumpf, S. A., & Smith, R. R. (2011). Manager support predicts turnover of professionals in India. *Career Development International, 16*(3), 293–312. doi:10.1108/13620431111140174

Uddin, M., Ali, K. B., & Khan, M. A. (2020). Perceived social support (PSS) and work-life balance (WLB) in a developing country: The moderating impact of work-life policy. *Iranian Journal of Management Studies.*

Ueda, Y. (2012). The relationship between work-life balance programs and employee satisfaction: Gender differences in the moderating effect of annual income. *Journal of Business Administration Research, 1*(1), 65. doi:10.5430/jbar.v1n1p65

Ugai, T. (2016). Evaluation of sustainable roof from various aspects and benefits of agriculture roofing in urban core. *Procedia: Social and Behavioral Sciences, 216*, 850–860. doi:10.1016/j.sbspro.2015.12.082

Uka, A., & Prendi, A. (2021). Motivation as an indicator of performance and productivity from the perspective of employees. *Management & Marketing. Challenges for the Knowledge Society, 16*(3), 268–285. doi:10.2478/mmcks-2021-0016

United Nations Global Impact. (2013). *Global Corporate Sustainability Report 2013 2.* www.unglobalcompact.org

Unruh, G., & Ettenson, R. (2010). Growing green. *Harvard Business Review, 88*(6), 94–100.

Ura, K., Alkire, S., Zangmo, T., & Wangdi, K. (2012). *An extensive analysis of GNH index.* Centre for Bhutan Studies. doi:10.35648/20.500.12413/11781/ii036

Vachon, S., & Klassen, R. D. (2006). Extending green practices across the supply chain: The impact of upstream and downstream integration. *International Journal of Operations & Production Management, 26*(7), 795–821. doi:10.1108/01443570610672248

Van Holt, T., Statler, M., Atz, U., Whelan, T., van Loggerenberg, M., & Cebulla, J. (2020). The cultural consensus of sustainability-driven innovation: Strategies for success. *Business Strategy and the Environment, 29*(2), 3399–3409. doi:10.1002/bse.2584

Velho, G. (1978). Observando o familiar. In A aventura sociológica: objetividade, paixao, improviso e método na pesquisa social. Zahar.

Verru, S. (2013). The impact of consumer behavior on Green Marketing: An Analysis. *International Journal of Advanced Trends in Computer Science and Engineering*, *2*(1), 569–575.

Vezzulli, L., Colwell, R. R., & Pruzzo, C. (2013). Ocean warming and spread of pathogenic vibrios in the aquatic environment. *Microbial Ecology*, *65*(4), 817–825. doi:10.100700248-012-0163-2 PMID:23280498

Vijayaraghavan, K. (2016). Green roofs: A critical review on the role of components, benefits, limitations, and trends. *Renewable & Sustainable Energy Reviews*, *57*, 740–752. doi:10.1016/j.rser.2015.12.119

Vyas, A., & Shrivastava, D. (2017). Factors affecting work life balance-a review. *Pacific Business Review International*, *9*(7), 194–200.

W. S. (2007). JThe Anthropocene: Are Human Beings Now Overwhelming the Forces of Nature? *Ambio*, *36*, 614–621. doi:10.1579/0044-7447(2007)36[614:TAAHNO]2.0.CO;2 PMID:18240674

Walcott, S. M. (2011, Spring). Making the Most of Geographic Disadvantage: Modernizing Bhutan, Education about. *Asia*, *16*, 1.

Walker, B. (2019). *Finding Resilience: Change and Uncertainty in Nature and Society*. CSIRO Publishing.

Walster, E., Berscheid, E., & Walster, G. W. (1976). New directions in equity research. In Advances in experimental social psychology (Vol. 9, pp. 1–42). Academic Press.

Wangai, M. M. (2012). *Factors affecting job satisfaction among secondary school teachers of Mwatate district, Kenya. Project paper presented as a partial fulfillment for the requirement of the award of certificate in senior management course*. GTI EMBU.

Wang, F., Shi, C., Dong, J., & Nie, H. (2021, December). Association between ambient temperature and atopic dermatitis in Lanzhou, China: A time series analysis. *Environmental Science and Pollution Research International*, *28*(47), 67487–67495. doi:10.100711356-021-15198-2 PMID:34254239

Wang, H., Ma, B., & Bai, R. (2019). How Does Green Product Knowledge Effectively Promote Green Purchase Intention? *Sustainability (Basel)*, *11*(4), 1193. doi:10.3390u11041193

Wang, L. C., Chen, S. S., Tsang, D. C. W., Poon, C.-S., & Shih, K. (2016). Recycling contaminated wood into eco-friendly particleboard using green cement and carbon dioxide curing. *Journal of Cleaner Production*, *137*, 861–870. doi:10.1016/j.jclepro.2016.07.180

Wangmo, T. (2016). *Starting Strong on the SDGs in Asia: Readiness in Bhutan*. IGES Discussion Paper, Institute for Global Environmental Strategies.

Wang, S.-T. (2014). Consumer characteristics and social influence factors on green purchasing intentions. *Marketing Intelligence & Planning*, *32*(7), 738–753. doi:10.1108/MIP-12-2012-0146

Wang, Y., Hu, H., Dai, W., & Burns, K. (2021). Evaluation of industrial green development and industrial green competitiveness: Evidence from Chinese urban agglomerations. *Ecological Indicators*, *124*, 107371. doi:10.1016/j.ecolind.2021.107371

Wannimayake, W., & Randiwela, P. (2008). Consumer attractiveness towards Green Products of FMCG sector: An empirical study. Oxford Business and Economics Conference Program, 1-19.

Wasik, J. F. (1996). *Green marketing and management: A global perspective*. Blackwell Publishers Inc.

Wei, X., Lo, C. K. Y., Jung, S., & Choi, T.-M. (2021). From co-consumption to co-production: A systematic review and research synthesis of collaborative consumption practices. *Journal of Business Research*, *129*(May), 282–294. doi:10.1016/j.jbusres.2021.02.027

Welz, G. (2019). 'Environmental Orientations' and the Anthropology of the Anthropocene. *Anthropological Journal on European Cultures*, *27*(1), 40–44. doi:10.3167/ajec.2018.270107

Westerlund, J. (2005). New simple tests for panel cointegration. *Econometric Reviews*, *24*(3), 297–316. doi:10.1080/07474930500243019

Westerlund, J. (2007). Testing for error correction in panel data. *Oxford Bulletin of Economics and Statistics*, *69*(6), 709–748. doi:10.1111/j.1468-0084.2007.00477.x

White, C., & Maniam, B. (2020). Flexible working arrangements, worklife balance, and working women. *Journal of Business and Accounting*, *13*(1), 59–73.

Wiering, M., Liefferink, D., Boezeman, D., Kaufmann, M., Crabbé, A., & Kurstjens, N. (2020). The Wicked Problem the Water Framework Directive Cannot Solve. The Governance Approach in Dealing with Pollution of Nutrients in Surface Water in the Netherlands, Flanders, Lower Saxony, Denmark, and Ireland. *Water (Basel)*, *12*(5), 1240. doi:10.3390/w12051240

Wijayathunga, L. A. D. H. R., & Rathish, D. (2023). Non-financial Rewards and Job Satisfaction among State University Academics of Sri Lanka. *Sri Lankan Journal of Human Resource Management*, *13*(1), 27–39. doi:10.4038ljhrm.v13i1.5676

Williams, N., Rayner, J. & Raynor, K. (2010). Green roofs for a wide brown land: Opportunities and barriers for rooftop greening in Australia. *Urban Forestry & Urban Greening*, *9*(3), 245–251. https://doi.org/.ufug.2010.01.005 doi:10.1016/j

Wirtenberg, J., Harmon, K. D., Russell, W. G., & Fairfield, K. D. (2007). HR's role in building a sustainable enterprise. *Human Resource Planning*, *30*, 10–20.

Wise, S., Bond, S., & Meikle, A. (2003). *Work-life balance-literature and research review*. DTI and Fair Play.

Wolf, M. (2016). Rethinking urban epidemiology: Natures, networks and materialities. *International Journal of Urban and Regional Research*, *40*(5), 958–982. doi:10.1111/1468-2427.12381 PMID:32336869

Wood, J., Oh, J., Park, J., & Kim, W. (2020). The relationship between work engagement and work–life balance in organizations: A review of the empirical research. *Human Resource Development Review*, *19*(3), 240–262. doi:10.1177/1534484320917560

Workman, A., Jones, P. J., Wheeler, A. J., Campbell, S. L., Williamson, G. J., Lucani, C., Bowman, D. M. J. S., Cooling, N., & Johnston, F. H. (2021). Environmental Hazards and Behavior Change: User Perspectives on the Usability and Effectiveness of the AirRater Smartphone App. *International Journal of Environmental Research and Public Health*, *18*(7), 3591. doi:10.3390/ijerph18073591 PMID:33808395

World Economic Forum. (2021). *The Global eco-weakening: how consumers are driving sustainability, climate breakthroughs: The Road to COP26 and Beyond*. Author.

Wu, W., & Jim, C. Y. (2021). Urban green infrastructure and sustainable urbanization: A comprehensive review of the roles of green roofs. *Landscape and Urban Planning*, *212*, 104097. doi:10.1016/j.landurbplan.2021.104097

Wysocki, J. (2021). Innovative green initiatives in the manufacturing SME sector in Poland. *Sustainability (Basel)*, *13*(4), 2386. doi:10.3390u13042386

Xu, L. (2009). View on work-family linkage and work-family conflict model. *International Journal of Business and Management, 4*(12), 229–233. doi:10.5539/ijbm.v4n12p229

Yaacob, M., & Zakaria, A. (2011). Customers' awareness, perception and future prospects of green products. *The Journal of Commerce, 3*(2), 1–10.

Yadav, H., Soni, U., & Kumar, G. (2023). Analysing challenges to smart waste management for a sustainable circular economy in developing countries: A fuzzy DEMATEL study. *Smart and Sustainable Built Environment, 12*(2), 361–384. doi:10.1108/SASBE-06-2021-0097

Yadav, R. K., & Dabhade, N. (2013). A case study with overview of job satisfaction. *Indian Journal of Economics and Development, 1*(8), 136–146.

Yadav, R., & Pathak, G. S. (2016). Young consumers' intention towards buying green products in a developing nation: Extending the theory of planned behavior. *Journal of Cleaner Production, 135*, 732–739. doi:10.1016/j.jclepro.2016.06.120

Yang, Y., Lili, W., & Hongchi, Z. (2022). Bibliometric research on the evolution of resilience theme from the perspective of Geographical Science. *2022 29th International Conference on Geoinformatics*, 1-9. 10.1109/Geoinformatics57846.2022.9963870

Yangka, D., Newman, P., Rauland, V., & Devereux, P. (2018). Sustainability in an emerging nation: The Bhutan case study. *Sustainability (Basel), 10*(5), 1622. Advance online publication. doi:10.3390u10051622

Yangka, D., Rauland, V., & Newman, P. (2019). Carbon neutral policy in action: The case of Bhutan. *Climate Policy, 19*(6), 672–687. doi:10.1080/14693062.2018.1551187

Yenipazarli, A., & Vakharia, A. J. (2015). Green, greener or brown: Choosing the right color of the product. *Annals of Operations Research, 250*(2), 537–567. doi:10.100710479-014-1781-5

Yilanci, V., Ozgur, O., & Gorus, M. S. (2019). The asymmetric effects of foreign direct investment on clean energy consumption in BRICS countries: A recently introduced hidden cointegration test. *Journal of Cleaner Production, 237*, 117786. doi:10.1016/j.jclepro.2019.117786

Younis, I., Naz, A., Shah, S. A. A., Nadeem, M., & Longsheng, C. (2021). Impact of stock market, renewable energy consumption and urbanization on environmental degradation: New evidence from BRICS countries. *Environmental Science and Pollution Research International, 28*(24), 31549–31565. doi:10.100711356-021-12731-1 PMID:33608780

Yu, M., Jin, H., Zhang, H., & Chong, A. Y. L. (2023). ICT, Financial Development and Renewable Energy Consumption. *Journal of Computer Information Systems, 63*(1), 190–203. doi:10.1080/08874417.2022.2049017

Zappelli, C., Barbulova, A., Apone, F., & Colucci, G. (2016). Effective Active Ingredients Obtained through Biotechnology. *Cosmetics, 3*(4), 39. doi:10.3390/cosmetics3040039

Zedeck, S. (1992). Introduction: Exploring the domain of work and family concerns. In Frontiers of Industrial and Organizational Psychology, Vol. 5; The Jossey-Bass Management Series and The Jossey-Bass Social and Behavioral Science Series. Work, families, and organizations (pp. 1–32). Jossey-Bass.

Zeng, C., & Chen, H. (2020). An exploration of the relationships between organizational dissent, employee burnout, and work-family balance: A cross-cultural comparison between China and Finland. *Communication Studies, 71*(4), 633–648. doi:10.1080/10510974.2020.1749864

Zhang, C. Y., Yu, B., Chen, J.-M., & Wei, Y.-M. (2021). Green transition pathways for cement industry in China. *Resources, Conservation and Recycling, 166*, 105355. doi:10.1016/j.resconrec.2020.105355

Zhang, K. S., Shen, P., Yang, L., Rao, M., Nie, S., & Wang, F. (2021). Development of high-ferrite cement: Toward green cement production. *Journal of Cleaner Production*, *327*, 129487. doi:10.1016/j.jclepro.2021.129487

Zhang, Y. J., Fan, J. L., & Chang, H. R. (2011). Impact of China's stock market development on energy consumption: An empirical analysis. *Energy Procedia*, *5*, 1927–1931. doi:10.1016/j.egypro.2011.03.331

Zhang, Z., & Sundaresan, S. (2010). Knowledge markets in firms: Knowledge sharing with trust and signalling. *Knowledge Management Research and Practice*, *2010*(8), 322–339. doi:10.1057/kmrp.2010.22

Zheng, C., Kashi, K., Fan, D., Molineux, J., & Ee, M. S. (2016). Impact of individual coping strategies and organisational work–life balance programmes on Australian employee well-being. *International Journal of Human Resource Management*, *27*(5), 501–526. doi:10.1080/09585192.2015.1020447

Zhu, Y., Zhang, J., & Duan, C. (2022). How does green finance affect the low-carbon economy? Capital allocation, green technology innovation and industry structure perspectives. *Economic Research*, 1-23. doi:10.1080/1331677X.2022.2110138

Zinbarg, R. E., Barlow, D. H., & Brown, T. A. (1997). Hierarchical structure and general factor saturation of the Anxiety Sensitivity Index: Evidence and implications. *Psychological Assessment*, *9*(3), 277–284. doi:10.1037/1040-3590.9.3.277

Zinsstag, J., Crump, L., Schelling, E., Hattendorf, J., Maidane, Y. O., Ali, K. O., Muhummed, A., Umer, A. A., Aliyi, F., Nooh, F., Abdikadir, M. I., Ali, S. M., Hartinger, S., Mäusezahl, D., de White, M. B. G., Cordon-Rosales, C., Castillo, D. A., McCracken, J., Abakar, F., ... Cissé, G. (2018). Climate change and One Health. *FEMS Microbiology Letters*, *365*(11), fny085. Advance online publication. doi:10.1093/femsle/fny085 PMID:29790983

About the Contributors

Sumita Dave is an accomplished academician with extensive experience in management education. Currently serving as a Deputy Pro Vice Chancellor, Dean of the Faculty of Management & Commerce and Director of Amity Business School at Amity University Chhattisgarh, she excels in academic leadership, curriculum design, and industry collaboration. With a notable track record as Principal and Professor at various institutions, she has significantly contributed to program development and faculty training. She is also a registered supervisor for a Ph.D. at Amity University, Raipur. Six scholars have been awarded Ph.D. degrees under her supervision. Dr. Dave's research prowess shines through numerous publications and presentations in national and international conferences, focusing on areas such as retail, organizational transformation, and emotional intelligence. She has more than 90 publications to her credit in SCI and UGC CARE-listed journals, national and international journals with ISSNs with high impact factor. She has a notable record of establishing Centers of Excellence and has delivered numerous lectures, keynote addresses, and guest lectures at various management institutes. Her expertise in administration, teaching, and research positions her as a prominent figure in the field of management education.

Monica Sainy is a professor of Marketing at Amity Business School, Amity University, Chhattisgarh. Prior to this, she had worked with various reputed universities and institutes in Indore, namely Shri Vaishnav Vidyapeeth Vishwavidyalaya, Indore, as an Associate Professor of Marketing Management at the CH Institute of Management & Commerce, Acropolis Faculty of Management & Research, Chameli Devi Group of Institutes, and Laxmi Narayan College of Technology, Indore. As a faculty, She has been teaching marketing and general management subjects. She is also a registered supervisor for a Ph.D. at Devi Ahilya Vishwavidyalya, Indore, and Amity University, Chhattisgarh, Raipur. Three scholars have been awarded Ph.D. degrees under her supervision, and two scholars have submitted their theses under her guidance. She also completed a minor research project in MPCOST on E-Learning: Issues, Challenges, and Implementation at Schools in M.P." and published three edited books entitled "Innovation, Invention, and Industrialization: Emerging Issues (2017) and Diverse, Divergent, and Innovative Research: Contemporary Issues," published by Excel Books, New Delhi, and "Digital Transformation for Sustainability: Opportunities and Challenges (2019)," published by Excellent Publishing House, New Delhi. She had more than 50 publications to her credit in SCI and UGC CARE-listed journals, national and international journals with ISSNs with high impact factors, research papers in edited books with ISBN numbers, and conference proceedings at the national and international levels. She has conducted several research-based workshops, case-writing workshops, seminars, and conferences on various themes. She is also a reviewer in "The Asian Journal of Business and Management (AJBM) having an ISSN of 2321–2802, and was also nominated as Editor-in-Chief in "The Innovare Journal of Business Management and Research."

Disha Sharma is an Assistant Professor of Finance and Taxation at Amity Business School, Amity University in Raipur, India. She has a Ph.D. in Commerce, an M. Com, an MTTM, and an MA in English Literature also holds a BBA in Management. Dr. Sharma has received several academic achievements, including the UGC-NET (Commerce), and CG-SET (Commerce). Her research work includes a thesis awarded on "An Analytical Study of Women Participation in Mahatma Gandhi National Rural Employment Guarantee Act (With Special Reference to Bilaspur District)" and completed project work with UNICEF and ICSSR. She is also a lifetime member of the Indian Commerce Association and Indian Accounting Association. She has completed certification courses on Digital Analytics and Fundamentals of R and an online course on Understanding Open Educational Resources. She has also completed an Information Technology Training Course and attended various refresher courses and orientation programs. Dr. Sharma has expertise in diverse topics such as systematic literature review, bibliometric analysis, fintech innovation, financial empowerment, entrepreneurship, public policy, role of foreign direct investment in retail market, and many more. She is passionate about quality education, higher learning, and research, and has been working tirelessly to create opportunities for students and faculty members to enhance their knowledge and skills. Dr Sharma has an impressive research profile, having presented papers in various national and international seminars and webinars. Dr. Sharma is also an experienced editor and has edited a book titled 'Economic Growth, Development and Social Challenges in the new normal' and an upcoming edited book to be published by IGI Global. She has also published numerous papers and chapters in various journals of international repute. Overall, Dr. Disha Sharma is a dedicated academician and researcher with a diverse educational background and an impressive range of qualifications, achievements, and work experience.

Anupam Singh is a Lecturer in Digital Marketing & Analytics. Prior to joining the University of Stirling, Dr Singh was a post-doc research fellow at the Nicolaus Copernicus University, Poland. He earned a Ph.D. from the National Institute of Technology (NIT), Bhopal, India. He is a recipient of a prestigious junior research fellowship (JRF) and Senior Research Fellowship (SRF) award from the University Grant Commission (UGC), New Delhi, India. He also received a Faculty Research Fellowship from the Indian Institute of Technology (IIT), Delhi, India (SFRF-2020). Dr Singh is a researcher specializing in the study of algorithmic consumption in digital environments. His research focuses on understanding how algorithms influence behavior, decision-making processes, and consumption patterns in the digital realm. Another aspect of his research focuses on the application of big data and machine learning techniques to analyze the motivations and barriers to sustainable consumption. His research attempts to answer how individuals' pro-social and pro-environmental behavior can make our planet more resilient and sustainable.

* * *

Pınar Avcı is a lecturer at Tekirdağ Namık Kemal University, Turkey. She received her doctoral degree in 2019. Her research interests are finance, financial institutions, international finance and capital market.

Ayşegül Aytaç Emin, PhD, is a Lecturer at Social Sciences University of Ankara. Her main research interests include macroeconomic analyses and financial economics with a special focus on Islamic finance, banking and financial crises. She holds a BS degree in Economics and an MA degree in Economics from Bilkent University. From 2014 to 2017 she worked in Turkey Economic Policy Research Foundation of Turkey (TEPAV, Turkey) as a research associate. During the Turkey's G20 Presidency in 2015, she worked in TOBB (The Union of Chambers and Commodity Exchanges of Turkey) as a General Content Coordinator for Business 20 and she was also a Research Associate in the Think 20 process. She holds a PhD degree in Economics in Hacettepe University. She has several published articles, sectoral projects and policy notes within the scope of her research interest.

Esha Bharadwaj is currently working as a researcher in the Department of Economics, School of Humanities and Social Sciences, Jain (Deemed-to-be University), Bangalore, India.

Chandan Chavadi is the Dean & Professor of the Presidency Business School, Presidency College, Bengaluru. A PhD from Karnatak University, Dharwad, and an MBA (Mktg.) He holds a primary degree in B.E. (E&C). He has two years of corporate experience before his moving to academics. He has been in academics for the last 21 years. He has 32 papers & 2 book reviews to his credit, published in reputed journals and magazines such as ABDC-C journals, Web of Science, Scopus, UGC Care list and other indexed journals. Under Google scholar indices, he has total citations of 90, h-index of 6 & i10-index of 4. His paper has been accepted for publication in the IIM –A Vikalpa. Five of his research papers were recently published in the IIM Kozhikode Society And Management Review journal, IIM-Shillong Journal of Management Science, Business Perspective & Research journal of K J Somaiya Institute of Management, the MDI journal "Vision", and IIM-Lucknow journal "Metamorphosis". He is the recipient of Labdhi Bhandari Best Paper Award for the 7th IIM-A International Marketing Conference held on 11th to 13th Jan 2017. He is a recognized PhD Guide in Management for Bangalore City University.

Ashish Kumar is currently working as a researcher in the Department of Economics, School of Humanities and Social Sciences, Jain (Deemed-to-be University), Bangalore, India.

Praveen Kumar, Assistant Professor at the University School of Business in Chandigarh University, with 5.5 years of experience in teaching and Research. He has cleared the Junior Research Fellowship (UGC-JRF) which is conducted by University Grant Commission (UGC). He has awarded his Ph.D. in Management with the research topic "Consumers Perception and Purchase Intention towards Organic Food Products". He has attended many international and national conferences, workshops, seminars, and faculty development programs. He has published many Scopus-indexed research papers and book chapters in reputed Journals in the field of Marketing, Finance, and Human Resource Management. He has also written many book chapters in different areas of Management and published national and International Patents in various fields. He has practical experience and knowledge of Statistical software viz. SPSS, AMOS- SEM, PLS. He has been invited as a keynote speaker in various workshops, and seminars related to Research methodology at different colleges and Universities.

Susobhan Maiti is currently working as an Assistant Professor in the Department of Economics, School of Humanities and Social Science (SHSS) at JAIN (Deemed-to-be University), Bangalore, India. He has published many research papers in the area of industry, efficiency, and productivity in both national and international journals. He received ICSSR Doctoral fellowship for the year 2017-2018 from Ministry of Human Resource Development (MHRD) presently Ministry of Education, Government of India.

Pratik Mukherjee is currently working as a researcher in the Department of Economics, School of Humanities and Social Sciences, Jain (Deemed-to-be University), Bangalore, India.

Soni Rathi is presently a Research scholar at the University School of Business in Chandigarh University, Punjab. Her current research interests include work life balance, job stress, work life conflict, job satisfaction, employees' creativity, Artificial intelligence, human resource management and generational differences. She has attended many international and national conferences, workshops, and faculty development programs. She has practical experience and knowledge of Statistical software viz. Smart PLS 3 & 4, SPSS v.25, Vos viewer.

Carlos Rodríguez-Maillard is a 20+ years working experience in industries as Logistics, GBS, Shared Service Centers, building products, Banking and Teaching. Experience in Quote to Cash Process strategy, information management, Business Strategy, and Customer Centricity. Professor in Intelligence Organizations, User Centric Design and Project Management in EGADE Business School (Tecnologico de Monterrey) Fluent in Spanish (mother tongue), English and advanced Portuguese.

Shobhit Sagar, an esteemed individual, acquired a Bachelor of Commerce Degree from Dr. Bhimrao Ambedkar University in 2014. Continuing his academic journey, he accomplished a Master of Commerce degree with a specialization in International Business from Dayalbagh Educational Institute in 2016. Demonstrating his dedication to scholarly pursuits, he successfully qualified the UGC NET examination in November 2017. In pursuit of further knowledge, Dr. Shobhit Sagar completed a Master of Philosophy in Accountancy & Law from the same esteemed institute in the year 2018. His commitment to scholarly research was recognized when he was awarded a Ph.D. degree in the year 2022. His doctoral thesis, titled "Impact of Monetary and Credit Policy on the Financial Performance of Commercial Banks in India during the Tenure of RBI Governor Dr. Raghuram Rajan," showcases his expertise in the field. currently working as Assistant Professor in Mewar Institute of Management Ghaziabad U.P. India.

Disha Sharma is a Finance and Tax Faculty, Presently working as an Assistant Professor at Amity Business School, Amity University Chhattisgarh. She has qualified UGC-NET & SET, did Masters in three subjects, and completed her PhD from Pt. Ravi Shankar Shukla University. She has six years of teaching experience. She has Published 6 Scopus and 4 peer-reviewed papers. She has also written 6 book chapters in International and Scopus-indexed books. She also has experience as Co-Principal Investigator in ICSSR Project and as a Principal Director in a project under UNICEF. She has presented many research papers at national and international conferences. She is a lifetime member of the Indian Commerce Association and Indian Accounting Association.

Urvashi Shrivastava is Professor and Head, Department of Management, Bhilai Institute of Technology, Durg.

Sushma Singh is a Research Scholar at Bhilai Institute of Technology, Durg in the area of Management.

Suvam Sivam is currently working as a researcher in the Department of Economics, School of Humanities and Social Sciences, Jain (Deemed-to-be University), Bangalore, India.

Shriya Tripathi is a Management professional currently pursuing Ph.D from Durg University. She completed her M. Com (Business Management) from Narsee Monjee College, Mumbai and her BMS (Bachelor in Management Studies) from K.P.B Hinduja College of Commerce, Mumbai. She has more than 4 years of experience in academia, research, Soft skills training, and corporate. She also takes into credit 7 research papers at various National and International conferences. She has demonstrated her passion and commitment to teaching young aspiring minds in academia by successfully qualifying for the UGC NET and SET. She has conducted various trainings for educational and corporate organizations. Currently working as an Assistant Professor at Ganpat University, her extensive expertise lies in the domain of Marketing, Entrepreneurship, and Soft Skills.

Omar Vargas-González is Professor and Head of Systems and Computing Department at Tecnologico Nacional de Mexico Campus Ciudad Guzman, professor at Telematic Engineering at Centro Universitario del Sur Universidad de Guadalajara with a master degree in Computer Systems. Has been trained in Innovation and Multidisciplinary Entrepreneurship at Arizona State University (2018) and a Generation of Ecosystems of Innovation, Entrepreneurship and Sustainability for Jalisco course by Harvard University T.H. Chan School of Health. At present conduct research on diverse fields such as Entrepreneurship, Economy, Statistics, Mathematics and Information and Computer Sciences. Has collaborated in the publication of over 20 scientific articles and conducted diverse Innovation and Technological Development projects.

Index

A

Advertisement 84, 157-158, 164, 166, 170-172
Advertising 76, 78, 81-83, 85-88, 90-91, 158, 164, 168, 170, 172, 174, 205, 215, 228
Agricultural Waste 93
Anthropocene 105, 107-108, 113-114, 116-119, 122-123, 126

B

Bhutan 175-194
BRICS 18, 24, 27-29, 31-34, 37-38, 40-44
Buddhist Economy 179
Building Materials 94, 102-104, 195

C

Carbon Dioxide (CO2) 16, 42, 93
Carbon Emission 33, 35-38, 189-190, 206
Causality 15, 17-18, 21-22, 25-26, 28, 33-39, 41-43, 54-55, 107
Clean-Energy Consumption 15, 19, 27, 32
Climate 1-2, 15, 33, 44-45, 47-48, 59, 64-75, 78-79, 87, 94, 106-107, 110-125, 127-130, 133-135, 137-138, 145, 147-148, 151-153, 158, 172, 184, 188, 190, 193-194, 196, 199, 209-210, 230
CO2 15-20, 23-29, 33-38, 40, 42-46, 89, 93-94, 163, 195-199, 202, 204-205
Concrete 93-104, 136, 197-202, 204-206, 208-210
Confirmatory Factor Analysis 212, 223, 226
Construction 52, 93-94, 98-99, 101-104, 106, 112, 118, 128, 130, 135, 137, 140-145, 147, 184, 186, 195, 197-209
Consumer 11-12, 77, 80, 85, 90-92, 114, 118, 120, 123, 130, 157, 160-163, 166, 171, 173-174, 214-215, 226, 229-230, 236
Control 34-35, 43, 65, 86, 96-97, 109, 115, 140, 143-144, 186, 197, 243, 252
Corporate Social Responsibility 1, 51, 60, 147-149
Culture 1, 3, 5-9, 79, 81, 109, 112, 125, 140-142, 144, 149, 177-178, 180, 183, 185-186, 188, 190, 230, 232-237, 239-240, 243, 247, 254, 257

D

Digital 13, 72, 138, 232, 234-237, 239-240, 262
Digitalization 232, 234-236, 239-240
Disposal Management 93

E

ECBC Code 127
Eco Friendly 76
Economic Growth 15-19, 23, 25-31, 33-38, 40-46, 48, 50, 135, 142, 188, 213, 230
Economy and Environment 47
Ecosystem 48, 64, 68, 76-78, 85, 128-129, 133, 135, 137
Emission 16, 29, 33-38, 45, 89, 95, 141, 144, 189-190, 195, 197-199, 201, 206, 210
Employee Engagement 1, 3-6, 8, 10-13, 254
Employee Retention 14, 241-242, 249-252, 254-256, 260, 262
Employees 2, 4-13, 51, 112, 115, 142, 164, 232-235, 241-247, 249-260, 262
Environment 2, 4, 10-12, 15-16, 23, 26, 31, 33-35, 38, 44-45, 47-48, 66, 68, 71-72, 74-75, 77, 82, 85-86, 88-91, 94, 106, 110-111, 113, 118, 127-130, 132, 134-135, 137-138, 140-141, 143, 145, 147-150, 153-155, 157-158, 160-165, 169-174, 176-180, 184-185, 190, 195-196, 199, 201, 204, 209-210, 213-215, 226, 231, 233, 244, 246, 249, 253, 258
Equity Market Performance 15, 19, 32
Exploratory Factor Analysis 212, 220, 227

F

Financial 5, 8, 11, 17-18, 27-32, 37, 48, 50-52, 57,

59-61, 77, 112, 117, 145, 148, 175, 184-186, 192, 201, 228-229, 232-236, 239-240, 246, 255

G

GNH 175-178, 180, 182-186, 188-193
Green Advertising 76, 78, 87, 158, 174, 215, 228
Green Cement 195-196, 198-203, 205-211
Green Energy 16, 43, 127-128, 131, 159, 173
Green HRM 1-13
Green Innovations 47, 105, 107-109, 112, 116, 135, 147
Green Knowledge 117, 134-135, 137-138, 145, 152
Green Marketing 76-83, 85-92, 151, 157-160, 162-167, 172-174, 225, 228-230
Green Marketing Mix 76-77, 79, 90-91, 163-164, 225, 229
Green Supply Chain 212, 229-230
Green Technology 41, 110, 113, 126-128, 132-135, 137-138, 145, 205
Green Urban Areas 134, 140, 142
Gross National Happiness 175-180, 189-193

I

ICT 15-19, 23-28, 30-32, 169, 230, 235
Indian Buildings 127
Industrial Waste 89, 93-94, 196, 200-201
Industry 12, 34, 44, 59, 66, 91, 93-94, 97-98, 101, 111, 118, 126, 131, 138, 147, 149, 151, 180, 183, 185-189, 191, 195-197, 199, 201-209, 211, 233, 235, 239, 254-255
Infosys 232, 236
Islamic Finance 47-49, 51-52, 54, 56-61

J

Job Satisfaction 5, 9, 241-243, 246-248, 250-262

M

Marketers 84, 157-158, 160, 164, 170-173, 213
Marketing 77-92, 114, 119, 123-124, 151-152, 157-167, 169-170, 172-174, 225-226, 228-230, 256, 260
Multi-Dimensional Construct 212

O

Ordinary Portland Cement (OPC) 104
Organizational Green Innovation 134-135, 137, 139-140, 142, 145, 147

P

Panel Data Analysis 15, 26, 32, 42, 45
Performance 1-6, 8-13, 15, 19, 32, 49, 77, 90, 96, 99, 104, 109, 111, 122-123, 142, 148-152, 154, 158, 168, 195, 203-206, 208-209, 213, 225, 228-229, 232, 234-237, 239-240, 242, 244, 246-247, 249-250, 253, 255-257, 259-260, 262

Q

Quaternary Concrete 93-94, 98, 100-102

S

SDG 47-51, 56-62, 175, 189, 191
Solar Cell 127, 130
Sustainability 1-8, 10-14, 27, 29, 33, 36, 42-43, 61, 65, 73, 77, 79-81, 85, 88-90, 93, 95, 104, 110-111, 115, 117-118, 122-123, 125-126, 128, 130, 135-136, 139-140, 142, 145-148, 151-155, 159, 163, 173-174, 176, 178-179, 190, 194-197, 199-202, 206-210, 213, 215, 228-230, 239-240, 242, 254, 257, 259
Sustainable Development 2-4, 25, 29, 43, 47-52, 57-62, 94, 110, 113-114, 134-135, 137-138, 148-150, 158, 172-173, 175, 180, 183-186, 188-193, 195, 206-208, 210, 226, 228
Sustainable Development Goals 43, 47-52, 57-62, 137-138, 150, 175, 184, 188-190, 193, 207, 226
Sustainable Global Organisation 1
Sustainable Marketing 76, 86, 90

T

The Healthcare Sector 64
Tourism 28, 45, 122, 145, 176, 183, 185-188, 191-193
Transformation 48, 106-107, 112-113, 115, 131, 145, 227, 232, 234, 236, 239-240

U

Unidimensional Construct 212
Urban Anthropology 105-109, 112, 114-118, 122
Urban Planning 105-107, 109-113, 115-119, 123, 152-153, 155
Urban Socio-Ecology 105, 107-111, 113, 115-118

V

Vector-Borne Diseases 64, 66, 69, 71-73

W

Water-Borne Diseases 64, 68
Work-Life Balance 176, 241-244, 246, 249-262

Z

Z-Generation 157-158, 168-173

www.ingramcontent.com/pod-product-compliance
Lightning Source LLC
LaVergne TN
LVHW060421300926
849162LV00006B/1491